# Ending the Fossil Fuel Era

# Ending the Fossil Fuel Era

edited by Thomas Princen, Jack P. Manno, and Pamela L. Martin

The MIT Press
Cambridge, Massachusetts
London, England

MIT Press books may be purchased at special quantity discounts for business or sales promotional use. For information, please email special_sales@mitpress.mit.edu.

This book was set in Sabon LT Std by Toppan Best-set Premedia Limited, Hong Kong. Printed and bound in the United States of America.

Library of Congress Cataloging-in-Publication Data

Ending the fossil fuel era / edited by Thomas Princen, Jack P. Manno and Pamela L. Martin.
    pages cm
  Includes bibliographical references and index.
  ISBN 978-0-262-02880-6 (hardcover : alk. paper) – ISBN 978-0-262-52733-0 (pbk. : alk. paper)
1. Fossil fuels.  2. Energy security.  3. Energy–Governmental policy.  4. Environmental degradation.  I. Princen, Thomas, 1951-  II. Manno, Jack.  III. Martin, Pamela, 1971-
    TP318.E54   2015
    553.2–dc23
                                                                            2014034211

10  9  8  7  6  5  4  3  2  1

# Contents

# Preface

Throughout our academic careers, we coeditors have tackled issues of global environmental politics from the perspective of those who seek social and ethical transformation. At times we have put these efforts under the rubric of sustainability or sufficiency or decommoditization or *buen vivir* (the good life). All aim at building good lives while living lightly on the earth. In Rio in 1992 at the Earth Summit, the UN Conference on Environment and Development, Jack and Tom participated in the work of nongovernmental organizations (NGOs) and documented their role in global environmental politics. In the early 1990s, Pam worked in Latin America, investigating how transnational networks for social change also changed world politics, all centering on oil and the Amazon. By 2010, the beginning of this project on the fossil fuel era, all three of us had turned to questions of diminishing energy resources and a post–fossil fuel future. Tom saw a localizing trend in the Global North—a shift in attention and action from the global, the abstract, the placeless to the local, the concrete, the place based. Jack and Pam both worked with Indigenous peoples (*Indigenous* as a capitalized term refers to groups of peoples, like *European peoples* or *North American nations,* with a common identity that involves historic claims to sovereignty and nationhood). Jack worked with Onondaga Nation in New York State as they strategized to prevent hydrofracking on their ancestral territory. Pam encountered conflict in the Amazonian rain forest as she came to know the people and their place, and the politics of the Yasuní National Park and the oil beneath it.

When the three of us came together for the long talks that eventually became shared writing, we realized that from our respective vantage points, each of us saw that a fundamental shift, at once biophysical and social, moral and spiritual, is underway. With every extreme weather event, every economic bubble bursting, every excuse for inaction on a host of critical environmental and social issues, the call for a concomitant cultural shift is gaining momentum. This book is our contribution to accelerating that momentum.

With some trepidation, we admit, we began to advance an argument that not so long ago would have been considered extreme: imagining and building a case for keeping fossil fuels in the ground. We no longer see anything at all extreme in this argument, this possibility, this hope. What is extreme is extreme extraction, beyond anything remotely sustainable ecologically, let alone just; extreme wealth and power for the few, which is hard to imagine is socially sustainable; and extreme weather, which will absorb increasing amounts of resources, capital, and attention in the coming years and decades. As we conceptualized and sought empirical grounding for the possibility of deliberately keeping fossil fuels in the ground, we gradually came to believe that a transition out of fossil fuels will occur one way or another. So we asked, What might that shift look like? What possibilities are there for positive transition? What would be the cultural shift that parallels the energy shift? These are among the questions that prompted this study.

The more we delved into this topic, the more we wondered why so few scholars in the environmental sciences deal explicitly and directly with fossil fuels. Certainly there is plenty of work being done on the impacts, from health and ecosystem effects to cleanup and efficiencies. Looking at our own practices and work environments, we suspect the reason is that most of us enter this field because we like green plants and blue water, free-roaming animals and wide-open spaces. We steer away from gooey and sooty substances and the noxious smells they give off. And we steer away from that messiest of all human activities—politics.

We coeditors came to realize that there is a price to be paid for such neglect: the very substances most implicated in environmental degradation and the very actors who so effectively convert concentrated physical power to concentrated economic and political power get a free pass. Put differently, the fossil fuel complex—that network of independent and national oil companies and their enablers in finance and government—can hide in the shadows, pull the levers, write the rules of the game, displace many true costs, all while others fret about the consequences and seek fixes. What we coeditors came to realize in this project is that an environmental science of transformation, of transition out of that which is demonstrably unsustainable and unjust, requires going to the source—physically, culturally, ethically, even spiritually. It requires conceiving of a politics of extractive resistance, of exiting the industry before compelled by circumstances, of imagining the good life after fossil fuels.

So this book is about transitional politics. Insofar as modern industrial society is only beginning the transition away from fossil fuels, barely showing an awareness of it, little that we offer here is definitive. Rather, our hope with this book is to provoke a conversation and offer some language and some examples of arguments made with the language. Few wish to speak of the end of the fossil fuel era, let alone the end of material growth. Here we speak of it, and speak differently from the prevailing discourse of bounteous growth (conflating economic, material, and so many other forms of growth), efficiency, consumer prerogative, technological proliferation, pollution cleanup, commercial diffusion, and financial mastery.

Philosopher Richard Rorty once said, to paraphrase, that fundamental cultural change occurs not when people argue well but when they speak differently.[1] In this book, we attempt to speak differently, to create a language of positive transition out of fossil fuels. We presume that fundamental cultural change occurs when relevant cultures and their languages change—the organizational culture of ExxonMobil, for instance; the industry culture of oil and gas; the sectoral culture of energy (dominated as it is by fossil fuels); the high-finance culture of economic policy (which dominates the fossil fuel industry); the economic culture of growth (which derives in part from the history of fossil fuel growth and in part is a necessary condition for growth); the consumer culture of goods seen as good so more goods must be better (until recently supported by cheap energy and costless waste deposition).

What do we mean by language here, and how does it engage a politics? First, language is more than words, grammar, and syntax. It is concepts and ideas, principles and norms, metaphors and stories, all that help steer societal change in a particular direction, here away from fossil fuels and toward a sustainable world. That steering, then, is the politics, the influencing, the changing of images of the possible and definitions of the good life. It is framing that escapes the dominant frames of empire, machine, laboratory, commerce, consumption, freedom, comfort, speed, and power and makes normal living within our means, with attention to all peoples, not just the powerful and the privileged. For ending the fossil fuel era—that period when fossil fuels dominate all other energy sources worldwide—there is a distinctive politics that we try to capture through, yes, argument, but also story. A major task of this book is to articulate those politics. In a nutshell, they are the politics of resistance, exit, and the imaginative.

Until recently, resistance politics played out only in isolated cases—the coal mining operation that ignores safety warnings, the offshore oil rig that hurries the drilling, the natural gas fertilizer plant that inexplicably explodes. As subjects of academic study, policy analysis, and policymaking, these politics have paled against those of climate change, conservation, ecological modernization, and other topics of mainstream environmental debate. The resisters were indeed isolated, working against both negligent, secretive operators and a growth-manic, cheaper-is-better commercial norm. What is different today, we have found, is that resisters have collaborators and colleagues, some next door, others on the other side of the globe, all connected by media that make operating in the shadows increasingly difficult. And part of what they are communicating is that the commercial norm is no longer hegemonic, that their lives and those of future peoples matter as much as anyone else's, and certainly more than an incremental bump in an earnings ratio, more than a check to a political campaign. Their politics is globalization from below. It is local—and hence the charge of not in my backyard (NIMBY)—and global—and hence the reality of not on Planet Earth (NOPE). Their politics is bringing to an end that which destroys both the local and the global; it is ending the fossil fuel era. But it is more.

Ending the fossil fuel era is also about beginning a new one. While this book is not a blueprint for the future (despite what various readers of earlier drafts wanted), we came to realize in the course of this study that an effective politics, one that aspires to effect a positive transition—peaceful, democratic, just, and ecologically sustainable—is one that sees opportunities in dramatic change (and we believe that with 80 to 90 percent reductions in greenhouse gases, change will be dramatic). A full range of actors can capitalize on those opportunities, from governments to investors, Indigenous people to transnational corporations. So throughout these pages, we flag signs of positive transition and offer language for this imaginative politics of ending the fossil fuel era. And we entertain the possibility that the fossil fuel industry itself can play a positive role, with companies designing their own exit.

So this book is about ending an era that will end one way or another no matter what anyone does. Our intuition tells us that the world will be better off accelerating that end. By the end of an era, we must stress, we mean the end of fossil fuel dominance, not of fossil fuel use entirely—not, as some early readers and listeners insisted on believing, stopping cold all fossil fuel use and watching people scrounge for morsels of food and shiver in the dark. Following climate science, ending fossil fuel dominance means, again, some 80 to 90 percent reductions worldwide, especially in the Global North, the primary focus of transition in this study. Whatever the physical amount, the political effect clearly will be dramatic. Better to start stopping now, we argue, better to follow the lead of some early pioneers and carefully, openly, and explicitly end the fossil fuel era.

As a brief overview of the structure of the book, in part 1, we set up "the problem," and then take three conceptual cuts at that problem—biophysical, cultural, and ethical. Part 2 is a collection of case studies that illuminates the struggles of and the possibilities for resisting extractive, exploitative modes of resource use, suggesting in the process how to delegitimize and start stopping fossil fuel use. In part 3 we conclude in three very different ways, imagining paths out of the fossil fuel era. There, in chapter 11, Manno and Martin elucidate the potential of Indigenous thought contributing to both fossil fuel resistance and to a politics of the good life. In chapter 12, Princen and Santana posit corporate strategies to promote fossil fuel exit from within the industry. And in chapter 13, Princen, Manno, and Martin draw on previous chapters to develop several themes for a fossil fuel politics of resistance and exit, of imagination and restoration, of realism and reality.

In this work we raise many questions and provide all too few answers. We hope, instead, to provoke constructive thought and useful language for a fundamental shift, for easing a fossil fuel–dependent world out of its dilemma and into the next era.

## Acknowledgments

In this study, spanning some five years of writing papers, conducting workshops, interviewing, and observing, we owe debts of gratitude to many people. Perhaps foremost are those working on the ground to keep fossil fuels in the ground. Many of these people must remain anonymous, but otherwise they are noted in the pages that follow, especially in part 2. For specific comments on draft chapters, we thank Kristin Bartenstein, Raymond De Young, Paul Hirsch, Gert Jan Kramer, Seth Peabody, Nicole Seymour, Adele Santana, Andrea Parker, and members of the Works in Progress seminar of the Rachel Carson Center on Environment and Society at the Ludwig Maximilians University in Munich and of the Sustainability Ethics seminar at the University of Michigan in Ann Arbor. We are deeply appreciative of the leadership of Clay Morgan, senior editor at MIT Press, now retired, in the building of the field of global environmental politics. We thank him for his continued support and masterly editorship of this book, one of his last at MIT Press. We also thank his successor, Beth Clevenger who, with Miranda Martin, Bev Miller, and Marcy Ross, expertly shepherded the manuscript to completion. For research assistance we thank Kenneth Fahey and Dominique De Wit.

We thank Coastal Carolina University for hosting a workshop in 2012 on its campus and the International Studies Association for supporting another at its 2013 annual meeting in San Francisco. And Tom Princen gives a special nod to a gentleman (in the truest sense of the term) who, across a table, over many cups of coffee, made "normal" much of the thinking that Tom put into this book—Raymond De Young. Pamela Martin thanks Alberto Acosta for his constant support and commentaries on the good life and the road ahead toward it.

## Note

1. Richard Rorty, *Philosophy and the Mirror of Nature* (Princeton, NJ: Princeton University Press, 1979).

# Part 1

## The Fossil Fuel Problem

# 1
## The Problem

Thomas Princen, Jack P. Manno, and Pamela L. Martin

*Fossil fuels—can't live without them, the lifeblood of modern industrial civilization.*

*Fossil fuels—can't live with them, the fire in the oven destined to bake civilization beyond recognition.*

Two existential positions, poles apart. Will the twain ever meet? Will their opposition, the crux of the contemporary energy dilemma, ever be resolved? Or will we, the industrialized and consumerized denizens of a material system thoroughly out of sync with the self-correcting biophysical system in which this system is embedded, just have to play it out? So far, those who cannot imagine life without fossil fuels have the upper hand. Notions of progress and technological determinism and the magic of the market sustain a belief system that says this world, this industrial, growth-oriented, consumer-serving, fossil fuel–driven world, is the best of all possible worlds. There is no reason to exchange it for an uncertain one—no reason to reorganize, to build a world without fossil fuels. Why? Because the next energy transition, like previous transitions—from human power to animal power, animal to wood, wood to coal, and coal to oil—will make life better for all.[1] As happened before, there will be less drudgery, greater convenience, higher speeds, and more consumer choice—material progress forever. The bridge to this improved state is new technologies, including technologies to extract and burn—albeit "cleanly"—every last bit of affordable oil, coal, and gas. This is the dominant worldview, which we dub the "industrial progressive" view.[2]

Those who see the world and the coming transition differently are in a distinct minority. There are the doom-and-gloomers—scientists, popular writers, and filmmakers—who get people's attention but, like the progressive faithful, can't imagine the good life being other than some green and clean variation of business as usual. There are the rejectionists—back-to-the-landers, localists, survivalists—who are hunkering down, getting ready for collapse. And then there are those, a distinct minority within this minority, who dare to imagine a different order, a different transition, indeed a different politics, one that is peaceful, democratic, just, and biophysically sustainable. To the extent this minority is even noticed, they are readily dismissed as idealists, not realists.[3]

In this book, we imagine a different kind of realism, one that starts with the sheer brevity of the fossil fuel era, defined as that period when fossil fuels have dominated all other energy sources, which began in the United States and worldwide only in the 1890s.[4] Until recently, that era can be characterized by two unassailable facts: easy-to-get, cheap, high-density energy and ever-increasing amounts of such energy.

To put numbers to these two facts, humans have extracted and burned roughly 1 trillion barrels of conventional oil to date. There's another 1 trillion available. And there's some 4 to 5 trillion, some say as much as 18 trillion barrel equivalents, in other fossil fuels.[5] The first trillion has been enough to disperse toxic substances to every corner of the globe, erode soil on a global scale, permanently deplete underground aquifers, and disrupt the climate and acidify the oceans, all this possibly irretrievably and with huge human costs. Extracting, refining, combusting, and dispersing the by-products of another trillion or more will only compound those effects (not just add to them or continue them) and will almost certainly be catastrophic.

To use more and more fossil fuels would thus be irrational, to say the least. Morally it would be a global crime of negligence. Politically it would be a policy failure to dwarf all others. The response, rational and moral, is to drastically reduce such use. The question industrial societies face at this historical juncture is how to navigate the transition and how to do so given that the fossil fuel era will end and that these fuels will be rationed, although on the current path, not soon enough to avert catastrophic environmental and social impacts. And because the pollutants *will* be emitted, soils degraded, water sources depleted if the oil, coal, and natural gas are taken out of the ground, a reasonable premise given the history of fossil fuel management (see chapter 3), the only feasible means of stopping such use is to leave these fuels in the ground.

This is the realism of this book—what we call "twenty-first century" realism. It is not the dominant view by far, not in mainstream policy debates, not in the media, not even in academe, where one would think that such trends are well discussed. But those who live in the energy world—oil companies and power utilities, for instance—and those who study the trends—energy analysts and historians, for instance—seem to agree. The former president of Shell Oil Company US put it this way: "The resources are there. The question is: do we *want* to continue to use these fossil fuels at current—or increasing—rates until they are eventually exhausted? The answer, unequivocally, is no. The economic, social, and environmental costs of such an approach are becoming ever clearer and ever higher."[6] Resource geographer Vaclav Smil puts it this way: "Ours is an overwhelmingly fossil-fueled society. ... This grand solar subsidy, this still-intensifying depletion of an energy stock whose beginnings go back hundreds of millions of years, cannot last, and the transition to a non-fossil future is an imperative process of self-preservation for modern high-energy civilization."[7]

For all the concern about fossil fuels, the predominant approach to both ground-level air pollution and high-level climate change is to *manage* emissions. In climate, it is to reduce a couple of centuries of history to one chemical element, carbon, to, in effect, seek end-of-pipe solutions when the real problem is upstream, in a global infrastructure and power structure that is extremely adept at laying new pipes. So we ask, "What exactly is `the problem'"? We start with climate change in part because it has dominated environmental discourse for a couple of decades now. But we must emphasize that "the problem" of fossil fuels is much more than climate change; climate change is but one symptom of a larger systemic problem. That problem can be described as a system of extraction, expansion, and exploitation that even if it were ethically and politically justifiable, would still require another planet or two to continue.[8] It is, in short, a system of excess that a focus on fossil fuels affords and, we assert, warrants much more examination.

**The Problem: Extraction, Not Emissions; Fossil Fuels, Not Carbon**

Climate change deliberations have centered on two primary realms of activity—state action and science—and associated actors—diplomats and scientists. If this were the sole locus of action, then the implicit theory of social change would be something like this:

*Climate change is a global problem. Like global security and global trade, it must be managed globally. Global managers are of two sorts—those with the authority—states—and those with the requisite knowledge—scientists. Only global managers can work at the global level. Only they can perceive the problem (through their instruments) requiring, as it does, vast data sets, sophisticated modeling, and the funds to support the science. Only they can marshal the resources to tackle such a gargantuan problem. Only they can reach the agreements that overcome the global collective action problem and arrange the incentives so their respective publics behave correctly.*

Curiously missing from this global management formulation of the problem and the rightful actors who would solve it are those actors who organize to pull fossil fuels out of the ground— private and state oil, gas, and coal companies and the industrial development arms of governments. Missing are the complex networks of actors who accomplish the remarkable transformation of raw materials to usable products (for example, shippers, refiners, manufacturers, distributors, petrochemical companies), ensure the flow of such materials (domestic security and court systems, international security forces), and finance it all (bankers, investors, consumers). Our experience is that the great bulk of the attention in academe is on the first set of actors. Certainly nearly all the funding goes to the science, some to the economics and intergovernmental relations, but next to nothing to an understanding of the political economy of extraction and combustion.[9]

In this book, we formulate "the problem" differently. We hold that the central problem is not about what is done *after* extraction and combustion; it is about extraction itself. Put differently, it is not about carbon but about "fossil fuels.

A carbon focus is reductionist, possibly the greatest and most dangerous reductionism of all time: a 150-year history of complex geologic, political, economic, and military security issues all reduced to one element—carbon. This chemical framing implies that the problem arises after a chemical transformation, after fuels are burned. It effectively absolves of responsibility all those who organize to extract, process, and distribute, state and nonstate. It leaves unquestioned the legal requirements to extract created by the selling of fossil fuel reserves in futures markets and the widespread use of reserves for collateral in financial transactions. So constructed, extraction is called "production," and the burden of harm and responsibility for amelioration falls on governments and consumers rather than extractors. This is a situation presumed to be given—normal or inevitable or desirable. Finally, "carbon" portrays the global ecological predicament as completely one-dimensional: solve the climate change problem—that is, deal with carbon—and everything else follows.

To focus on fossil fuels, by contrast, is to ask about the status of oil, coal, and natural gas in the ground and how and why these complex hydrocarbons come out of the ground. It does not take such how-and-why questions as self-evident (consumers want the energy; producers get it). A fossil fuel focus directs attention, analytic and eventually political attention, upstream to a whole set of decisions, incentives, and structures that conspire to bring to the surface hydrocarbons that otherwise sit safely and permanently in the ground. It forces one to consider that once fossil fuels are extracted, their by-products—petrochemical endocrine disrupters, sulfur dioxide, atmospheric greenhouse gases—inevitably and unavoidably move into people's bloodstream, into ecosystems, and into the atmosphere and oceans.

So the difference, both analytic and rhetorical, between "carbon" and "fossil fuels" is the difference between reductionism and complex systems, between global management and deliberative decision making, between management and elimination, between end of pipe and prevention, between cleanup and abstention, between technocracy and democracy. To question extraction is to consider deliberately limiting an otherwise valuable resource, rationing and setting priorities for its uses. It is to take renewable energy, conservation, equity, and environmental justice seriously and create the institutions, local to global, capable of doing so. It is to delegitimize excess rates of use and make special modest rates of use (chapter 3). It is to ask what the implicit ethic of fossil fuels has been and what a twenty-first-century ethic of fossil fuel use might be (chapter 4). It is to view fossil fuel resistance and abolition as more than NIMBY ("not in my backyard"). It is to deliberately choose a post–fossil fuel world (chapter 12). All told, it is to imagine *a politics that challenges excess and disavows exploitation by confronting extraction.*

In this book, we choose *fossil fuels* as our primary analytic lens in part because, like any analytic premise, our intuition tells us that end-of-pipe and cleanup approaches do not work, that complex networks of fossil fuel actors have had a free ride, presumed innocent by analytic omission. We do so because to question extraction, not just emission, is to ask about the meaning of fossil fuels and how that meaning varies across cultures and subcultures (chapters 3 and 11). It is to ask what would be a politics of resistance, on the one hand, and imagination (of a post–fossil fuel world) on the other (part 2). It is to entertain what, in mainstream policy and academic debate, heretofore has been unthinkable: *keeping fossil fuels in the ground.*

## Beyond End of Pipe

If the predominant approach to ground-level air pollution, high-level climate change, persistent toxic substances, and a host of other environmental ills is to manage fossil fuel emissions, how could one go beyond end of pipe?

First, consider the implications of end of pipe. From an *economic perspective*, end of pipe effectively says the problem occurs at the end of a long chain of production and consumption decisions. Pollutants emerge after the goods are produced, as a by-product, an unfortunate yet unavoidable side effect. All previous steps in the production-consumption-disposal chain are indeed about "goods," that which can be presumed beneficial or benign. Exploration, testing, drilling, transporting, securing, processing, manufacturing, distributing, advertising, and lobbying are all given—given, that is, by a combination of entrepreneurial spirit, extraordinary risk taking, technological innovation, capital investment, managerial choice, and, once all this is instituted, consumer demand and political imperative. So construed, emissions are merely an unfortunate and inconvenient side effect. But because the goods come from all that is given and captured by those actors so engaged and the bads come from the emissions that nobody wants, a society (read government or taxpayers) is obligated to ameliorate the bads. This division of goods and bads is terribly convenient for those who actually make the key decisions and reap so much of the rewards. It is an ethic in its own right, one that says producing goods is inherently good. But in the larger scheme of things, in that system that incorporates both the extraction and the disposal over geologically long periods of time—the only system that could legitimately be called an ecosystem—it is hardly ethical when downstream vulnerable populations now and in the future face the bads. And it is hardly realist if the interests of all these populations are of account.

From a *political economy perspective*, what we know to be the natural order of things is actually deliberately constructed, an order whose rules have evolved over time to suit well those who benefit most. In the fossil fuel order, certainly all players—rule makers and rule followers—benefit in some fashion (investors gain returns on their investments, risk-taking employees are paid well, consumers have cheap and abundant energy), but such benefits are time constrained, as are all mining operations. Boom times may last years, even a century or two, but the bust is perfectly predictable (chapters 2 and 3). The fossil fuel binge may be hugely profitable, amazingly stimulating, not to mention convenient and fun; it may transport, heat, and feed billions, but it is not sustainable. It will end. The only question is how: how peacefully and democratically, how planfully and equitably. This, once again, is twenty-first-century realism.

End of pipe is also problematic *from a material flow and systems integrity perspective.* Material systems, whether ecosystems, hydrologic systems, agricultural systems, or financial systems, have integrity when, among other things, negative feedback loops kick in to modulate positive feedback loops. In a mining economy, positive feedback during boom times is ubiquitous and immediate (return on investment, low consumer prices, political favors), whereas the negatives are delayed and often diffuse (years of safe digging and drilling are followed by mine disasters and oil spills; decades of economic growth fueled by fossil fuel combustion are followed by decades and centuries of respiratory, endocrine, oceanic, and climate disruption). The fossil fuel mining system is thus unstable. It grows rapidly, even exponentially, then collapses.[10] All the wishful thinking about new energy sources doesn't change the fact that the energy density and ease of extraction and transport of fossil fuels cannot be replicated (chapter 2).[11] Although all players lose in the collapse, those with the fewest options and the least decision-making authority (generally the poorest) lose the most (part 2). This takes us to an ethical, and no less realist, perspective on end-of-pipe solutions and the industrial progressive view.

Boom-and-bust mining operations papered with end-of-pipe fixes are inherently inequitable. By taking the system as given and hence perpetuating a system that will ultimately fail, proponents of new sources and end-of-pipe management, cleanup, and technical fixes are, at best, delaying the inevitable. At worst, they are exhibiting ever-increasing complexity, creating new and unnecessary problems (deepwater drilling, stratospheric aerosols for global cooling), and making ever more difficult the transition away from fossil fuel dominance.[12] What's more, it is becoming apparent that the bottom fifth or two of the world's population will bear the brunt of the adapting. In the process, fossil fuel proponents invoke "consumer sovereignty" (we just produce what consumers want), absolving themselves of responsibility for the plethora of harms—personal, societal, and environmental—these fuels and their fixes cause.

All this leads to the uncomfortable conclusion (for fossil fuel proponents) that the only realistic means of stopping fossil fuel emissions is to leave fossil fuels in the ground. The only safe place for fossil fuels is in place, where they lie, where they are solid or liquid (or, for natural gas, geologically well contained already), where their chemistry is mostly of complex chains, not simple molecules like carbon dioxide that find their way out of the tiniest crevices, that lubricate tectonic plates perpetually under stress, that react readily with water to acidify the oceans and float into high places filtering and reflecting sunlight, heating beyond livability the habitats below. Again, this is realism in a biophysical and political sense. This, we argue further in chapter 13, is twenty-first-century realism.

## A Politics of Urgent Transition: Putting Power Front and Center

Global management schemes may derive their legitimacy from their very rationality and scientific soundness (including economic calculations), creating the impression of being realist, but their appeal—to environmentalists and oil companies alike, it seems—derives from a different source: they are essentially apolitical; they don't attempt to rewrite the constitutional rules of the game—namely, that extraction proceeds full speed ahead. It is not a politics of urgency, of fundamental transition. It is a politics of accommodation. Instead of confronting extraction and the power of the fossil fuel complex—independent and national oil companies and their enablers in finance and government—the global managers create their own politics, a tame politics, one that ruffles few feathers but keeps everyone pointed in the same direction. It is a politics that diverts attention from that which the fossil fuel complex knows all too well—the politics of ensuring access and stable prices—which is to say, the politics of total extraction (chapter 4).[13] For that politics, everything else—distributing carbon credits, assigning liabilities for oil spills—is a convenient distraction, a great diversion.[14]

A new politics is needed, one of urgency, of transitioning out of fossil fuels, of confronting extremely powerful actors, of being "realist" across cultures and generations.[15] A normative shift as monumental as any other in human history will define the politics of the end of the fossil fuel era. The project of leaving fossil fuels in the ground, is one approach, one that presumes that the fossil fuel era will come to an end and, with it, the fossil fuel industry—private and state. The consequences, however, will persist for generations, and vast quantities of oil, coal, and natural gas will remain in the ground. The central question is: With great cause-effect time lags between extraction and combustion on the one hand, impacts on the other, will fossil fuel use drop soon enough?

More specifically, what will be, or can be:

1. The exit strategies of the fossil fuel industry (explicit or implicit; intended or inadvertent)?
2. The resistance strategies of those who are dumped on (directly or indirectly)?
3. The adaptation strategies of publics and societies as a whole, from the local to the global, both to declining energy availability and to rising defensive expenditures?

This book explores these three questions. It does so as a futuring exercise (What, given the trends, will happen?) and as a normative exercise (What should happen given a goal such as a peaceful, democratic, just, and ecologically sustainable transition?). Where the preponderance of work is descriptive (What are the trends?) or critical (What is the distribution?) and key actors are reluctant to engage issues of power (concentrated and highly unequal power), we feel this is precisely where a realist research agenda should be located.[16]

So if the "natural" exit and adaptation strategies are not fast enough to avert catastrophic change, the normative question is: How can fossil fuel exit and societal adaptation be accelerated? The positive questions (regarding knowledge of trends and power disparities) and normative (regarding acceleration) combine into a politics of urgent transition.

For such a politics, the current state of affairs can be summed up thus: if fossil fuel use could be presumed net beneficial in its first century, it cannot in its second and third centuries. Current technologies, market demand, and geopolitical strategic imperatives are sufficient to drive the extraction and burning of catastrophic amounts of fossil fuels.[17] Given this and the state of global management schemes to date, it is reasonable to assume that when fossil fuels are extracted, their by-products *will* enter people's bloodstream, water supplies, the oceans, and the atmosphere. Consequently, a normative shift on the order of abolition, industrialization, democratization, international peace, suffrage, and civil rights will be needed for an early exit. Moral entrepreneurs will have to find leverage points in current material systems even if they cannot imagine, let alone offer, a plan for the post–fossil fuel era.[18] That normative shift will define the politics of the end of the fossil fuel era.

If accelerating the end of the fossil fuel era is ultimately a moral question, and we are convinced it is, then the ultimate strategy for bringing the fossil fuel era to an early end may well be delegitimization, a topic we explore in different ways throughout this book and conclude with in chapter 13. For now, suffice it to say that by delegitimization, we do not mean a vilification of the fossil fuel industry. The industry has a century and more of vilification starting with charges against Rockefeller's Standard Oil (the "Octopus") and continuing to today (a former Shell president titled his book *Why We Hate the Oil Companies*). Nor do we mean a repudiation of the industry's antidemocratic, antienvironmental tactics.

Rather, by *delegitimization,* we mean the reconceptualization and revalorization of fossil fuels or, to be precise, humans' relations with fossil fuels. We mean a shift from fossil fuels as a constructive substance to a destructive substance, from necessity to indulgence, even addiction, from a "good" to a "bad," from lifeblood (of modern society) to poison (of a potentially sustainable society). We mean a shift from fossil fuels as that which is normal to that which is abnormal. In other words, fossil fuels will make a *moral transition in parallel to the material transition.* Much as slavery went from universal institution to universal abomination and as tobacco went from medicinal and cool to lethal and disgusting, the delegitimization of fossil fuels will flip the valence of these otherwise wondrous, free-for-the-taking complex hydrocarbons. And rather than pin blame on "big bad oil and coal companies" or, even worse, on "all of us" because we all use fossil fuels, delegitimization simply recognizes that a substance once deemed net beneficial can become net detrimental. All it takes is a bit of evidence (in the case of fossil fuels, a mountain of evidence already exists), some incisive critics, effective communication, and, for the "moral entrepreneurs," a whole lot of persistence and willingness to themselves be vilified.[19] It would start with the simple observation that there are some things humans cannot handle. Their level of understanding, their susceptibility to convenience or power, their inability to organize globally and for the long term all mitigate against having such things as ozone-depleting substances, lead, drift nets, land mines, rhino horns, and, someday perhaps, nuclear weapons and nuclear power plants.

This, then, would be the essential politics of early fossil fuel exit. It would acknowledge that there is little to gain prosecuting agents of economic and political development of a bygone era (which persists to today), of a period in human history that, through the keep-them-in-the-ground lens, looks almost naive (How could they believe this would last and have no ill effect?). Delegitimization says that now, with today's accumulated knowledge and with a complexity of threats that jeopardize human existence, it is time to banish the offending substance and the practices and impacts that go with it. It is time to fess up to the impossibility of marginal improvement: just as more humane shackles didn't address slavery as an institution, nor filters on cigarettes the industrial tobacco culture, more efficient cars and high-technology cleanup technology won't address the system-jeopardizing properties—physical and social—of fossil fuels (box 1.1).[20]

**Box 1.1**
Technology to the Rescue: Or Not

"Energy independence has always been a race between depletion and technologies to produce more and use energy more efficiently. Depletion was winning for decades, and now [with hydraulic fracturing, offshore platforms, and the like] technology is starting to overtake its lead."

—Bill White, former US deputy energy secretary under President Bill Clinton and former mayor of Houston[a]

From Houston to Washington, Caracas to Riyadh, this is the dominant view of the role of technology in energy production, especially high-capital, megaproject production. It is a powerful argument, with history on its side, not to mention a wished-for future with its bright new technologies. The history of oil development shows how each new technology (e.g., carbide drill bits, floating platforms) expanded the range, depth, and extent of production dramatically. With unlimited human ingenuity and virtually unlimited fossil fuels in the ground (between five and eighteen times what has been extracted so far), it will go on for a very, very long time.

The argument, indeed the rhetoric, is compelling. And a lot of people and quite a few countries have gotten fabulously wealthy following such logic.

But like many other utopian dreams, it neglects a few things. First, geologically and historically, the fossil fuel era has been brief, just a hundred years or so (even less from a cultural perspective). The game—the game of endless extraction, combustion and emission—has hardly played itself out. Positive feedback has overwhelmed the negative so far (mostly), but systems theory tells us the checks eventually kick in. Social theory tells us that when inequities intensify and people's dignity is denied, social structures change.

Second, technologies never exist independent of resources. The steam engine required wood and coal, the Internet cheap electricity (still generated mostly from coal). Technologies do not produce energy any more than they produce education, health care, or national security. They do not create energy, only convert it to usable forms. The inexorable decline in energy return on investment (chapter 2) reveals this fact: for all the incentives for maintaining high energy returns, no technology or combination of

**Box 1.1**
(continued)

technologies has yet succeeded in reversing the decline. The energy industry is effectively betting that before parity is reached (a unit invested returns only that unit), such a technology will come along. To the extent political leaders and publics go along with such a global bet, it may be the mother of all social bets.

Third, technologies are constrained by more than energy sources. Financial capital is limited, as any banker or businessperson will attest. Because financial capital flows to its best use—its highest short-term monetary return—at some point drilling for water to grow food and supply drinking water, not drilling for oil and gas, will be the "best use" for scarce capital. Social capital is also limited, especially when it rests on the legitimacy of the governance system and its leaders. To date social capital has flowed to fossil fuels: some 85 percent of energy consumed worldwide is fossil fuels. But this is true of any addiction. Eventually the body and the body politic degrade. Unquestioned faith in the net beneficence of fossil fuels controlled by a powerful few shifts to a faith in powering one's self under one's own control. Social capital shifts to uses that ensure security and equity.

The boosters' unbounded faith that technologies will overcome all is not future grounded, not biophysically, not socially. And while fossil fuels have been highly profitable for some, they are not for the many, not in places like the Niger Delta and Louisiana's Cancer Alley anyway, not for those who must live through the coming years and decades of ecological and health disruption.

**Source**

a. White quoted in Clifford Krauss and Eric Lipton, "U.S. Inches toward Goal of Energy Independence," *New York Times*, March 23, 2012, A21.

Arguably, such delegitimization has begun. The belief in the net beneficence and fair distribution of fossil fuels is declining, along with a decline in trust in major institutions, domestic and international.[21] Every new act of local resistance contributes to a new normative belief one that says that the game is illegitimate, that it benefits a powerful few and their clients while fobbing the costs off on others in space and time. While such local acts of resistance are quickly dismissed as NIMBY by industrial progressives, from the perspective of global threat and globalization from below, they are part of a larger project of delegitimization (box 1.2).[22] We offer several examples of such local (and national and international) acts in part 2.

**Box 1.2**
Silent Flows, Silent Agreements

Nebraska is a deeply conservative state, just about as middle America as one can get. It is, writes psychologist Mary Pipher, "generally not friendly to environmentalists."[a] But when TransCanada, the oil and gas company, announced in 2008 plans to build the Keystone XL pipeline across Nebraska to carry crude oil from Alberta tar sands, a deeply felt resistance emerged. The BP oil spill in the Gulf of Mexico in 2010 and the bursting of a tar sands oil pipeline in Battle Creek, Michigan, that same year (which cost over $720 million and closed parts of the Kalamazoo River for years), prompted questions about reliability and impact on the land and water supplies, especially when the proposed pipeline would cross Nebraska's fertile Sand Hills region and go over the Ogallala Aquifer.

What may have been the clincher, though, was TransCanada's tactics: threats to take property by eminent domain if landowners weren't cooperative and contract requirements prohibiting the landowner from discussing terms publicly. Coercion and gag orders don't sit well with Nebraskan ranchers and farmers, or city dwellers and Native people. But "our government wasn't helping," says Pipher, "so they realized they needed to save themselves."[b] An alliance of disparate groups, rural and urban, left and right, organized, transforming "our feelings of sorrow, fear, anger and helplessness into something stronger and more durable."

At the core of such opposition are two issues. One is the very legitimacy of a contract, of an economic relationship, of an industrial practice that depends on forced removal and silencing. In Canada, for the very same Alberta tar sands oil to flow to coastal ports for export, the federal government has threatened to revoke the charitable status of environmental groups. Public Safety Canada has classified environmentalists as a potential source of domestic terrorism, put on the same list as white supremacists. A major pipeline company, Enbridge, has offered First Nations, which by law must be consulted about pipelines crossing their lands, a 10 percent "take" in one of its projects. Coercion and bribery may be standard practice around the world, but they signify an inability, possibly an increasing inability, to gain broad-based political acceptance.

The other issue is the very meaning of an energy source: Is it BTUs and dollars, economic growth and jobs (however temporary)? Or is it a community's sense of place (chapters 11 and 13), a place where, as Pipher puts it, "ordinary heroes [decide] that money couldn't buy everything and that some things were sacred"?[c]

**Sources**

a. Mary Pipher, "Lighting a Spark on the High Plains," *New York Times*, April 18, 2013, A25. See also Elisabeth Rosenthal, "Canada Seeks Alternatives to Transport Oil Reserves," *New York Times*, June 14, 2012, A14.

b. Pipher, "Lighting a Spark on the High Plains," A25.

c. Ibid.

So we sense that a perceptual shift has begun, one that will probably not be fully intelligible for years or decades to come, especially to the extent a moral shift accompanies it. That shift is from fossil fuels as amenity to fossil fuels as existential threat, from fossil fuels as the essential lifeblood for life as we know it to fossil fuels as a toxic threat to a way of life, to livelihood, to the good life, to life for millions and millions (chapter 12). Those who were first to make the perceptual shift were climate and other environmental scientists and their followers in the academic, policy, and activist communities, always a tiny minority yet at times vocal and influential. The general public began to shift, prompting a backlash, but as far as we can tell, the backlash has not reversed the public's shift. Perhaps the most notable shift has been among those who, irrespective of ideology or religious belief, are making changes in practice. Water masters, forest and range managers, irrigators, farmers, timber companies, the fishing industry, landscapers, weather forecasters: all are changing the way they do things. Even oil companies that explore, prepare sites, drill, and pump with time horizons of decades must factor in climate change (not to mention political change) to protect their investments and deliver a product. To be effective, indeed profitable, they must plan for long-term changes in water availability, temperature, pests, storms, sea level, and so forth. It is simply bad policy and bad business to do otherwise.

These shifts among scientists, policy elites, resource managers, and businesses are arguably pragmatic, not moral. The source of the moral shift is likely to come initially from those who suffer the most from such changes: those who will be driven from their ancestral lands onto marginal lands or into urban slums, those whose water has become unfit to drink, their air unfit to breathe (part 2). As these peoples, not all from the Global South, gain a voice, they may well drive the moral shift. While much of the politics is aimed at oppressive regimes and rapacious corporations, a logical material focal point will be fossil fuels. Many may hope to share in the spoils of fossil fuel exploitation, but that aspiration will be blunted by the broad-scale social learning in recent decades of the true costs locally and nationally of such exploitation. "The hopes of people watching new pipelines built through their communities or seeing the impressive installations of offshore platforms can be palpably felt," says political scientist Terry Lynn Karl and relief official Ian Gary in reference to oil development in Africa. "They believe that oil will bring jobs, food, schools, healthcare, agricultural support, and housing." In reality, write Karl and Gary, "When taken as a group, all 'rich' less developed countries dependent on oil exports have seen the living standards of their populations drop—and drop dramatically." What's more, "countries that depend upon oil exports, over time, are among the most economically troubled, the most authoritarian, and the most conflict-ridden states in the world today."[23]

Not so long ago, people North and South had little reason to believe that oil wealth brought anything but great prosperity. Now the evidence of the "resource curse" (a decrease in social well-being as oil wealth increases) is abundant and spreads globally, and the literature is extensive, albeit still contested (chapter 3).[24] If specific acts of resistance can be labeled NIMBY, the knowledge base for such actions is anything but localist.[25]

Thus, a growing delegitimization of fossil fuels is a realistic scenario—not as a response to the abstract claims of the scientific community but as a response to existential threats perceived by peoples everywhere (some of which are indeed environmentally induced). The fossil fuel boosters can no longer be believed; their promises ring increasingly hollow as their depredations become increasingly visible (chapter 3).

So the changes among previously marginalized peoples may well be a harbinger of things to come—not just parochial protectionism, not gated localism, not NIMBY, but something approximating NOPE (not on Planet Earth) and KIIG (keep it in the ground). People the world over are saying, "Enough." What the environmental scientists and others started, yet cannot finish with their top-down, expert-led, managerialist plans and technological fixes, will be augmented and accelerated by moral commitments. What will they be? In a nutshell, following the scenario of this study (neither doom and gloom nor rejectionist but, rather, transitional), they will start with the observation that business as usual, mining all available fossil fuels and imposing huge reverberating, time-lagged effects would, once their by-products are unavoidably released, render the planet uninhabitable for humans and other creatures. Thus, the physical dimension of an ethic of future fossil fuel use is that although exhumation can continue technically and economically for a long time, life cannot; fossil fuel use must slow dramatically well before technical limits compel it. The social dimension is the understanding that while fossil fuel–dependent societies cannot stop cold, their overarching policy must be to *start stopping now* (chapter 4). And the personal and community dimension is that local action matters for *both the locality and the world's localities* (part 2). Put differently, although there is unlikely to ever be effective global management of a phase-out of fossil fuels, local action may well add up to global effect; it would be ethical for both amenity and existential reasons.

To delegitimize a substance (or a process like exploring and drilling) as opposed to condemning an actor or charging a system or blaming everyone puts the focus on the offending substance or, more specifically, on its use. Fossil fuels are perfectly "natural"; ancient uses were, for all we can tell, compatible with other substances and processes. In a strategy of delegitimization, the burden shifts from the contest of interest groups (e.g., environmentalists versus industrialists) to a contest over the politics of the good life. Industrialist progressives have enacted one vision of the good life. Its efficacy in the twentieth century can be debated (and probably will be for a long time to come), but the politics of delegitimization is about now and the future, including the distant future. What is more, as a normative theorizing exercise, it is about creating the good given the biophysical trends underway, about co-creating the good life without expansionism, extractivism, or displacement (chapters 11 and 12). It is about confronting excess.

So the politics of delegitimization of fossil fuels does not investigate motive or assign blame. It takes as given the threatening nature of fossil fuels when used at excessive levels, a poison at heavy doses, continuously injected into bodies and bodies of water and the atmosphere. So perceived, a politics of delegitimization becomes a politics of creation— let's create an economy, a healthy economy, without fossil fuel dependency. This stands in contrast to the prevailing politics, that of marginal improvement—let's manage the ill effects but don't dare question the injections. The industry will spend itself one way or another. A creative politics of early exit and accelerated end, promoted by a variety of actors including industry itself, aims to soften the industry's unavoidable business decline and society's overall material descent, aiming all along for well-being ascent. This, once again, is the scenario of this study—that of a peaceful, democratic, just, and biophysically sustainable transition.

Finally, delegitimization may seem like a distant prospect. But in some sense it began a half-century ago with, ironically enough, a founder of the Organization of Petroleum Exporting Countries, Juan Pablo Perez Alfonso, Venezuela's minister of mines and hydrocarbons. Toward the end of his life, he took to calling petroleum the "excrement of the devil."[26] Now, in the twenty-first century, it is time to leave the excrement in the ground, well covered, out of sight, untouched by human hands or machines.

## Better Than the Alternatives

The case for delegitimization and accelerating the end of the fossil fuel era rests on two key premises. The first, as noted, is the net detriment presumption. In the first century of the fossil fuel era, the costs of extraction and burning—e.g., miner injuries and deaths, city dwellers' lung disease, land and waters polluted—were well known yet seemingly overwhelmed by the benefits: steel production, increased agricultural yields, railroad and automobile transport, and buildings lighted, heated, and cooled. Now a fuller picture of the present costs of past fossil fuel use—toxic buildup, freshwater drawdown, soil depletion, extreme weather—is coming to light both scientifically and experientially (land managers, for example, see, touch, and work with the physical changes). Projecting into the future leads to a convincing set of scenarios if business as usual continues, all with catastrophic consequences for human and other populations.

At some point, the net benefit presumption falls away. Only in retrospect will we know with confidence when the switch point occurred. Some of us think it has already occurred. Others see it coming. And yet others debate it or, most commonly, ignore it or never give it a thought. What is nearly certain, however, is that when it happens, the calculation will be superfluous. Most important, it is hardly prudent to wait for such a determination.

Thus, the net detriment presumption of this study simply states that global society is at or near the switch point. The ethical position is that it is wise to act now *as if* net benefits are a thing of the past. (Notice that in this formulation, "valuing future generations" is not needed to make the case to act now.) The politics are those of precaution, of acting without conclusive scientific or economic evidence. They are a politics that accepts the incommensurability of life-support risks and economic risks, rejecting specious arguments that a fossil fuels phase-out would "hurt the economy" and that extreme weather, soil, and freshwater can be handled with more economic growth (which, to date, results in more material and energy growth, more emissions, more impacts). They are a politics that confronts the concentrated power of business as usual, power that reinforces narrowly short-term benefits, especially financial, while it disregards the long-term ecological and societal costs.

The second key premise in making the case for ending the fossil fuel era is the presumption that the alternatives—that is, the *alternatives to deliberately keeping fossil fuels in the ground*—are infeasible, unduly risky, or unjust. Those alternatives fall into four categories, a proper critique of which could occupy an entire chapter, if not book. We coeditors take the position that debates around such alternatives will continue for a long time, probably until material circumstances render them moot. A precautionary approach, once again, says it is better to act now. That means it is better to begin investing now in the fossil fuel phase-out and to begin adjusting to having less and less available concentrated energy than to bet that the alternatives to keep it in the ground will pan out.

Put differently, not using a destructive substance can be done now, with certainty; hoping for a cure while continuing to use the destructive substance is foolhardy. It is especially foolhardy when life-support systems are at risk, especially when great cause-and-effect time lags reveal how contemporary decision making—that which has narrow time horizons due in part to the financialization of the economy, in part to the pervasive commercialization of major institutions, from governments to media to academe—is not up to the task. And it is unjust when some populations are harmed far more than others. This position amounts to a no-regrets policy: if an alternative to KIIG emerges as the answer to fossil fuel dependency, then those unused fossil fuels are likely to be even more valuable in the new, "saved" energy regime. The four alternatives to keep-it-in-the-ground are global agreement, renewables, technological solution, and pricing.

## Global Agreement

For some two decades, would-be global managers from the scientific, academic, nongovernmental organization (NGO), and diplomatic communities put great stock in controlling greenhouse gasses through treaty making. Buoyed by successes in other environmental realms, most notably the Montreal Protocol on Ozone-Depleting Substances, they marshaled powerful arguments—scientific, economic, and political (in the rationalist sense). For all the power of those rationalist arguments, their proponents did not reckon with the power behind fossil fuel dominance: the power of concentrated economic and political influence that comes from concentrated energy, the power of world domination that comes from a fossil-fueled military, the power of industrial production, and the very tempting idea of endlessly growing material abundance. In short, they did not account for a culture of abundance and, now, excess. Cultural transformation was not in the data sets or the computer models; they were the words that could not be spoken by diplomats. So the fossil fuel culture (chapter 3) prevailed just as it did in the apparent precedents (including the Montreal Protocol), its legitimacy unchallenged.

Maybe a global treaty will be written one day. Diplomatic persistence does pay off sometimes. And maybe, after its every provision is gamed for short-term strategic advantage and the accessible sources are exhausted, the fossil fuel complex will give up and the culture of fossil fuel dominance will fade away. A more likely scenario is that dramatic biophysical events will compel local and regional adaptations (such as are occurring already in drought- and storm-prone areas) that lead to national action and, in turn, international agreement. In the process, fossil fuel legitimacy will be slowly upended. Our project is about upending now; it assumes humans, other species, and life-support systems don't have that kind of time. It is about ending the legitimacy of an increasingly illegitimate substance, about weaning human populations off a substance they cannot handle, about undermining the power of an extremely powerful transnational coalition of industrial interests that to date have received a pass from the would-be global managers.

## Renewables

The promotion of renewables as a means of pushing out fossil fuels has relied on at least three assumptions. One is swap in, swap out: if enough windmills are put up, a coal-fired plant comes down. Without an explicit swap-out mechanism and a declining ceiling on fossil fuel use, energy production diversifies, but fossil fuel use does not necessarily diminish. Just because renewables can replace fossil fuels does not mean they automatically will.

The second assumption is that renewables are not themselves a product of fossil fuel dependency, constructed as they are with fossil fuels and reliant on fossil fuel inputs to install and maintain (chapter 2). A more realistic assumption is that renewables will indeed replace fossil fuels (because there is no alternative), but not at a level of energy availability and reliability even close to that enjoyed by modern societies for the past century or so. What's more, because the fossil fuel era was a function of vast, easy-to-get energy and cheapness, to presume that renewables will somehow underprice fossil fuels in the near future is to presume the cheapness of fossil fuels itself can be replaced. But the fossil fuel era is a one-time event in human history, never to be repeated, its fuels coming from the concentrated sunlight of accumulated biomass over hundreds of millions of years in quantities that no amount of solar panels and windmills can replicate.

Finally, the "push-out" assumption has its own implicit politics or, rather, lack of politics. It assumes that a technical innovation—efficient converters at a good price—can beat a form of economic and political power humans had never previously known and, as argued in chapter 3, still don't really know or appreciate. Put differently, to push out a powerful industrial network, one intermeshed with nearly all aspects of modern life, from food production to warfare, requires its own source of power, political power; no technology or pricing scheme can muster that (box 1.3).

**Box 1.3**
Fossil Fuel Replacement: An Impossibility Theorem

- Fossil fuels provide 88 percent of the world's energy.[a]
- Fossil fuel infrastructure occupies an area the size of Belgium.
- If biofuels were to replace fossil fuels, the required area would be roughly the size of the United States and India.[b]
- $38 trillion in oil and gas infrastructure is needed by 2035 to meet industry and agency projections of increased energy demands.[c]
- 7.3 to 10 calories of energy input are used to produce 1 calorie of food energy.[d]
- Direct fuel subsidies to agriculture in United States: $2.4 billion.[f]
- Proven fossil fuel reserves, owned by private companies, state companies, and governments, exceed the planet's remaining carbon budget (to keep within a 2 degree C temperature increase) by a factor of *five*.[g]

Fossil fuels, these numbers suggest, not only have impacts; their replacement is virtually impossible. A plausible scenario, the one that informs this study, is that the high-consuming Global North will be consuming drastically less total energy. The sooner those reductions are made, the better off life-support systems will be.

**Sources**

a. Vaclav Smil, "Global Energy: The Latest Infatuations," *American Scientist* 99 (2011): 217.

b. Vaclav Smil, *Energy Transitions: History, Requirements, Prospects* (Santa Barbara, CA: Praeger, 2011), 117.

c. International Energy Agency, *World Energy Outlook 2011* (Paris: OECD/IEA, 2011), 69.d. Roni A. Neff, Cindy L. Parker, Frederick L. Kirschenmann, Jennifer Tinch, and Robert S. Lawrence, "Peak Oil, Food Systems, and Public Health," *American Journal of Public Health* 101 (2011): 1589.

**Box 1.3**
(continued)

e. Ibid.

f. Carbon Tracker Initiative, *Unburnable Carbon: Are the World's Financial Markets Carrying a Carbon Bubble?* (March 2012), http://www.carbontracker.org/carbonbubble.

g. Paul R. Epstein and Jesse Selber, eds., "Oil: A Life Cycle Analysis of its Health and Environmental Impacts" (Boston: Center for Health and the Global Environment, Harvard Medical School, 2002), 4.

## Technological Solutions

Two technologies stand out at present; others are sure to emerge.

The first is geoengineering—aerosols in space or iron filings in the ocean, for example. We dismiss it quickly, not because it is not a serious issue or that some actor, a rogue state or terrorist group, for instance, couldn't attempt such planetary manipulation. Rather, we dismiss it because it is not a policy, not a long-term plan, only a desperate stop-gap measure. It largely ignores the history of engineering complex systems like rivers and local weather and downplays the likelihood of unintended consequences. It fundamentally violates two core principles of scientific research: experiments should take place in highly controlled settings—a laboratory—not in households or communities, let alone over the entire planet; and, if individuals are experimented on, it is with informed consent, what cannot happen on the planetary level. If such an intervention was subject to public approval through free and open debate, we presume it would be blocked. What's more, if the choice were geoengineering or phasing out fossil fuels, we presume an informed public would eventually opt for phase-out.

The second potential technological savior is carbon capture and storage (CCS), which seems to be the favorite of the fossil fuel industry and, by extension, the financial and manufacturing industries as well. Because the oil and gas industry has for decades injected carbon dioxide into the ground (to push up stubborn deposits), it assumes the technology exists to put offending greenhouse gases, most notably carbon dioxide, back where the parent hydrocarbons came from. In so doing, proponents largely overlook three things: scale, incentives, and permanence.

The scale of enhanced recovery by using carbon dioxide injections is nowhere near what would be needed for a significant capture of carbon dioxide, enough, say, to stabilize climate. Energy analyst Vaclav Smil made some preliminary calculations regarding requisite infrastructure. He concluded "that in order to sequester just a fifth of current $CO_2$ emissions we would have to create an entirely new worldwide absorption-gathering-compression-transportation-storage industry whose annual throughput would have to be about 70 percent larger than the annual volume now handled by the global crude oil industry."[27]

It further assumes that because the oil and gas industry built out the infrastructure for its business over the past century, the infrastructure for underground storage can be similarly built out—by someone, at some attractive price. But CCS is not a business. It is a cleanup operation. Governments—that is, taxpayers—would have to foot the bill as they do for waste problems with a commons feature (e.g., solid and biological wastes). The incentives are to do just the opposite, delay, which is what has happened for at least a decade after the initial boosterism of CCS. Alternatively, governments would have to reach a global agreement to price carbon dioxide and create the business. This takes us back to the first issue, global agreement.

Finally, permanence. CCS is highly doubtful not because of the injection technology, but because no one seems to know, or to have asked, if the offending gases will actually stay in the ground. One German study began its life cycle analysis of CCS by assuming, "due to the lack of data," "no underground storage leakage."[28] Another German study, Tom Morton notes in chapter 9, concluded that after a number of CCS pilot projects have been abandoned or postponed, CCS technology will not play a role in the German energy sector for the next twenty years. The technology has proved "too technologically demanding and too expensive to implement."[29] In its 2005 report, the Intergovernmental Panel on Climate Change (IPCC) states that "if continuous leakage of $CO_2$ occurs it could at least in part offset the benefits of CCS for mitigating climate change."[30] Then in its 2013 report it states, "The effects of CDR [carbon dioxide removal] methods are in general slow on account of long time scales required by relevant carbon cycle processes, and thus may not present an option for rapid mitigation of climate change during the next century. ... The level of scientific knowledge upon which CDR methods can be evaluated is low, and uncertainties are very large."[31]

To be a solution, even a significant piece of a solution, let alone a reason to keep extracting fossil fuels, the captured carbon dioxide would have to stay in the ground. Oil, coal, and gas can stay in the ground with great certainty. But these studies suggest that there is great uncertainty as to whether carbon dioxide can actually be kept underground;

proponents, including governments and industry, just assume so. The oil and gas industry has had no reason to know; its carbon dioxide injections have been for one purpose, increasing recovery rates, not cleaning up combustion by-products. We know of no independent, long-term analysis of the permanence of gaseous sequestration in the earth's continually shifting crust. We thus remain skeptical, as any scientist should, of a technology and a policy that rests on an untested, possibly untestable geologic assumption.

In short, CCS is the ultimate end-of-pipe solution, one that presumes all activities up the value chain—extraction, refining, consuming—are legitimate and that cleanup is a relatively minor matter. We presume, by contrast, that the cleanup of fossil fuels' detritus—its carbon dioxide, methane, persistent organic pollutants—would be a gargantuan task, an energy-consuming one no less, likely dwarfing all of history's previous public projects combined, from the space race, to dam and road building, to public sanitation. We presume that as weather extremes intensify and sea levels rise and freshwater supplies disappear, publics everywhere will have better uses for scarce capital, financial and social.

**Pricing**

On many occasions, we editors have heard statements of the following sort, often from economists: "We know what the answer [to climate change] is—put a price on carbon." Putting aside the smugness of such statements and the supreme reductionism (it is all about carbon and price), the operative term is *put*. That is, this solution presumes that someone will convince some set of influential actors to act exactly opposite to what has heretofore been in their best interests—namely, promoting economic growth, industrial expansion, cheap consumer products. To say that all it takes is good pricing is a superb instance of technocratic conceit: erase the politics—the influencing, shaping, organizing, negotiating, strategizing, in short, what Hannah Arendt argued is part and parcel of the human condition—and the technical fix will work its magic.[32]

To be clear, we editors wish carbon would be priced, and significantly. But it is too risky, not to mention politically naive, to presume that politics will melt away once the fix is offered, its airtight rationality explained, and somehow significant publics come to understand it and support it. Instead, what we get is another politics—business as usual—and the perpetuation of the fossil fuel order. A carbon tax may eventually get implemented globally (the only scale at which it would be effective), but we presume that only after some essential preconditions are established. Among them are the delegitimization of fossil fuels, the articulation of a twenty-first-century good life without fossil fuels, and overt challenges of the exploitative, the extractive, the excessive.

In sum, constructing a politics of ending the fossil fuel era with such notions as delegitimization, making fossil fuels special (chapter 3), resisting extractive pressures (part 2), and seeking a twenty-first-century "good life" (chapters 11 and 13) may be the least efficient, most cumbersome, unlikeliest approach to the global environmental problematic. It is just, we continue to argue in the coming pages, better than the alternatives.

## Envisioning a Post–Fossil Fuel Era

To imagine deliberately leaving fossil fuels in the ground, much less the end of the fossil fuel era, is difficult. No matter how much environmental science we absorb, how much geologic and ecological perspective we attain, how much ethical commitment we muster, it is hard to escape industrial progressivism. It just seems as if all this will continue, with adjustments, an efficiency here, some greening up there. But there is just too much knowledge piled up to say otherwise, and not just scientific knowledge, but political and strategic knowledge. Environmental scientist Vaclav Smil, even-tempered and apolitical in his prognostications, sums up his analysis of the coming energy transition:

A world without fossil fuel combustion may be highly desirable, and eventually it will be inevitable, and our collective determination could accelerate its arrival—but making it a reality will demand great determination, extraordinary commitment, substantial expense, and uncommon patience as the process of a new epochal energy transition unfolds across decades.[33]

If Smil's is a reasonable perspective on the transition—energetic and technological—grounded as it is in the history of previous energy transitions, then this book's contribution is to play out the political dimension. It does so in part by making possible the imagining of a post–fossil fuel order. And it does so in part by engaging the normative—that is, the "shoulds" of policy analysis and policy prescription.

Envisioning a post–fossil fuel order can lead to the retrospective question: How is it that fossil fuels have become like the water in which we swim, ubiquitous yet almost unnoticed? But the prospective question is different: Under what conditions will the fossil fuel era, that period little more than a century long of fossil fuel dominance, come to an end? Unlike conventional approaches to "energy transition," our focus is not on new sources and new technologies. Nor is it on the impacts. Rather, our focus, once again, is on the politics, broadly construed where power is central, of deliberately keeping fossil fuels in the ground. What's more, in this prospective mode, we ask the normative question: What can be done to accelerate the end of the fossil fuel era given the huge cause-and-effect time lags, the short-term concentration of benefits among the few and the long-term dispersion of costs across the many, the great difficulties in assigning responsibility, and the increasing likelihood of catastrophic outcomes if industrial societies continue on this path?

This study, then, is an exercise in reframing the problem, documenting an emerging social movement, and positing alternative paths. It is, in short, a normative theorizing project.[34] At the same time, it does not presume to create a theory of fossil fuel phase-out. Such a theory, should it be inductively developed, will have to await a richer and more definitive set of cases where successes and failures are evident and key variables allow comparative purchase.[35] Such a theory, like so many others in the social sciences, would indeed be retrospective and descriptive, not prospective and normative. That is, it would only say what is and how it is, not what is likely to come or what should come given a set of objectives. This study is precisely a forward-looking, anticipatory exercise, one that presumes a societal objective of peaceful, democratic, just, and biophysically sustainable transition. If there is theory building here, it is hypothesis generating, not hypothesis testing. Because few in the relevant disciplines have asked questions of transition, of exiting fossil fuels, of pursuing the good life without fossil fuels, there is no theory to test.

More important, however, this study asks what decision makers should do given a set of conditions. Among those conditions are the end of cheap energy, the rise in defensive environmental expenditures (e.g., to adapt to freshwater decline and climate disruption), and concomitant social changes—changes from the nature of the corporation to the structure of the economy, changes in the meaning of nation and community, of progress and the good life. The "should" in normative theorizing derives not, in the first instance, from the authors' personal ideologies but from a common conviction that these changes are profound and, on the whole, unprecedented. What is more, we authors presume we are not alone in wishing the transition to be peaceful, democratic, just, and ecologically sustainable. On top of all this, we put front and center these conditions and these normative positions. (For more on methodological commitment, as well as the role of case studies, see the introduction to part 2.)

So we conclude this introduction by offering three observations to facilitate the determination and commitment Smil deems necessary. This is an envisioning exercise, drawing in particular on the futuring work of Donella Meadows and designed to help articulate a politics of transition.[36]

First, the fossil fuel era, which began worldwide in the 1890s when fossil fuels surpassed wood as the dominant energy source, is only about six generations old. Most individuals experience five generations in their lifetime: one's parents and grandparents, one's children and grandchildren, and oneself. Many of us alive now have personally known people who were alive before the fossil fuel era. It was not that long ago. And this is only from an energetic perspective. From a cultural perspective, one that includes societal perceptions, values, and sources of power, it would be even shorter (chapter 3).

So the fossil fuel era has not been that long, it is not that permanent, and its continuation is not inevitable. In fact, the purported inevitability of continued fossil fuel use may be a defining characteristic of twentieth-century energy politics, but its declining use may well define twenty-first century politics.

Second, the initial stage of an energy source's use is one where benefits are highlighted and costs unknown or shaded (displaced in time and space; chapter 3). So we can expect that fossil fuels have the same quality, only on a far grander scale than anything before. Coal's depredations—from miners' bodies to the lungs of those who have asthma, from decimated communities in Appalachia to Mongolia—are well known. Coal's early exit is virtually a no-brainer. No wonder their anti–climate change activism has been so vehement.[37] Oil, arguably the most consequential energy source of all time, is widely deemed essential (and thus the rush for alternative liquid fuels), but it too will eventually fade out. The fossil fuel era will likely draw to a close well before conventional analysis and decision making would indicate. And just as global fossil fuel production will decline as all wells and oil fields do, the industry will decline too (chapter 12). Just because no one in the industry or anyone dependent on it (virtually everyone) wants to talk about it doesn't make it otherwise. To envision a post–fossil fuel world is to accept that fossil fuel production and the fossil fuel industries *will* decline.

Third, implicit in the envisioning of a post–fossil fuel era is a theory of social change captured by this aphorism, one relevant, we assert, not just to political revolutions but also to broad-scale, fundamental social change—that is, transition: before the revolution, everyone says it's impossible; after the revolution, everyone says it was inevitable. The apparent impossibility of ending the fossil fuel era is based on a belief that might go something like this:

*Fossil fuels are central to peoples' lives the world over. They comprise 85 percent of all energy consumed and 95 percent of all industrial output. Even if heating and transportation could be dramatically curtailed, food consumption cannot: modern, vertically integrated, globally distributed food systems depend on fossil fuel–based fertilizers, pesticides, and transport. What is more, the fossil fuel industry, with representation at the highest levels of government and allies in finance, auto manufacture, and petrochemicals, is so powerful, so tightly articulated with the seats of power, that no amount of arguing and pleading will effect fundamental change.*

*The fossil fuel industry is king of the mountain, and it shows no signs of stepping down. Once the mountain is completely mined, the fossil fuel industry will move to the next one. The mighty have always prevailed, and they always will. Better to direct one's efforts at protecting what's left of that mountain, distributing the proceeds of its removal to those in need, and finding a place for the tailings. Or to finding new mountains.*

This, to our read, is the dominant thinking. But if fossil fuels, and the power that goes with them, can't last, then this thinking can't last either. It will change. The moral imperative, given mounting costs to vulnerable peoples in the short term (e.g., downwind, downstream populations and those dependent on glacier-fed rivers) and to everyone in the long term, as well as the great cause-and-effect time lags, is to start changing that thinking now. An essential step is to imagine the shift.

One way to imagine the shift is to heed the voices of Indigenous and other traditional peoples, many of whom have resisted high-energy ways of life and warned of the consequences, often couching their stories in terms of unleashing evil or chaos from deep underground (chapter 11). Twenty-first-century efforts to live well while living lightly on the earth, with considerably less fossil energy, will need traditional knowledge and place-based values drawn from long experiences of dependence on local foods, fiber, materials, and sources of energy. All of us had ancestors who were Indigenous to some place and ecosystems, and we will all have Indigenous ancestors if we are to survive long into the future. Like the fossil fuel era itself, the globalization era, with its dependence on far-flung resources transported globally, will also draw to a close. Finding our way back home will be part of the essential work of building good lives informed by an understanding of how good lives and good minds depend ultimately on healthy land, air, water, and energy.

Another way to do the imagining is, ironically, by examining the fossil fuel industry itself (chapter 12). Preliminary evidence suggests that serious people in the oil and gas industries, along with the automobile and petrochemical industries, know this game cannot go on.[38] "Energy executives know that the existing supply capacity from traditional sources is about tapped out," writes former Shell Oil Company president John Hofmeister.[39] They know the easy stuff is effectively gone.[40] Now they are factoring into their financial calculations the effects of a changing climate where the very tundra their trucks depend on is melting, the rigs they thought were secure are breaking down.[41] What they say publicly is different, of course. Their jobs, their way of life, their personal and professional identity, their future: all are on the line. They seem to pray that a miracle technology will come along to keep the game going a little while longer.[42] This difficulty is perfectly understandable. And yet people in equally entrenched positions (witness, slavery and smoking; see box 1.4) have made such shifts.[43]

**Box 1.4**
From Smoking to Fossil Fuels

A smoking analogy to fossil fuels may seem a stretch. But when antitobacco advocates began their campaign a few decades ago, their analogy to hard drugs seemed a stretch. Once thought to be healthful and cool, smoking, we now know, is deadly and disgusting. Nevertheless, there is still a long ways to go in making smoking abnormal.

Two medical doctors with the nonprofit health advocacy group National Jewish Health continue to work on doing just that. They wrote a public service advertisement that appeared in major American newspapers. Here, as an exercise in envisioning a post–fossil fuel world, is an edited version of their commentary with fossil fuel inserts in brackets; the doctors' language is verbatim:

All Americans [humans] … smokers and nonsmokers [high consumers and low consumers] alike, pay the price for smoking [using fossil fuels]. …

Despite … staggering costs and health effects, and despite the fact that nicotine [fossil fuel] addiction is hard to overcome without help, the budgets for proven smoking-cessation [conservation and alternative energy sources] have been cut in recent years. At the same time, the astoundingly profitable tobacco [fossil fuel] industry continues to devote massive

**Box 1.4**
(continued)

resources to maintaining its profits by promoting tobacco [fossil fuel] use. … The facts show we simply cannot afford tobacco [fossil fuel use] any more. …

Innovative and bolder steps toward the eradication of tobacco [fossil fuel] use are needed by governmental agencies. …

Can America [the world] develop a culture in which smoking [fossil fuel extraction and consumption] is not acceptable? We are moving in that direction, but with high hurdles and formidable opponents. Achieving this goal is not easy, but we cannot afford the alternative.

**Sources**

Allan M. Brandt, *The Cigarette Century: The Rise, Fall and Deadly Persistence of the Product that Defined America* (New York: Basic Books, 2007); David Tinkelman and Michael Salem, "We Cannot Afford Tobacco," paid advertisement, *New York Times*, [2012].

In short, a deliberate policy, state led or not, of leaving fossil fuels in the ground is at once preposterous and perfectly sensible. Stranger things have happened. How it would happen, at what rate, with what local effects, is still anyone's guess. That fossil fuels will be in the ground and stay there when the fossil fuel era ends is beyond doubt.[44] The question is whether enough will stay to relieve dumped-on populations, reverse water and land degradation, stabilize climate, and avert social calamity. The question we raise in this book is whether a politics of fossil fuel delegitimation can overcome the inertia of impact denial and fossil fuel continuation.

To summarize, the argument for KIIG rests on three pillars, each biophysically grounded. First is the premise that net beneficence can no longer be presumed. Looking back a century or more to the heyday of fossil fuels when coal fueled industrialization and oil-powered high-speed transport and military conquest, the benefits were many and huge, even transformational—steel, lighting, residential heating, shipping, troop transport, and so much more. Although there were always downsides—mine deaths, London "fogs," oil-sullied beaches—the upside was overwhelming. The mere fact that, to our knowledge, the question of net benefits arose only recently is evidence enough that as a normative question—a question of what has been culturally normal regarding fossil fuel use—fossil fuels have indeed been presumed net beneficial. As further evidence, though, consider the outcry from industry that routinely accompanies proposed restraints on fossil fuel consumption, perhaps especially in the United States, whether those be to mitigate pollution or restrict drilling or even remove taxpayer subsidies of the fossil fuel industry and exempt the industry from environmental regulations (see table 3.1 in chapter 3). Elsewhere the vicious response to rising fuel prices, often regardless of the reason (e.g., the removal of government price controls), is a politician's worst nightmare (along with price rises for food and water).

In short, for at least a century, fossil fuels have been at least implicitly presumed to be net beneficial. Now, looking forward, a plausible scenario for continued fossil fuel use is widespread societal collapse and conflict as island nations disappear, major coastal cities fail to cope with repeated flooding, and inland communities are weakened by extreme heat, winds, and drought. If and when that scenario plays out (and more and more scientists seem to think it already is), at some point net beneficence can no longer be presumed. In fact, only the few who can buy their way out could possibly harbor such an idea. What position, we ask, is preposterous, then? And at what point in this historical progression from presumed net beneficence to obvious net detriment does preposterousness shift? We coeditors take the position that societies worldwide are at or near the shift. The transition is imminent. Prudence requires the rapid elimination of fossil fuel dominance and dependency (not to be confused with eliminating all use of fossil fuels). And if our argument is correct, that will not happen with better prices, improved technologies, and emissions management. It's time to start stopping, to accelerate the end of the fossil fuel era.

### Notes

1. Vaclav Smil, *Energy Transitions: History, Requirements, Prospects* (Santa Barbara, CA: Praeger, 2010), especially chap. 2.

2. For discussion of the hold a notion of progress has on modern life, see Christopher Lasch, *The True and Only Heaven: Progress and Its Critics* (New York: Norton, 1991); Michael Greer, "Progress vs. Apocalypse: The Stories We Tell Ourselves," in *The Energy Reader: Overdevelopment and the Delusion of Endless Growth*, eds. Tom Butler, Daniel Lerch, and George Wuerthner (Sausalito, CA: Foundation for Deep Ecology, 2012), 95–101.

For a popular and apparently influential expression of the industrial progressive view, see any number of articles on energy and the environment by *New York Times* columnist Thomas Friedman. A February 26, 2012, piece is illustrative: "When [ethanol] is combined with improved vehicle fuel economy ... it will inevitably drive down demand for gasoline and create more surplus crude [for the United States] to export. ... All this is good news, but it will come true at scale only if these oil and gas resources can be extracted in an environmentally sustainable manner."

3. A director of the American Petroleum Institute represented this "idealist, not realist" perspective well regarding undue concerns about offshore drilling, this after the 2010 BP spill in the Gulf of Mexico: "Halting offshore development must be evaluated with a *realistic* understanding of the energy need to keep our economy moving. ... [It] won't reduce our need for oil and natural gas; it will only outsource it" [emphasis added]. Letter to the editor, May 29, 2010, *New York Times*, A18.

4. Smil, *Energy Transitions*, 63, 108.

5. A. R. Brandt and A. E. Farrell, "Scraping the Bottom of the Barrel: Greenhouse Gas Emission Consequences of a Transition to Low-Quality and Synthetic Petroleum Resources," *Climate Change* 84 (2007): 241–263; C. Hall, P. Tharakan, J. Hallock, C. Cleveland, and M. Jefferson, "Hydrocarbons and the Evolution of Human Culture," *Nature* 426 (2003): 318–322; International Energy Agency, *World Energy Outlook 2010* (Paris: OECD/IEA, 2010); US Joint Forces Command, *Joint Operating Environment (JOE) Report* (2010); Robert L. Hirsch, "The Inevitable Peaking of World Oil Production," *Atlantic Council Bulletin* 16, no. 3 (2005): 1–9.

6. John Hofmeister, *Why We Hate the Oil Companies: Straight Talk from an Energy Insider* (New York: Palgrave Macmillan, 2010), 48.

7. Smil, *Energy Transitions*, 148–149.

8. Mathis Wackernagel, Niels B. Schulz, Diana Deumling, Alejandro Callejas Linares, Martin Jenkins, Valerie Kapos, Chad Monfreda, et al., "Tracking the Ecological Overshoot of the Human Economy," *Proceedings of the National Academy of Sciences* 99 (2002): 9266–9271.

9. For a concise and insightful critique of the "scientific separatism" and "technological determinism" that underlie this theory of social change, see Sheila Jasanoff, "The Essential Parallel between Science and Democracy," *D.C. Science*, February 17, 2009. Also see Bruno Latour, *Politics of Nature: How to Bring the Sciences into Democracy*, trans. Catherine Porter (Cambridge, MA: Harvard University Press, 2004).

10. M. King Hubbert, "Exponential Growth as a Transient Phenomenon in Human History," in *Valuing the Earth: Economics, Ecology, Ethics*, ed. Herman Daly and Kenneth Townsend (Cambridge, MA: MIT Press, 1996), 113–126; R. D. Kerr, "World Oil Crunch Looming?" *Science* 322 (2008): 21; Center for German Army Transformation, Group for Future Studies, "Implications of Resource Scarcity on (National) Security," unofficial translation, July 2010, http://www.theoildrum.com/node/6912.

11. Hall et al., "Hydrocarbons and the Evolution of Human Culture" 2003; Smil, *Energy Transitions*.

12. On increasing complexity and diminishing marginal returns on social investments in ancient civilizations, see Joseph A. Tainter, *The Collapse of Complex Societies* (Cambridge: Cambridge University Press, 1988).

13. On the ethics of total extraction, see Thomas Princen, "A Sustainability Ethic," in *Handbook of Global Environmental Politics*, ed. Peter Dauvergne (Cheltenham, UK: Edward Elgar, 2012), 466–479, and chapter 4, this volume.

14. Certainly there are critical analyses of these global management politics, but little, to our read, that takes the next step, that is, the prescriptive step, the normative step, precisely the purpose of this book. See, for instance, Pratap Chatterjee and Matthias Finger, *The Earth Brokers: Power, Politics and World Development* (London: Routledge, 1994).

15. To be clear, the "extremely powerful actors" are not just the oil companies and petro states. They are that larger edifice with a revolving door in the middle, the one where the players are extractors and financiers one moment, rule makers the next, and then back again.

16. For readers still resistant to the idea of futuring and normative theorizing, consider that this is precisely what the foreign policy, security, and business communities, both academic and practitioner, routinely do.

17. On the inertia of current patterns of production and consumption, see Smil, *Energy Transitions*.

18. On moral entrepreneurs, see Ethan A. Nadelmann, "Global Prohibition Regimes: The Evolution of Norms in International Society," *International Organization* 44 (1990): 479–526.

19. Ibid.

20. For compelling evidence of the inability of global economic actors to actually reduce total impact at the same time they claim "sustainability," see Peter Dauvergne and Jane Lister, *Eco-Business: A Big-Brand Takeover of Sustainability* (Cambridge, MA: MIT Press, 2013).

21. Business journalist Eduardo Porter, after reviewing a range of public opinion surveys in the United States, concludes that "Americans are losing trust in a broad range of institutions, including Congress, the presidency, public schools, labor unions and the church." "The Spreading Scourge of Corporate Corruption," *New York Times*, July 11, 2012, B1, B5. Internationally, others find declining trust in most Western countries. David Brooks, "The Segmentation Century," *New York Times*, June 1, 2012.

22. For evidence of this rising voice, see Rob Nixon, *Slow Violence and the Environmentalism of the Poor* (Cambridge, MA: Harvard University Press, 2011); Jerry Mander and Victoria Tauli-Corpuz, eds., *Paradigm Wars: Indigenous Peoples' Resistance to Globalization* (San Francisco: Sierra Club Books, 2006); and chapter 11, this volume.

23. Terry Lynn Karl and Ian Gary, "Bottom of the Barrel: Africa's Oil Boom and the Poor" (Baltimore, MD: Catholic Relief Services, 2003); all quotes on p. 18.

24. Mahmoud A. El-Gamal and Amy Myers Jaffe, *Oil, Dollars, Debt, and Crises: The Global Curse of Black Gold* (Cambridge: Cambridge University Press, 2010); M. Humphreys, J. Sachs, and J. Stiglitz, eds., *Escaping the Resource Curse* (New York: Columbia University Press, 2007); Terry Lynn Karl, *The Paradox of Plenty: Oil Booms and Petro-States* (Berkeley: University of California Press, 1997); Karl and Gary, "Bottom of the Barrel"; Michael Klare, *Resource Wars: The New Landscape of Global Conflict* (New York: Holt, 2002); Michael L. Ross, *The Oil Curse: How Petroleum Wealth Shapes the Development of Nations* (Princeton, NJ: Princeton University Press, 2012).

25 On the difference between localism (about geographically limited scales such as that of cities and communities) and localization (about a process of social change where attention is directed away from the abstract and global to the concrete and local), see Raymond De Young and Thomas Princen, *The Localization Reader: Adaptations for the Coming Downshift* (Cambridge, MA: MIT Press, 2012).

26. Daniel Yergin, *The Prize: The Epic Quest for Oil, Money, and Power* (New York: Simon and Schuster, 1991), 525.

27. Vaclav Smil, "Global Energy: The Latest Infatuations," *American Scientist* 99 (2011): 219.

28. Peter Viebahn, Joachim Nitsch, Manfred Fischedick, Andrea Esken, Dietmar Schüwer, Nikolaus Supersberger, Ulrich Zuberbühler, et al., "Comparison of Carbon Capture and Storage with Renewable Energy Technologies Regarding Structural, Economic, and ecological aspects in Germany," *International Journal of Greenhouse Gas Control* (2007), doi:10.1016/S1750-5836(07)00024-2.

29. See DIW Berlin, CCS Technologie ist für die Energiewende gestorben, February 8, 2012, http://www.diw.de/de/diw_01.c.392660.de/themen_nachrichten/ccs_technologie_ist_fuer_die_energiewende_gestorben.html.

30. Intergovernmental Panel on Climate Change, 2005. In B. Metz, O. Davidson, H. C. de Conninck, M. Loos, and L. A. Meyer, eds., *IPCC Special Report on Carbon Dioxide Capture and Storage. Prepared by Working Group III of the IPCC,* Cambridge University Press, Cambridge, United Kingdom and New York, NY, USA. Cited in Viebahn et al., 2007, p. 6.

31. Fifth Assessment Report of the Intergovernmental Panel on Climate Change, 2013, in Thomas F. Stocker, Dahe Qin, Gian-Kasper Plattner, Melinda M. B. Tignor, Simon K. Allen, Judith Boschung, Alexander Nauels et al., eds., *Climate Change 2013: The Physical Science Basis* (20013), 6, https://www.ipcc.ch/report/ar5/wg1.

32. Hannah Arendt, *The Human Condition* (Chicago: University of Chicago Press, 1959); Leslie Paul Thiele, *Indra's Net and the Midas Touch: Living Sustainably in a Connected World* (Cambridge, MA: MIT Press, 2011).

33. Smil, *Energy Transitions*, 142.

34. Paul Wapner, "The Resurgence and Metamorphosis of Normative International Relations: Principled Commitment and Scholarship in a New Millennium," in *Principled World Politics: The Challenge of Normative International Relations,*

ed. Paul Wapner and Lester Edwin J. Ruiz (Lanham, MD: Rowman & Littlefield, 2000), 4–5.

35. Paul F. Steinberg and Stacy D. VanDeveer, eds., *Comparative Environmental Politics: Theory, Practice, and Prospects* (Cambridge, MA: MIT Press, 2012).

36. Donella H. Meadows, "Envisioning a Sustainable World" (paper for the Third Biennial Meeting of the International Society for Ecological Economics, October 24–28, 1994, San Jose, Costa Rica). For histories of some of the early futurists or, scenario builders, including Meadows, the Shell Scenarios Group, the Hudson Institute, and the Rand Corporation, see Art Kleiner, *The Age of Heretics: Heroes, Outlaws and the Forerunners of Corporate Change* (New York: Currency Doubleday, 1996).

37. Barbara Freese, *Coal: A Human History* (New York: Penguin, 2003).

38. This evidence comes in part from informal interviews with high-level officials in the automobile and oil industries, as well as perusal of the mainstream press. See also George Mobus, "Industry Leaders Seem to be Showing More Openness to Energy Descent Issue," accessed May 4, 2010, http://www.the oildrum.com/node/6419; reports of the Shell scenarios groups, http://www.shell.com/scenarios.

39. Hofmeister, *Why We Hate the Oil Companies*, 168. To be sure, Hofmeister goes on to argue that in the United States, we have "more energy than we'll ever need" (211).

40. Hirsch, "The Inevitable Peaking of World Oil Production"; Hubbert, "Exponential Growth as a Transient Phenomenon in Human History; Kerr, "World Oil Crunch Looming?"; Center for German Army Transformation, "Implications of Resource Scarcity on (National) Security."

41. Steve Coll, *Private Empire: ExxonMobil and American Power* (New York: Penguin Press, 2012).

42. Hofmeister, *Why We Hate the Oil Companies,* may be the exception in his willingness to confront declining easy energy. But conversations with high-level officials in two of the world's largest private oil companies and readings elsewhere suggest he is not alone, even at the top of major fossil fuel businesses.

43. Regarding slavery, see Adam Hochschild, *Bury the Chains: Prophets and Rebels in the Fight to Free an Empire's Slaves* (Boston: Houghton Mifflin, 2005); and regarding smoking, Allan M. Brandt, *The Cigarette Century: The Rise, Fall and Deadly Persistence of the Product That Defined America* (Basic Books, 2007).

44. For geologic, energetic, and economic reasons, no well or seam is completely mined out. In fact, worldwide, only some 30 to 50 percent of oil reserves are normally recovered. E. Tzimas, A. Georgakaki, C. Garcia Cortes, and S. D. Peteves, "Enhanced Oil Recovery Using Carbon Dioxide in the European Energy System" (Peten, Netherlands: Institute for Energy, December 2005).

# 2

## The Biophysical: The Decline in Energy Returned on Energy Invested, Net Energy, and Marginal Benefits

Jack P. Manno and Stephen B. Balogh

To understand the logic of leaving oil in the ground we need to understand the good, the bad, and the ugly of fossil fuels.

The good: Compared with other sources of energy, oil has many favorable qualities. It is energy dense, its potential energy is highly concentrated yet stable, and it can be readily and economically transported to where it is needed. Until recently it was relatively easy and cost-effective to get it out of the ground. Before fossil fuels, most of human history was powered by much less efficient solar energy, either directly, as wind or running water; released in wood fires, or as photosynthesized by plants feeding human and animal labor.[1] The growth and development of modern economies has been fueled by a concurrent and equivalent growth in the supply of fossil fuels, especially oil.[2] The few alternatives to oil—for example, biofuels, natural gas, electrification—are either very energy intensive to produce (biofuels) or require massive investment in infrastructure and replacement of capital equipment. Or, like natural gas, the supposed "clean fossil fuel" or "bridge" fossil fuel, they have their own climate-related problems.[3]

The bad: the combustion of oil and other fossil fuels releases carbon dioxide that has been sequestered from the earth's atmosphere for millennia. Burning it produces many chemicals (e.g., sulfur and nitrogen oxides, ozone) and airborne particulates that are harmful to human health and ecosystems. As time goes on and the sources of easy-to-get, high-quality "light" crude declines, the available inventory becomes increasingly expensive in both energy and financial terms. The oil that remains is of increasingly poor quality, is found in environmentally and socially hostile locations, and requires increasing investments in capital equipment and energy to produce. In addition, spills and other impacts from the production of oil have despoiled countless hectares of land and vast areas of ocean.

The ugly: Finally, there are wide and varying social effects from local, regional, and global competition over access to oil—wars, conflict, socio-economic inequality, manipulation, corruption, despotism, and more.[4]

In this chapter, we articulate a biophysical logic that complements the social, political, and ethical arguments against the unfettered extraction of fossil fuels. We believe that making a sound and rational decision about the future of energy requires approaching this issue from a broad systems perspective.

According to conventional economic logic, there are two and only two reasons to keep fossil fuels in the ground: when access is technologically unfeasible (too deep, too remote, too diffuse, too difficult to refine) and, related, when the cost to the developer of getting fossil fuels out of the ground and to the market is greater than the price at which the fuels are currently valued or anticipated to be worth in the foreseeable future. There is nothing new about keeping fossil fuels in the ground. They will stay underground as long as getting them out is too difficult or too costly. Prices and technology are related: rising prices create incentives for investments in technological advances that overcome the technical and cost barriers to extraction. Fossil fuels, by this logic, will then flow out of the ground and into the pipelines, flooding markets and reducing prices, and thus reducing the price of practically everything, and thereby fueling consumption-based prosperity for all.

What's new about the concepts and cases presented in this book is the suggestion that there should be more to the extraction decision than just feasibility and profitability. Burning fossil fuels has social and environmental consequences that must also be taken into account. These are real costs spread throughout society and borne mostly by those who never profit directly from fossil fuels. The authors of this book and the activists highlighted in several of the chapters reject the suggestion that the only decisions in which civil society and political representatives need be involved are those related to managing fossil fuels and their by-products and unintended consequences only after they've been taken from the ground. Decisions considered to be subjects of debate are limited to conserving energy, keeping prices affordable, siting infrastructure, encouraging discovery and development of new sources, and cleaning up the messes—but never the decision about whether to extract in the first place. The concepts and cases in this book demonstrate that the decision to keep fossil fuels in the ground can and should be a social one, influenced by social movements that raise broad ethical, political, and economic concerns. Given the environmental and social consequences of fossil fuel dependence, energy decisions should be determined by a logic other than one based solely on industry profits and government revenues.

There is an alternative argument—a scientific argument—for making decisions that include keeping fossil fuels in the ground. The logic depends largely on three concepts drawn from the emerging fields of biophysical and ecological economics that form a foundation for analyzing net social and economic costs and benefits: they are energy return on energy invested, or simply energy return on investment (EROI); welfare return per unit of energy invested or simply welfare returned on energy investment (WROI); and the threshold hypothesis.[5]

### Energy Return on Investment

The new alternative logic of keeping fossil fuels in the ground begins with a ratio known as EROI. It takes energy to get energy. A single barrel of oil (or its energy equivalent) invested in obtaining oil might return one hundred barrels, fifty, twenty, one, or less. Alternative energy projects can be analyzed in the same terms: how much energy measured in terms of barrels of oil equivalent is produced per one unit of oil or equivalent consumed in production. While differences in quality must be accounted for—for example, many renewable energy technologies produce electricity directly, while fossil fuels must be combusted to be converted to electricity (usually at about a 30 to 40 percent conversion efficiency)—EROI provides an alternative method to traditional cost benefit analyses. Briefly, economic rationality does not always equal energetic or environmental rationality.

Each activity in the life cycle of energy consumes energy: finding new reservoirs, setting up and maintaining infrastructure, drilling, pumping, transporting, storing, marketing, and, finally, consuming. EROI represents the amount of energy obtained from an energy-producing activity divided by the energy used to make that amount of energy available for its end use.[6] Traditionally the boundary for calculating most EROI analyses is set at the point of production or extraction (e.g., wellhead for oil or natural gas, the mine mouth for coal). Alternatively, the boundary for the EROI of fuels can be expanded to include refining and delivery, or even the infrastructure needed to use the energy source (roadways, vehicles, insurance, and so on). It is theoretically possible to extend the boundaries even further to include the environmental and human health burdens of fuel use as an intrinsic cost of consuming that fuel. Doing so would reduce the EROI by increasing the denominator of the ratio. There has never been complete agreement about where to set the boundaries, although there have been attempts more recently to create a protocol for doing so.[7]

EROI is a ratio; a related term, *net energy*, refers to the remainder from subtracting energy input from energy output. When EROI approaches 1:1, net energy gain approaches zero. EROI analysts regularly debate what level of total EROI from the energy industry is necessary to support schools, the arts, transportation, and all the other amenities of society and community. Clearly it must be considerably higher than 1:1.[8]

In the early twentieth century, the EROI for oil exploration and extraction was high. US oil producers found over 1,000 barrels of oil for every barrel of oil's worth of energy they invested in the discovery process, and they extracted nearly twenty-four barrels for every barrel's worth of energy consumed.[9] Oil production reached its peak in the United States in 1970. Since then, much of the easy-to-access oil has been extracted, and producers have been forced to develop poorer-quality, lower-EROI resources. Thus, the estimated EROI for US oil (and associated gas) production fell to 11:1 by 2007. This includes only the direct fuel consumed on site and the indirect energy embedded in the equipment and material. It does not include the energy used in transportation or processing or the energy costs of dealing with the unintended consequences of its use. The decline in EROI for oil and gas production has not been restricted to the United States. Researchers report falling EROI for oil and gas in the North Sea in Norway, China, and at the global level.[10] A 2011 synthesis of twenty papers on the topic of EROI in the journal *Sustainability* concluded that "the EROI of essentially all fossil fuels are declining, in many cases, sharply."[11]

The decline in EROI and net energy of fossil fuels results from the fact that most of the readily available high-quality reservoirs are already in production, and finding new supplies requires increasingly intense drilling effort.[12] Newer discoveries tend to be in deep water, remote and hostile environments, at great depths, bound in shale or sand, or otherwise in conditions requiring considerably more energy to bring to market than has been the norm in the petroleum age.[13] As in all other endeavors, we tend to take the easiest-to get fuels first, the proverbial low-hanging fruit that one picks before climbing to the tops of the tree or investing in specialty tools for picking fruit from the upper branches.

There are energy and economic opportunity costs to the general trend of a declining EROI for our major fuels. Every barrel of oil that needs to be reinvested in energy extraction is unavailable to be used in another sector of the economy. A society reliant on lower-EROI fuels must divert an increased amount of human effort, money, and energy resources to maintaining the flow of energy to its economy. This leaves less energy and capital to invest in maintaining or developing infrastructure and, ultimately, less to spend on discretionary goods and services.[14]

Despite a general decline in the EROI of fossil fuels, this synthesis of EROI studies concluded that "traditional fossil fuels almost universally have a higher, often much higher, EROI than most substitutes" —substitutes such as solar, wind, biofuels, and other alternative and renewable sources of energy.[15] However, the EROI analyses presented do not include the energy it takes to respond to the effects of a warming climate, of air and water pollution or the energy needed to deal with the health effects that can be attributed to fossil fuel consumption.

Total net energy represents the productive energy available to heat and power all the economic, social, and cultural activities of modern, industrial daily life. Fossil fuels provide dense concentrations of energy described mathematically in the higher EROI values for fossil fuels compared to other energy sources. Until recently the EROI of most renewable alternatives did not come close to producing the net energy of fossil fuels. But gradually the EROI for oil and gas has declined while some of the alternatives have increased.[16] Still, the divide in most cases is wide, meaning that society must either develop higher EROI alternatives or find a means of maintaining or improving our quality of life at lower net energy levels (see box 2.1).[17]

**Box 2.1**
Deepwater Drilling: How Complexity Begets Complexity

> If there is an inevitability in the oil and gas business, it is not more and bigger finds but more complexity in making those finds. From an energetic perspective, increasing complexity requires increasing energy inputs just to get the same amount of usable energy. Energy return on energy invested declines. From a societal perspective, such complexity drains financial, physical, and human capital from other uses (e.g., growing food). And at some point, the social benefits of such fossil fuel extraction no longer exceed the opportunity cost—that is, the costs to higher social and environmental values—of that extraction.
>
> So how does such complexity come about? Offshore rigs experience hurricane winds exceeding 100 mph, giant waves 80 feet high or more, powerful ocean currents that can change direction, and undersea mudslides. They

**Box 2.1**
(continued)

can be located so far offshore as to prevent a rapid response should a fire break out on board. Below 3,000 feet, temperatures are just above freezing, which can harden natural gas into hydrates—crystal-like structures that can clog pipelines. Some drilling is now at depths over 9,000 feet, with even deeper ones planned. Divers can't go to such depths, so remote-controlled submarines are required, which have an assortment of technical issues.

"The industry has entered a new domain of vastly increased complexity and increased risks," said Robert Bea, a professor at the University of California, Berkeley. The real question is what risks, to what, to whom and for how long?

### Source

Michael Wines, "Risk-Taking Rises to New Levels as Oil Rigs in Gulf Drill Deeper," *New York Times,* August 30, 2010, A1, A10 (quote on A10).

As EROI approaches 1:1, there are financial and thermodynamic limits to the amount of oil that will be extracted ultimately from the earth's crust regardless of the total volume of oil that actually remains in the ground. At an EROI of 1:1, the energy breakeven point, there is no net energy delivered to society, only energy moved from one application to another. It does not make sense energetically to invest in a barrel of oil to get a barrel of oil out of the ground. This breakeven point, however, may lie some decades into the future. If sufficient resources are available and current levels of use is maintained, it will take only three to four decades to consume as much oil as has been consumed in the first century of the oil era.

Oil fields that do not have a net energy profit will never be exploited without major subsidies, much of which would, in effect, be energy debt—and one that most likely would never be paid. These subsidies come in the form of tax breaks, direct investment, military commitment, skewed investment decisions, the nonexistent price of dumping carbon into the atmosphere, narrowed range of corporate and personal liability for damages, and so on. The costs of these subsidies are dispersed among all, especially the poor, while the profits concentrate with the rich. One can imagine oil at great depth or present in too small an amount to be energetically profitable. But even when the fossil fuels industry can demonstrate a net energy surplus from its activities, the surplus could disappear when analysts expand their boundaries to include the energy costs borne by local communities and the broader society for cleaning up spills, repairing roads, building temporary shelters and schools, and, most of all, recovering from the weather-related disasters exacerbated by climate change. Fundamentally, *a business profit does not equal an energy profit.*

Depending on how comprehensively the boundaries for the EROI calculation are drawn, society may soon or may have already reached the point when further exploitation of fossil fuel deposits will no longer provide an energy surplus to humans. This phenomenon may be directly related to what is commonly referred to as "peak oil," the time when oil production and consumption no longer continue to grow and gradually decline. Peak oil is a geological phenomenon mainly, but it may also be exacerbated by economic barriers to higher oil prices.[18] At that point, there will remain large amounts of hydrocarbons deeply or diffusely within the earth's crust. Peak oil (or peak other fossil fuels) does not mean the world has run out of oil. It can mean the time has come when energetic rationality becomes economic rationality and the only logical place for fossil fuels is in the ground. These rationalities will be heeded only when the subsidies mentioned above come to an end.

When the EROI of a nonrenewable fuel declines toward that of renewable alternatives or when the consumption of energy exceeds a threshold after which the marginal benefits of energy consumption turn negative, then leaving fossil fuels in the ground makes economic, social, and environmental sense. Economic logic would then align with the environmental, social, and spiritual arguments presented in the rest of this book, making a compelling case for deciding to keep much of the remaining fossil fuels in the ground and to deliberately hasten the end of the fossil fuel era. Continuing to drill ever deeper beneath the ocean floor, in ever more remote or environmentally sensitive areas, using ever more energy-intense and ecologically disruptive methods, just to find highly subsidized fossil fuels with low or even negative net energy, makes no economic, ecological, or ethical sense.

Thus, what may appear to be a radical choice—keeping fossil fuels in the ground—becomes rational when three things happen: (1) the useful energy obtained from the fuel is actually less than the energy used in searching, testing, digging, pumping, processing, transporting, managing, selling, buying, and burning fuels along with the energy used to deal with negative side effects (environmental, health, infrastructure, climate related) of the fossil fuel cycle; (2) the benefits measured in terms of human and ecological well-being are less than the damages measured in human and ecological harm; and (3) the gains and losses are measured at the scale of the whole earth.

That is, if profit and loss are measured only at the scale of an energy company, energy expended and energy gained are measured at the scale of the refinery, and human well-being is measured only in terms of dollars and cents, then leaving fossil fuels in the ground appears not to be a rational choice. Nevertheless, even at these restricted scales, declining EROI and increasing negative side effects of the fossil fuel cycle will eventually lead to keeping some fossil fuels in the ground (as already happens now; recovery rates are generally less than 40 percent).

Unfortunately this is most likely to happen only after it is too late to matter. The fundamental problem lies in the fact that that which makes financial sense in the fossil fuel business and consumer sense at the pump often makes no sense for the long-term functioning of human and ecological systems.

**Welfare Return on Energy Consumption and the Threshold Hypothesis**

In addition to EROI, we propose another measure we are calling welfare returned on energy investment (WROI). It would be a measure of the efficiency of energy consumption in terms of human benefits. Not all energy consumption is equal. Some produces clear benefits, and some is destructive and reduces human welfare and well-being. WROI is a useful concept that places emphasis on the underlying purposes of energy use and consumption: to improve quality of life. Hypothetically, high WROI would result from lower fossil-energy-intense projects that improve human well-being, such as community organizing, solar thermal energy, organic agriculture, repair, maintenance, and reuse and sharing of tools and consumption goods—in other words, the whole panoply of institutions and behaviors of sustainability, learning how to live well while minimizing energy and material consumption or throughput. As EROI for fossil fuels declines, WROI necessarily declines for all products and activities that consume them. The result is increasing economic incentives (from an ecological economic perspective) for low-energy lifeways and livelihoods and deliberate choices to keep fossil fuels in the ground. At present there are few analyses of WROI because of the difficulty of quantifying welfare and well-being.

In one of the early issues of the journal *Ecological Economics*, the Chilean economist Manfred Max-Neef offered a threshold hypothesis.[19] He argued that "for every society there seems to be a period in which economic growth (as conventionally measured) brings about an improvement in the quality of life, but only up to a point—the threshold point— beyond which, if there is more economic growth, quality of life may begin to deteriorate." Valentina Niccolucci, Frederico M. Pulselli, and Enzo Tiezzi reviewed measurements of the index of sustainable economic welfare and genuine progress indicator, two indicators commonly offered as alternatives to gross domestic product (GDP) as a measure of a nation's progress.[20] Analyzing the relation between GDP and a nation's ecological footprint, the authors concluded that "increase in economic wealth often results in worse, not better, conditions for people because the welfare related to a given GDP is 'polluted' and diminished by environmental stress and social pressures." This is consistent with Herman Daly's concept of uneconomic growth, defined as occurring when "increases in production come at an expense in resources and well-being that is worth more than the items made."[21] Martinez and Ebenhack found that improvements in the human development index (HDI) became saturated at moderate levels of energy consumption for both oil-importing and oil-exporting nations.[22] A study by Lambert and others showed that indexes of well-being (e.g., HDI, percent children underweight, health expenditures, gender inequality index, literacy rate, and access to improved water) were well correlated with an energy index that included measures of energy quality (EROI at the societal level), availability (energy per capita), and distribution (Gini index).[23] This study provides support for the threshold hypothesis: improvements in well-being level off above 200 gigajoules per capita, 30:1 EROI at the societal level.

Given the tight correlation and logical link between economic output and total energy use, the same holds for the relation between increasing energy use and measures of well-being. Energy analyst Vaclav Smil has noted, "There are some remarkably uniform inflection bands beyond which the rate of gains declines sharply, and some clear saturation levels beyond which further increases of fuel and electricity consumption produce hardly any additional gains."[24]

The UN Development Programme (UNDP) estimates that it requires one tonne of oil equivalent (TOE)/per capita (approximately the per capita consumption of Honduras and Indonesia) as the minimum required to reach a fairly high state of national health and development.[25] At the same time, it has estimated that consumption of 1.194 to 1.672 TOE of commercial energy per person (the level of Jordan, Lebanon, and Mexico) is enough to meet essential physical needs plus high-quality education and social services.[26] On average, in 2005, the world's population consumed 1.778 TOE per capita annually.[27] In other words, the amount of energy required for a good life is remarkably close to the 2005 per capita energy consumption. What is more, there is at present enough energy to meet everyone's need for it. But the distribution of energy is highly unequal. The world's energy glutton, Qatar, consumed over 19 TOE per capita, while in the United States and Canada, annual consumption hovers around 8 TOE per capita. In general, as annual per capita energy use increases, measures of quality of life increase in step, up to a threshold point after which increases in quality of life are no longer evident. Smil, who analyzed data on infant mortality, female life expectancy, and a nation's proportion of undernourished people to analyze the relation of energy consumption and well-being, concluded that

there are pronounced gains as commercial energy use increases toward 30 and 40 GJ (1 GJ = 238 kg of oil equivalent) per capita, and clear inflections are evident at annual consumption levels of 50–60 GJ per capita; these inflections are followed by rapidly diminishing returns and finally by a zone of no additional gains accompanying primary commercial energy consumption above 100–110 GJ per capita. ...

These realities make it clear that a society concerned about equity, determined to extend a good quality of life to the largest possible number of its citizens and hence willing to channel its resources into the provision of adequate diets, good health care, and basic schooling could guarantee decent physical well-being with an annual per capita use (converted with today's prevailing efficiencies) of as little as 50 GJ (1.2 Tonnes of Oil Equivalent). A more satisfactory combination of infant mortalities below 20, female life expectancies above 75 years, and HDI above 0.8 requires annually about 60 GJ. But, once the physical quality of life reaches a satisfactory level, other concerns that contribute to the overall well-being of populations become prominent. ...

Actual US and Canadian per capita energy use is thus more than three times the high-level minimum of 110 GJ, and almost exactly twice as much as in Japan or the richest countries of the EU—yet it would be ludicrous to suggest that the American quality of life is twice as high. In fact, the US falls behind Europe and Japan in a number of important quality-of-life indicators.[28]

The message is that it is possible and certainly desirable to create conditions in which all people can obtain a high quality of life even while overall humanity consumed lower amounts of fossil fuels. It will require new energy opportunities, in particular, renewable energy, to be made available to parts of the world where people live in energy poverty and that the energy gluttons of the world considerably reduce their levels of energy consumption. In the context of such a strategy, decisions about keeping fossil fuels in the ground become logical, especially as well-being produced per unit of energy consumption declines.

Similarly, self-reported levels of happiness among the poor tend to rise with increased income while levels of emotional depression decline. The relation between self-reported state of happiness and personal income, however, largely disappears beyond moderate levels of income.[29] Up to a point, one can buy at least a chance at happiness, but that point may be well below what is taken for granted in affluent societies. Too much energy introduced into a system can overwhelm it. For an analogy, consider the problem of eutrophication of a lake or pond. An overabundance of fertilizing nutrients entering the water stimulates the growth of plankton and thus excessive photosynthesis and a decline in oxygen that leads to rapidly degrading conditions for most fish. The shore becomes awash in rotting algae and dead fish. Remember Lake Erie in the 1970s. Consider also how a starving person rapidly improves by increasing caloric intake. But the average North American who now takes in 3600 calories a day (the world average is 2700 calories) is not well served by adding another pound of steak to his or her daily diet. Consider that the energy that goes into producing, processing, storing, transporting, and preparing the average American meat-eating diet emits 8,800 pounds of carbon dioxide per day, just less than the average US car.[30] The WROI beyond these points would turn rapidly negative.

Many of the means for reducing energy use are also steps toward improving one's health: walking, riding bikes, driving less; eating more fruits and vegetables, preferably organic and locally grown. The Worldwatch Institute in its State of the World report points out how isolated, lonely lives expand energy consumption: "A one person household in the United States uses 17% more energy per person than a two person household."[31] Friendship, sharing meals and tools, conversation, skill sharing, and other community-building activities go a long way toward reducing individual consumption to achieve satisfaction and maintain health while significantly reducing ecological footprints.

These three concepts of energy return on energy invested, welfare return on energy consumed, and the threshold hypothesis articulate the critical point that although fossil fuels provide the energy to meet basic human needs powering much of value in modern life, they also power much that is destructive in contemporary civilization: violence, addiction, and wholesale environmental destruction that together decimate the quality of human lives. A high-quality, low-energy world may be difficult to attain, given where we are now, but it is possible to imagine and possibly achieve (see chapters 1 and 12). While conventional industry logic presumes that fossil fuels should never be kept in the ground when it is arguably narrowly profitable to get them out, a decision logic based on EROI, WROI, and the threshold hypothesis demonstrates how and why the balance between the constructive and destructive powers of fossil fuels is shifting. We need new ways to describe and enumerate this balance and make choices about when fossil fuels should be kept in the ground for the common good. And we need to do so based on existing knowledge of the human and environmental costs of extracting and burning fossil fuels.

As long as the industry, rather than the broader public, largely controls decisions about mining and extraction, the great majority of people employed in the field of fossil fuels will be considered the technical experts. They will be the ones asked how to get the fuels out of the ground. They will not be asked, and will not be inclined to consider, the net energy and net welfare effects of continuing this irrational game of total extraction (chapter 3). For every analyst now examining and comparing values such as EROI (and there are very few), there are many thousands, well funded and with considerable state support, engaged in overcoming obstacles to extraction. Each industry expert knows some aspect of fossil fuel extraction extremely well, and each is highly rewarded for that knowledge. Petroleum geologists scour the surface of the globe looking for the telltale signs of rock domes, or anticlines, under which oil could be trapped. They direct sound waves down into the earth looking for the sonar signal that can mean big rewards. Then other experts do the test drilling. Still others tell you the properties of the reservoir rock: the porosity (storage capacity) and permeability (ability to flow between pores) of the oil itself: the API gravity (the density of oil compared with water). The best reservoirs have high porosity, permeability, and low-density oil. Other experts determine the quality of the source (how much additional processing and refining will be required). Others compare costs, potential sales, investments required. Others know how to bring the goods safely to the surface. Still others know total financial costs, estimate revenues, and calculate whether a given well and source can be expected to turn a profit.

So many great minds are busy on all aspects of getting fossil fuels out of the ground and onto the market. From the industry perspective, a decision to keep fossil fuels in the ground is always a negative, a failure to find or extract or deliver. We venture to say that there are no experts in the industry looking at the benefits (social, environmental, psychological, or spiritual) of keeping fossil fuels in the ground, benefits for society as a whole, for the earth as a whole, for specific ecosystems—at least, none that show up on the industry's books.

The literature on EROI is full of warnings about the dire consequences of declining energy availability as the energy sector continues to consume increasing amounts of energy for its own production needs. They ask, "What ratio of EROI is necessary to support civilization?"[31] while defining *civilization* by the energy-intense norms of energy-gluttonous nations. While declining EROI will challenge all societies to learn to be frugal, simpler, and more cooperative, when one looks at the range of inverted-U relations between energy consumption and human well-being, the notion begins to dawn that society should be guided by the concept of an optimal level of energy use. We should ask what is needed to support personal well-being rather than push to maximize economic activity and thus energy consumption. It is entirely possible that wealthy industrial societies have exceeded a sustainable optimum of energy use, in particular those nations that can be considered energy gluttons. For societies still on the upward slope of the energy and well-being U curve, more available energy should lead to more social and individual well-being (i.e., increasing WROI). Given the generally increasing trends in EROI for windmills and solar panels and the potential benefits of increasing energy consumption, alternative and renewable sources may make sense for new energy development in those nations. For modern industrial societies that appear to be on the downward slope, where marginal cost of energy use exceeds marginal benefit, an energy diet would be beneficial. Like most addicts, people in modern industrial societies are unlikely to voluntarily choose to live less energy-intense lifestyles despite the best persuasive efforts to encourage simpler living. But in the face of less energy availability and higher prices associated with declining EROI, ultimately what we will have is less energy. This knowledge makes it possible to plan for this day. We must begin to design energy production and economic systems that will maximize well-being produced per unit of energy consumed while minimizing the impacts on ecosystems and human health. Decisions to leave fossil fuels in the ground would likely accelerate the decline in net energy available to society. Those who advocate post–fossil fuel policies must be honest that the consequences will mean leading lives that consume less energy. Scholars need to focus much more attention on the communities and conditions that lead to a higher quality of life with lower energy consumption, which equals improving the efficiency of all forms of energy, increasing the human welfare benefits provided per unit of energy or increasing WROI.

## Conclusion

The chapter advances an alternative logic to the dominant one of unending extraction. It is to underpin with physical and welfare analysis the many efforts to keep fossil fuels in the ground. In each of the cases in this book, those seeking to keep these fuels in the ground have been or will be accused of advocating a position that is likely to impoverish all, lead to economic decline, even collapse, and is unwarranted, if not immoral. That argument assumes that without growth in fossil fuel availability, there can be no economic growth, and without economic growth, there can be no prosperity. Clearly, declining energy availability will require much higher levels of human welfare produced per unit of energy consumed. Beginning to end the fossil fuel era will require consistently high levels of WROI in agriculture, housing, transportation, and industrial plants, all of which will depend on more cooperation, more recycling, more reuse, more community, more awareness of each individual's connection to and responsibility for the land, water, and ecosystems on which we all depend. In the end, this is also exactly what is needed for the long-term survival of humans on the planet.

## Notes

1. Vaclav Smil, *Energy in World History* (Boulder, CO: Westview Press, 1994).

2. Cutler J. Cleveland, Robert Costanza, Charles A. S. Hall, and Robert Kaufmann, "Energy and the US Economy: A Biophysical Perspective," *Science* 225 (1984): 890–897.

3. R. W. Howarth, R. Santoro, and A. Ingraffea, "Methane and the Greenhouse-Gas Footprint of Natural Gas from Shale Formations," *Climatic Change* 106 (2011): 679–690.

4. D. O'Rourke, and S. Connolly, "Just Oil? The Distribution of Environmental and Social Impacts of Oil Production and Consumption," *Annual Review of Environment and Resources* 28 (2003): 587–617; M. L. Ross, "Blood Barrels: Why Oil Wealth Fuels Conflict," *Foreign Affairs* (2008): 2–8.

5. Cleveland et al., "Energy and the US Economy; Jack P. Manno, "Looking for a Silver Lining: The Possible Positives of Declining Energy Return on Investment (EROI)," *Sustainability* 3 (2011): 2071–2079; Manfred Max-Neef, "Economic

Growth and Quality of Life: A Threshold Hypothesis," *Ecological Economics* 15 (1995): 115–118.

6. Charles A. S. Hall, Cutler J. Cleveland, and Robert Kaufmann, *Energy and Resource Quality: The Ecology of the Economic Process* (New York: Wiley Interscience, 1986).

7. David J. Murphy, Charles A. S. Hall, Michael Dale, and Cutler Cleveland, "Order from Chaos: A Preliminary Protocol for Determining the EROI of Fuels," *Sustainability* 3 (2011): 1888–1907.

8. Charles A. S. Hall, S. Balogh, and D. J. Murphy, "What Is the Minimum EROI That a Sustainable Society Must Have?" *Energies* 2 (2009): 25–47; Jessica G. Lambert, C. A. Hall, S. Balogh, A. Gupta, and M. Arnold, "Energy, EROI and Quality of Life," *Energy Policy* 64 (2014): 153–67; Jessica G. Lambert, C. A. Hall, and S. Balogh, "EROI of Different Fuels and the Implications for Society," *Energy Policy* 64 (2014): 141–152.

9. Megan C. Guilford, Charles A. S. Hall, Peter O'Connor, and Cutler J. Cleveland, "A New Long Term Assessment of Energy Return on Investment (EROI) for U.S. Oil and Gas Discovery and Production," *Sustainability* 3 (2011): 1866–1887.

10. Leena Grandell, Charles A. S. Hall, Mikael Höök, "Energy Return on Investment for Norwegian Oil and Gas from 1991 to 2008," *Sustainability* 3 (2011): 2050–2070; Yan Hu, Lianyong Feng, Charles A. S. Hall, and Dong Tian, "Analysis of the Energy Return on Investment (EROI) of the Huge Daqing Oil Field in China," *Sustainability* 3 (2011): 2323–2338; Nathan Gagnon, Charles A. S. Hall, and Lysle Brinker, "A Preliminary Investigation of Energy Return on Energy Investment for Global Oil and Gas Production," *Energies* 2 (2009): 490–503.

11. Charles A. S. Hall, "Synthesis to Special Issue on New Studies in EROI (Energy Return on Investment)," *Sustainability* 3 (2011): 2496–2499.

12. Charles A. S. Hall and Cutler J. Cleveland, "Petroleum Drilling and Production in the United States: Yield per Effort and Net Energy Analysis," *Science* 211 (1981): 576–579.

13. Charles A. S. Hall, P. Tharakan, J. Hallock, C. Cleveland, and M. Jefferson, "Hydrocarbons and the Evolution of Human Culture," *Nature* 426 (2003): 318–322.

14. Charles A. S. Hall, R. Powers, and W. Schoenberg, "Peak Oil, EROI, Investments and the Economy in an Uncertain Future," in *Biofuels, Solar and Wind as Renewable Energy Systems*, ed. David Pimental (Amsterdam: Springer Netherlands, 2008), 109–132.

15. Charles A. S. Hall and Cutler J. Cleveland, "Petroleum Drilling and Production in the United States: Yield per Effort and Net Energy Analysis," *Science* 211 (1981): 576–579.

16. Nathan Gagnon, Charles A. S. Hall, and Lysle Brinker, "A Preliminary Investigation of Energy Return on Energy Investment for Global Oil and Gas Production," *Energies* 2 (2009): 490–503; Leena Grandell, Charles A. S. Hall, and Mikael Höök, "Energy Return on Investment for Norwegian Oil and Gas from 1991 to 2008," *Sustainability* 3 (2011): 2050–2070; Megan C. Guilford, Charles

A. S. Hall, Peter O'Connor, and Cutler J. Cleveland, "A New Long Term Assessment of Energy Return on Investment (EROI) for U.S. Oil and Gas Discovery and Production," *Sustainability* 3 (2011): 1866–1887.

17. Jessica G. Lambert, C. A. Hall, and S. Balogh, "EROI of Different Fuels and the Implications for Society," *Energy Policy* 64 (2014): 141–152.

18. Gail Tverberg, "Oil Supply Limits and the Continuing Financial Crisis," *Energy* 37 (2012): 27–34.

19. Manfred Max-Neef, "Economic Growth and Quality of Life: A Threshold Hypothesis," *Ecological Economics* 15 (1995): 115–118.

20. Valentina Niccolucci, Frederico M. Pulselli, and Enzo Tiezzi, "Strengthening the Threshold Hypothesis: Economic and Biophysical Limits to Growth," *Ecological Economics* 60 (2007): 667–672.

21. Herman Daly, "Economics in a Full World," *Scientific American* 293 (2005): 100–107.

22. Daniel M. Martinez and B. W. Ebenhack, "Understanding the Role of Energy Consumption in Human Development through the Use of Saturation Phenomena," *Energy Policy* 36 (2008): 1430–1435.

23. Lambert et al., "Energy, EROI and Quality of Life."

24. Vaclav Smil, "Science, Energy, Ethics, and Civilization" in *Visions of Discovery: New Light on Physics, Cosmology, and Consciousness*, ed. Raymond Y. Chiao, Marvin L. Cohen, Anthony J. Leggett, William D. Phillips, Charles L. Harper, Jr. (Cambridge: Cambridge University Press, 2010), 709–729 (quote on p. 722).

25. UN Development Program, International Cooperation at a Crossroads: Aid, Trade and Security in an Unequal World, UN Development Report, accessed September 23, 2011, http://hdr.undp.org/en/reports/global/hdr2005.

26. Vaclav Smil, *Energy at the Crossroads Global: Perspectives and Uncertainties* (Cambridge, MA: MIT Press, 2010).

27. World Resources Institute, "Earth Trends: The Environmental Information Portal," accessed July 28, 2010, http://earthtrends.wri.org.

27. Smil, "Science, Energy, Ethics, and Civilization."

28. Robert E. Lane, *The Loss of Happiness in Market Democracies* (New Haven, CT: Yale University Press, 2000).

29. Robert Goodland, "Environmental Sustainability in Agriculture: Diet Matters," *Ecological Economics* 23 (1997): 189–200.

30. Worldwatch Institute, *State of the World 2004* (Washington, DC: Worldwatch, 2004).

31. Hall, Balogh, and Murphy, "What Is the Minimum EROI That a Society Must Have?"

# 3

## The Cultural: The Magic, the Vision, the Power

Thomas Princen

When sugar, once a rare substance used mostly for medicinal and decorative purposes, became mass produced and mass consumed, ingested for calories as much as sweetness, it became not just another food but an agent of cultural transformation. When meat, once produced by local farmers and processed and sold by the local butcher, became a product of the disassembly line, corporate packing, high finance, and marketing genius, it transformed not just eating patterns but entire landscapes and the way people related to those landscapes. When the automobile, for decades a plaything of daredevils and the rich, became a mode of everyday transport, even a domicile, the landscape again changed dramatically, and with it people's notion of speed, mobility, and freedom. Likewise, when coal, once a dirty, even diabolic substance from deep in the earth's bowels, fit for use only by society's lowest classes, became an acceptable heat source for the upper classes and the fuel of choice for industry, its status rose, affecting virtually all aspects of modern industrial life. And then, when petroleum, long used for rituals or as a medicine, could be converted into kerosene to chase away the darkness at home and power machines around the clock in the factory, followed by its use as a fuel for transport, heating, and electric power, not to mention weapons, fertilizer, and synthetics, spurring economic growth at unimaginable rates, societies the world over changed rapidly and dramatically.[1]

### Not Just a Fuel

Fossil fuels were never just energy sources, never just raw materials. They were, like sugar, meat, automobiles, and so many other commodities in the modern, consumerist, efficiency-driven, growth-oriented industrial world, agents of cultural change.[2] To understand the agency of a substance, especially a substance that pervades so many aspects of life to the point where it is the proverbial water in which we fish swim, essential but unnoticed, I argue in this chapter, is to understand the simultaneity of the blessing and the curse of fossil fuels—their wonder and everydayness, their revelations and disguises, their fertility and toxicity. If fossil fuels were "just another energy source," the transition from wood to coal and then to oil would have been a simple "swap out, swap in" affair. But those cultural changes were in fact momentous, and we have every reason to believe that the current transition—out of fossil fuels as the dominant energy source—will be as well.

So fossil fuels are not just energy sources. Nor is it the case that the extraction of oil, coal, natural gas, or any other mineral is just about the mined substance. From a utilitarian or economic perspective, all the attention is indeed on the uses and use values of the oil or coal, the costs of production, and concomitant prices and output levels. But from a systems perspective, one that accounts for the physical system of the landscape and the social system of the extractors, extraction is a social activity. It is one that indeed begins with exploration and investment and ends with consumption, emissions, disposal, and returns on investment, and yet is entwined in a system of perceptions and expectations, in a set of beliefs and values, in a story of the good life. Moreover, the beneficiaries of extraction, narrowly construed to include only the mineral and its revenues, are often distinct from those who live in and with these larger cultural systems. So as we will see in part 2, the gold in gold mining may be the focus of attention for the mining company and its investors, as well as the government's revenue collectors. But for those who live in and around the gold mines and expect to live there for generations, gold extraction is water contamination. In Ecuador, oil drilling is forest destruction. In Australia, uranium mining is the sacrilege of ancestral places. In Appalachia, mountaintop removal is the destruction of a rural way of life. Everywhere shale oil and gas are climate catastrophe. To focus on the commodity, as is commonly done in the business and policy arenas, is to reduce a systemwide intervention to a convenient fiction—that extraction is only about the mineral, the commodity, the trade, the revenues, only about the stuff in play.

This belief system is convenient for those who benefit most and are adept at getting away with displacing the costs onto others. Theirs is a belief system born of extraction and expansion. These are the other agents—individuals, organizations, institutions—that, along with the agency of the substance itself, form a two-pronged vector of cultural change, an arrow whose power is unrivaled in human history, whose effects are only now coming to light. What is known, indeed experienced and celebrated, is the fact that homes are heated and streets lit, food grown and rushed to store shelves, travelers propelled at the speed of birds, even of sound, and messages sent at the speed of light. It all started with a certain mystery and a certain magic.

**The Mystery, the Magic, the Myth**

From the earliest recorded uses, fossil fuels have had a certain magical quality, a mystery of origin and effect, with one result being a sense of endless abundance.[3] Oil in the Middle East, known for millennia from seepages, was believed to have miraculous properties. The continuous burning of oil seepages and escaping gases provided the basis of fire worship. Around Baku near the Caspian Sea, there were "eternal pillars of fire" worshipped by the Zoroastrians. In ancient Babylon, although bitumen was used as a mortar and a caulk and for paving roads, its primary use was medicinal: it "healed wounds, treated cataracts, provided a liniment for gout, ... drew together severed muscles, and relieved both rheumatism and fever."[4] In North America, before drilling derricks brought great quantities to the surface, Indians and Europeans alike believed oil had medicinal properties; it was used "to relieve everything from headaches, toothaches, and deafness to stomach upsets, worms, rheumatism, and dropsy."[5] When those drills did strike oil, oil under great pressure, it would erupt with "the sound of a cannon shot and then with a continuing and deafening roar," writes oil historian and consultant Daniel Yergin. The first gusher in Texas, Spindletop, was so violent "some people thought it was the end of the world."[6]

In the early years and decades of industrialization, fossil fuels inspired awe, even terror. The first coal miners in Newcastle in the sixteenth century had to be coaxed to go underground where, except for the occasional deep cave, no humans had ever gone before, where, they believed, demons and goblins haunted the mines, causing inexplicable violent explosions and death. As one journalist wrote in the early 1800s, a mine "would require all the fortitude of nature to refrain from fear. ... The immense depth, the innumerable windings and the dark solitary wastes of a coalmine are truly astonishing, and create a sensation of horror in the imagination."[7] Even the imprint of a fern so deep underground contradicted people's understanding of the world, one where plants grew above ground. For consumers of coal, the smell of sulfur in the smoke was troubling because sulfur, otherwise known as brimstone, came from the demonic underworld.[8] And lest such beliefs be dismissed as those of the uneducated and superstitious, chemist Robert Boyle, a founder of modern chemistry and the Royal Society, encouraged his colleagues to investigate scientifically whether the miners "ever really meet with subterranean daemons."[9]

With industrialization, the mysteries created a new magic. Oil, in the form of kerosene, could "push back the night," offering light as no candle could, liberating countless households from darkness, enabling homemakers to read and sew and perform other tasks otherwise limited to daylight hours. It was "a light fit for Kings and Royalists and not unsuitable for Republicans and Democrats," wrote one booster upon the discovery of kerosene. Industrialization itself was spurred by fossil fuel illumination, first coal gas and then kerosene, enabling factories to add night shifts and run around the clock, making efficient use of otherwise idle plant and equipment. And there was magic of a sort in the power a single individual could command directly. In 1900 a farmer with six large horses controlled 5 kilowatts of animate power; by 2000, with modern tractors, it was 300 kilowatts of power. In 1900, an engineer operating a locomotive commanded about 1,000 kilowatts of steam power; a century later an airline pilot commanded some 10,000 kilowatts (box 3.1).[10] To further illustrate the hidden energy, electric motors in modern buildings run almost continuously but are in the basement or on the rooftop or behind locked door. They keep vital services going but all virtually unseen, as if by magic.

In nineteenth-century Britain, with the steam engine and iron smelting, coal production soared. "Steam increased the demand for both coal and iron," writes historian Barbara Freese, "and also made coal and iron easier and cheaper to produce. Cheaper coal and iron made steam engines cheaper to build and to run, which, in turn, attracted more people to steam power, further increasing the demand for coal and iron, and so on."[11] Such were the self-reinforcing, positive feedback loops that drove industrialization. At the center was indeed coal, which, Freese writes,

had completely permeated society. It was not only directly present in the bellies of the steam engines, but indirectly present in the engine's iron cylinders and pistons, in the loom's iron frames, in the factories' iron girders, and later in the iron railroads, bridges, and steamships that would define the industrial age.

... And, because its energy had already been handily condensed over millions of years, coal concentrated the factories and workforces in urban areas instead of dispersing them throughout the countryside. In short, coal allowed the industrialization of Britain to gain a momentum that was nothing short of revolutionary.[12]

**Box 3.1**
Power at Hand

Physical power, which can translate to economic and political power, ultimately rests in people's hands. Here are examples of how power in the hands of operators has increased as the concentrated energy of fossil fuels has become more available via new technologies.

| 1800 | Coach driver (horse drawn) | 2.5 kW |
|------|----------------------------|--------|
| 1900 | Farmer (six large horses) | 5 kW |
| 1900 | Railroad engineer (steam locomotive) | 1,000 kW |
| 2000 | Farmer (modern tractor) | 300 kW |
| 2000 | Pilot (Boeing 737) | 10,000 kW |

**Sources**

R. Y. Chiano, M. L. Cohen, A. J. Leggett, W. D. Phillips, and C. L. Harper Jr., eds., *Visions of Discovery: New Light on Physics, Cosmology, and Consciousness* (Cambridge: Cambridge University Press, 2010); Vaclav Smil, "Science, Energy, Ethics, and Civilization," in *Visions of Discovery* 709–729.

As more and more coal was extracted and more uses discovered, the mysteries shifted from being perceived as mostly negative (the filth and dangers of deep mining and household burning) to positive, a force for good. Coal, long reviled as dirty and low class, a symbol of poverty and disappointment (a lump of coal in a Christmas stocking) or stinginess (the Grinch with a heart of coal), achieved a certain favor when timber supplies were exhausted, and even the wealthy heated with coal. No less a Romantic figure than Ralph Waldo Emerson could now wax eloquently about coal and its magical ability to create its own climate, even its own civilization:

Every basket [of coal] is power and civilization. For coal is a portable climate. It carries the heat of the tropics to Labrador and the polar circle; and it is the means of transporting itself whithersoever it is waned. Watt and Stephenson whispered in the ear of mankind their secret, that a half-ounce of coal will draw two tons a mile, and coal carries coal, by rail and by boat, to make Canada as warm as Calcutta; and with its comfort brings its industrial power.[13]

Comfort and power followed coal in America too but there the real excitement, the real magic, was with oil, first discovered in western Pennsylvania in the 1860s. "The enthusiasm for oil seemed to know no limits," writes historian Yergin, "and it became not only a source of illumination and lubrication, but also part of the popular culture. Americans danced to the 'American Petroleum Polka' and the 'Oil Fever Gallop," and they sang such songs as 'Famous Oil Firms' and 'Oil on the Brain.'"[14] Arguably that enthusiasm has never waned, only gone underground, as it were, in industry and politics, to surface when shortages arise or a war effort needs more supply or a spill riles up influential residents.

Thus it is that the early magic of fossil fuels became the magic of industrial transformation, of economic growth, of, indeed, "power and civilization." It was a "time of ingenuity and innovation, of deals and

**Figure 3.1**

Embossed postcard from a Texas oil field in 1905. Muscle power supplied by biofuels (presumably hay and oats) distributed oil at the beginning of the fossil fuel era. Soon horsepower was found almost entirely in automobiles and trucks. Oil became everyday, horses special. Perhaps a century or so later, as the fossil fuel era comes to an end, oil will help convert vehicles to biofuels (and build windmills and solar plants). Then muscle power, including human muscle power, and self-generated electrical power will be everyday, fossil fuels special. *Source:* American Gas & Oil Society, http://aoghs.org/wp-content/uploads/2013/01/Oilfield-Humble-TX-post-card-AOGHS.jpg.

**Figure 3.2**

Horse-drawn oil tanker in Germany, likely in the early 1900s. Even in the birth-place of the automobile, horses distributed oil (very likely shipped by sail power from Texas) in the early years of the fossil fuel era. In the last years of the fossil fuel era, maybe much of it will stay in the ground, unwanted and unneeded. *Source:* Photo courtesy Deutsches Museum, Munich.

frauds," writes Yergin, "of fortunes made and fortunes lost, fortunes never made, of grueling hard work and bitter disappointments, and of astonishing growth."[15] On that path of unending growth, each step was empirically and experientially grounded: producers and consumers alike could see it, experience it, feel it. The magic was sensuous in the sense of engaging all the senses, especially among motorists. "Driving in a motorcar, like all mechanical calisthenics, effects a brisk activization of the entire organism," a 1909 German promotion stated. The motorcar "stimulates the activities of the skin and lungs, in a pleasant manner and, in so doing, initiates an extremely advantageous unburdening of the internal organs, which are quite excessively gorged with blood."[16]

In fact, speed itself became one of the magical elixirs of the fossil-fueled industrial age, starting with the very first public railway, the Liverpool and Manchester Railway, in 1890. "I stood up, and with my bonnet off 'drank the air before me,' " declared a young actress. "When I closed my eyes this sensation was quite delightful, and strange beyond description; yet strange as it was, I had a perfect sense of security and not the slightest fear."[17] The automobile offered yet more. "Driving was … an adventure, producing above all the gratification of having success-fully overcome the fear," writes cultural analyst Wolfgang Sachs. "But then there was the fear's reward: breath-taking speed."[18]

The more that fossil fuels insinuated themselves into every aspect of modern life—from homemaking to manufacturing, from pumping water to growing food, from commuting to shopping, from policing the neighborhood to policing the world—the more they cast a spell over the body politic. No substance was more central, once again, than oil. Harold Ickes, President Roosevelt's interior secretary and so-called oil czar in World War II, probably put best the political centrality of oil in the United States midcentury: "There is no doubt about our absolute and complete dependence upon oil," he said. "We have passed from the stone age, to bronze, to iron, to the industrial age, and now to an age of oil. Without oil, American civilization as we know it could not exist."[19]

That sentiment has hardly changed in mainstream political debate a half century and more hence. It just acknowledges some downsides, accommodating the need to do it all cleaner and safer. The fossil fuel mantra of today might go like this:

*Fossil fuels are the lifeblood of society. They are essential to all that we know to be progress. If there are downsides—and certainly there are some—we must deal with them. In fact, it is the very wealth that fossil fuels bring us that is needed to manage the pollution.*

*Fossil fuels have been with us for a very long time and will be for a very long time to come. The reservoirs are vast, virtually unlimited. It is our right, indeed our duty, to use them. But use them wisely—efficiently, safely, cleanly.*

For the general public to come under this spell, little convincing was needed, yet much was applied. Marketers could not only turn a phrase, they could back it up with an endless cornucopia of goods, delivering speed, comfort, and prestige by the barrelful. Shoppers and drivers and party hosts need only live like, and keep up with, the Joneses, committing themselves to all that is new and improved, to the next big thing—and, not incidentally, refrain from questioning the magic, from pulling back the curtain.

So the original mystery of fossil fuels was in their source—mostly deep underground, and extent—vast pools and seams. The occasional seep, gas vent, or exposed coal seam revealed their existence for millennia, but their extent became common knowledge only once mine shafts were dug and geologists found or theorized vast stocks. And, experience told, those stocks were always yielding, whether as perpetual flares or as a result of digging and drilling and sounding into subterranean unknowns. Many thought the coal beds of Newcastle were inexhaustible, much like the forests of the colonies across the seas.[20] Resources did not just yield continuously; they yielded increasingly. The belief in endless extraction was consistent with a larger historical belief: that resources of all kinds "resurged." Trees could be coppiced, fish caught, corn harvested, and ores mined, and, in time, it all came back.[21] In seventeenth-century Britain, some people believed coal was alive, derived from "special seeds for its reproduction and growth underground."[22] Extraction only reinforced the belief—repeatedly, continuously, and at unimaginable scales—with years and decades of ever increasing oil, gas, and coal production.

For decades, indeed for several centuries, an expectation of more was rewarded, often handsomely. There were no guarantees, and there was plenty of competition, but to plan otherwise, to underinvest, to let a dry hole discourage exploration, to avoid exotic lands, was to be foolish. The storied history of oil has larger-than-life characters like John D. Rockefeller and his nemesis in Europe, Marcus Samuel, for a reason: they were fearless, they took huge risks (including with their own lives, not to mention with others' lives), and when they succeeded, they succeeded big time.

In short, a cultural myth of endless extraction could arise from a belief in "resurgence"—that resources always came back—or because the very geology of the stuff first disguised, then, with the full blossoming of the fossil fuel era, revealed so much. Either way (or both ways) people's sheer experience, not just desire or ignorance or self-promotion, was reinforcing: there has always been more, there always will be more. The evidence has been and continues to be overwhelming: with enough coaxing, the good earth always yields; it always has, always will. Booms do go bust but, say the petro-optimists, there is always another boom, somewhere, at some time. "My father was one of the pioneers in the oil industry," said J. Howard Pew of Sun Oil. "Periodically ever since I was a small boy, there has been an agitation predicting an oil shortage, and always in the succeeding years the production has been greater than ever before."[23]

In fact, one read of Daniel Yergin's landmark history of oil, *The Prize*, is that cries of depletion, "a chronic malady in the oil business,"[24] are as common as the discoveries that follow. "Earlier fears of shortage, at the beginning of the 1920s, in the mid-1940s, had ended in surplus and glut," Yergin writes, "because rising prices had stimulated new technology and the development of new areas."[25] This line of reasoning is very much alive today regarding the prospects for shale oil and gas, where, for example, in the United States, boosters are claiming that North America is becoming "the next Saudi Arabia." For them and consultants like Yergin, continuous fossil fuel abundance is all a matter of prices and technologies and the ability of governments to assure access—along with the steadfastness of industry and government in ignoring the fears and pessimism of the naysayers.

In the end, the magic and spell of fossil fuels fed into a dominant material myth of modern, industrial, consumerist society: growth. This myth has it that material growth (economic and political) is both good and necessary, as natural and healthy as a growing crop and a growing child, as inevitable and everlasting as the rising and setting sun, the waxing and waning tide. Never before had an energy source and associated technologies promised—and delivered—ever increasing, ever cheaper (with occasional exceptions) usable energy (chapter 2). Human and animal power, hydro and wind were always inherently constrained. Coal production, by contrast, increased steadily for centuries, oil for a century and a half. Availability for households and industry alike increased steadily as prices declined. Fossil fuels, it turns out, may be singularly, and maybe uniquely, well suited to creating and perpetuating such a myth of endless growth. ExxonMobil's chief executive got it right in 2006 when, criticizing President George W. Bush's statement that America is addicted to oil, he said, "To say that you're addicted to oil and natural gas seems to me to say you're addicted to economic growth."[26]

From an experiential perspective, generation after generation knew growth, and growth not just in energy. In fact, as described later in this chapter, the energy basis of a growing economy and growing political power was largely invisible, shaded to do the magic. Rather, they knew growth in goods: the number of household items one could buy, the miles one could drive, the size of a home one could have, the ever increasing standards of heating and cooling one could dial up. For Americans, Brits, and others, one could know the projected power of one's country by watching its military bases spread around the world, its navies protecting freedom of seas. A society of consumers consuming more and more hardly needed to be convinced that endless growth was possible and right. They didn't have to see or calculate the energy behind it all. They merely had to earn the wages and spend the cash.

But growth was not just possible, it was increasingly imperative. For the imperial projection of power, wood and sail power imposed inherent limits. Trade and colonization were one thing, but controlling the sea-lanes and becoming a truly global power were possible only when Britain adopted coal-fired boilers and steel-clad hulls (the steel smelted with coal). When King George II asked Matthew Boulton who, with James Watt, introduced the steam engine (powered by coal) what business he was in, Boulton reportedly said, "I am engaged, your Majesty, in the production of a commodity which is the desire of kings." When asked what he meant, Boulton replied: "Power, your Majesty."[27]

At the turn of the twentieth century, imperial power played out in a race for naval supremacy between Germany and Britain that led to oil-fired ships and troop transport. Britain won that race but, like all other military advantages, technological or tactical, its strategic advantage dissipated as others adopted the new, readily transported, highly concentrated fuel. Once adopted, though, no one could go back. Fossil fuel consumption, for reasons of national security and the projection of power, let alone the need for parity, ratchets up (box 3.3). France's head of petroleum, speaking of oil's role at the end of World War I and the Allies' victory over coal-dominated Germany, put it thus: "As oil had been the blood of war, so it would be the blood of the peace. At this hour, at the beginning of the peace, our civilian populations, our industries, our commerce, our farmers are all calling for more oil, always more oil, for more gasoline, always more gasoline." And then, as if to presage the century's future, he declared, "More oil, ever more oil!"[28]

If one country produced more and more oil and benefited most in terms of power from its use, it was the world's largest producer for most of the twentieth century, the one that supplied 80 percent of the Allies' oil in World War I (a quarter of that from Standard Oil of New Jersey alone) and 90 percent in World War II—the United States.[29] As one oilman wrote to President Calvin Coolidge in 1924, "A deficiency of oil is not only a serious war handicap to us but is an invitation to others to declare war against us," a sentiment Coolidge shared: "The supremacy of nations may be determined by the possession of available petroleum and its products."[30]

So fossil fuels and their growth were necessary to extend national power, defend the nation, and, eventually, provide jobs and keep the

**Box 3.2**
Our Daily Oil

---

In 2011 the United States consumed 18.8 million barrels of crude oil a day; worldwide consumption averaged 87.3 million barrels a day.[a] Even those who work with such numbers daily cannot really grasp what they mean in everyday terms. So to make those quantities more tangible, consider the following comparisons.

*US Consumption*

- 18.8 million barrels is equivalent to 789.6 million US gallons.[b]
- This is equal to about 1,196 Olympic-sized swimming pools.[c]
- An average bathroom faucet would need to run for 683 years to release those 18.8 million barrels.[c]
- This would mean a faucet turned on in 1330 would only now be running out.

*Worldwide Consumption*

- 87.3 million barrels is equivalent to 3.666 billion US gallons.[e]
- This is equal to about 5,555 Olympic-sized swimming pools.[f]
- An average bathroom faucet would need to run for 3,171 years to release those 87.3 million barrels.[g]
- This would mean a faucet turned on in 1158 BC would only now be running out.

**Sources**

a. US Energy Information Administration, "Oil: Crude and Petroleum Products," http://www.eia.gov/energyexplained/index.cfm?page=oil_use.

b. U.S. Energy Information Administration, "Frequently Asked Questions," http://www.eia.gov/tools/faqs.

c. Based on Fédération Internationale de Natation, "FINA Facilities Rules 200—2013," http://www.fina.org/project/index.php?option=com_content&task=view&id=51&Itemid=119.

d. Assuming a flow rate of 2.2 gallons per minute according to maximum federal standards: Portland Water Bureau, "Indoor Water Efficiency: Faucets Fact Sheet," http://www.portlandoregon.gov/water/article/305150.

e. US Energy Information Administration, "Frequently Asked Questions."

f. Based on Fédération Internationale de Natation, "FINA Facilities Rules 2009–2013."

g. Portland Water Bureau, "Indoor Water Efficiency."

economy growing. To this day, few seem to see the irony in growth for growth's sake, despite the best efforts of economists like Herman Daly. Even fewer seem to see the biophysical impossibility of it all, again despite the efforts of Daly, but also of energy analysts like Charles Hall and Cutler Cleveland, social analysts like Richard Heinberg, and, now, many others. Instead, throughout the twentieth century to today, consumers are told at every turn that such growth is to be expected, if not demanded, that it is virtually a birthright, ordained by those with the most sophisticated technologies, the most efficient markets. From the early days of marketing to the present, a drumbeat of growth messages has greeted Americans and others. On billboards and through the airwaves, in the news and on the big screen, at home and in school, the message has been clear: a growing economy, a growing population, a growing infrastructure of roads and bridges, electric power lines, and fiber-optic cables—all this is a sign of a healthy, strong society.[31] What's more, fossil fuels are here to stay, like it or not. In 2005, for instance, ExxonMobil "advanced a carefully designed, research-tested campaign," writes journalist Steve Coll, "to persuade political and media elites that while the oil industry should not necessarily be loved, it should be understood as inevitable."[32] Curiously, ExxonMobil's world energy forecast projected out twenty-five years showed just that. And for their target audience, "informed influentials," it seemed to be just what they wanted to hear—more growth, progress forever.[33]

With all this "progress"—commercial and strategic—no one had to point out that little of it, certainly little at the rates experienced in the twentieth century, would have been possible without cheap fossil fuels—cheap economically, cheap energetically, cheap environmentally (see chapter 1). But business historian Alfred Chandler saw just that in the postwar "economic miracles" of Germany and Japan: "The German and Japanese miracles were based on improved institutional arrangements and cheap oil."[34] Oil historian Yergin goes a step further: "In the boom years of the 1950s and the 1960s, economic growth throughout the industrial world was powered by cheap oil."[35] More recently, the global business and economic consulting firm McKinsey and Company found that economic growth depended greatly on material surplus, a condition now coming to an end:

During most of the 20th century, the prices of natural resources such as energy, food, water, and materials such as steel all fell, supporting economic growth in the process. But that benign era appears to have come to an end. ... The past decade alone has reversed a 100-year decline in resource prices as demand for these commodities have surged.[36]

So the myth of endless growth arose in part because of its obvious appeal (goods galore, bads somewhere else) and the fact that people have had direct, self-reinforcing experience with it, and in part because it has been a terribly convenient myth for those who benefit the most: industrialists, investors, geopolitical power brokers, and, for all of them, their political enablers. Now the growth myth appears to be a myth in two senses. One is that which explains and justifies, that situates a society in the larger world and makes sense of one's place in that world; the other is that which is patently false, that is constructed to advance an agenda but cannot sustain scrutiny. Fossil fuels could undergird a growth myth in the first sense when one boom followed another and costs could be ignored or displaced. But just as no crop or child can grow indefinitely, the costs of continuous growth can no longer be ignored and displaced. What goes around (e.g., greenhouse gasses, toxic substances) does eventually come around. At best, fossil fuels can legitimate their own exit (chapter 4). Now the fossil fuel era is coming to an end and, with it, the myth of endless growth on a finite planet. If that myth could be constructed, it can be deconstructed and reconstructed.

In sum, the beginnings of the fossil fuel era were, from a material perspective, the beginnings of the myth of endless growth. The economic and the political justifications would follow. The myth got a boost with each entrepreneurial venture, with each new technology in the value chain, with each new law and regulation favoring exploration and refining, with each victory at sea and on the battlefield, with each new item of consumption (especially those once a luxury and now a necessity), with each ratchet up in the standard of living. Never mind that the increases in fossil fuel production rates resulted from new technologies—the steam engine, the railroad, the pipeline—not endless resurgence of the fuel itself. The key point here is that from a cultural perspective and, hence, from the perspective of continuous resurgence and seemingly costless abundance, a belief emerged that resources are endless. This belief system, so appealing, so experiential, so logical (there is no limit to human ingenuity), who could escape its spell? Who would want to? This belief system would be cast as if in stone, but in reality in one-time, nonrenewing sources of energy, slippery as oil, dusty as coal.

That fossil-fueled myth—endless material growth—so grounded geologically and economically, so appealing, so sensuous, so powerful, has endured for a century and more. Only now can it be questioned and even still, in mainstream discourse, only with respect to obvious and personal matters—obesity, drug abuse, financial debt. To begin the essential questioning of fossil fuels themselves and their tight link to endless growth, I turn from the first prong in the double agency of cultural change, fossil fuels themselves, to the second, the specific actors who ginned up frontier expansion from mere settlement to global consumerism, people who had a vision and the wherewithal to enact it.

### Boosterist Vision, American Emporium

In mid- and late-nineteenth-century America, with the frontier closing to the west, speculators, newspaper editors, merchants, and chambers of commerce competed across the Midwest, from Cincinnati to Toledo, from St. Louis to Chicago to become the next great city, the successor to New York, which they assumed was about to succeed London, the successor to Rome. For all their competition, though, they had a shared vision, one of commercial greatness and expansion, what some called "emporium," a curious term that combines grand market and empire. It was a vision that inspired those who went west, settled the land, built its great cities, and extracted its resources. It was a vision that, above all, attracted investment, both public (for the telegraph, canals, railways) and private (for railroads, grain and meat conglomerates, timber companies).

And while their "rhetoric always inclined toward enthusiastic exaggeration and self-interested promotion," writes historian William Cronon, "not all was fantasy. The 'boosters,' as they came to be known, expounded serious theories of economic growth that dominated nineteenth-century thinking about frontier development."[37] Indeed, a central concept in the boosters' rhetoric was growth—population growth, industrial growth, economic growth—"assured by nature—or better yet—ordained by God [such that] only a fool could doubt its future promise," writes Cronon. "No mere human power could alter the forces that compelled [a city's] growth."[38]

Boosters were more than self-interested promoters, though, more than cheerleaders. Hard driving, intelligent, and, above all, confident, they collected data and extrapolated growth trends of the past into the future, deeming such growth not only natural and ordained by God but as inevitable; it was "high destiny," material progress forever. And, perhaps foreshadowing the global flow of oil in the coming century, "the imperial metaphor which cropped up most frequently in booster prose, whether applied to New York or Chicago [was] 'tributary' [which] conjured up the image of a great river. ... Like Rome, Chicago's imperial future would arise from the wealth flowing into its coffers from the territory around it."[39] Far from contradicting the new nation's democratic aspirations, this triumphant vision of "America as a commercial empire allowed boosters and others to believe that the flow of 'tribute' among its various parts enriched all and impoverished none," writes Cronon. "Commercial 'conquest' yielded happy results for conqueror and conquered alike."[40]

What kind of a society were these boosters and their followers creating? What was its cultural character? Boosters insisted it was all about promise, progress, and prosperity, about individual opportunity and free choice, about epic grandeur and democratic freedom. Or, as one booster put it, it was all about "mutual concord, self-sustained; unlimited expansion: perpetual buoyancy, and perpetual life!"[41] A French political scientist was less buoyant. In 1891 Emile Boutmy wrote:

> The striking and peculiar characteristic of American society is, that it is not so much a democracy as a huge commercial company for the discovery, cultivation, and capitalization of its enormous territory. ... The United States are primarily a commercial society ... and only secondarily a nation.[42]

Boutmy wrote just as fossil fuels in the United States and worldwide were overtaking biofuels (wood) as the dominant source of energy. He could not possibly have imagined what would be the next frontier, given that this one was, geographically, closing. Who could guess that what Boutmy saw as a "huge commercial company" would become even bigger, in total and per capita, even more commercial, even more extensive, even more imperial? And who could guess that it wasn't a new frontier in the sense of virgin lands and untapped resources that enabled it, but vast amounts of nonrenewing fuels, dense and easily transported? As it turned out, of course, geographically, the "next frontier" would be the entire world, but its "perpetual life" would be the fossil remains of long ago life, its tributaries rivers of oil (and trainloads of coal and tanks of natural gas).[43]

If "emporium" captured the essence of late–nineteenth-century America, an America run largely on wood, with canals and carriages still mostly horse drawn and the railroad connecting the continent only in 1869, then perhaps "consumerized republic," as historian Liz Cohen calls it, captures twentieth-century America, run on fossil fuels, with the automobile, the filling station and interstate highways, department stores and shopping malls its defining icons.[44] This "next frontier," no less a product of boosterist dreams, of confident expansion, of capitalist investment, grew "perpetually" not because there were wide open spaces and limitless forests but because all that dense energy could power unimaginable levels of consumption. Shopping, commuting, eating out, traveling at superhuman speeds, heating and cooling to create "a portable climate," collecting big data: all this and more has depended hugely on cheap and abundant energy. The commercial society became a supercharged society, drunk on cheap energy, heedless to full costs, all too willing to let markets and technologies, indeed growth, not to mention the occasional military intervention, solve its problems.

Cronon sums up the nineteenth-century boosterist project of the great emporium thus:

The boosters expressed what many Americans believed—or wanted to believe—about the expansion and progress of the United States and its Great West. They offered seemingly rational arguments to reinforce the visionary faith that sustained many who lived and invested in the region. As a group, they present a strikingly consistent picture of how the western landscape would be absorbed into a commercial system.[45]

Now, in this hyperenergized consumerist and commercialized society, at once local and global, Americans and so many others want to believe that such expansion and progress can continue (box 3.3).

**Box 3.3**
Boosters in Brazil

Boosters come in for the rush, for the lucky strike, for the big find. When it's over, and it always is in the extractive industries, they are nowhere to be seen. Someone else reports the end, picks up the pieces, and cleans up the mess.

And yet everyone seems to want to believe the boosters. After all, they are so sure, so confident, so optimistic, indeed, so believable. They promise, and sometimes deliver, so much. Until it's over. Brazil, only one in a long line, is a recent case in point.

"God has been very generous with the Brazilian people, who have long waited for the chance to be respected in the world as we are today."
—President Luiz Inacio Lula da Silva after Petrobras, Brazil's national oil company, sold $70 billion in shares at a higher-than-expected price, 2010

"No other place on the planet is seeing this kind of investment. ... This decade is our chance to rise."
—Márcio Mello, former geologist for Petrobras then chief executive of HRT, a new Brazilian oil company, 2011

Brazil will become an oil power by the end of the decade, with production in line with that of Iran.
—Pedro Cordeiro, energy consultant, Bain & Company, 2011

**Box 3.3**
(continued)

> "[Petrobras] is now facing soaring debt, major projects mired in delays and older fields, once prodigious, that are yielding less oil. ... The undersea bounty in its grasp also remains devilishly complex to exploit."
> —Simon Romero, reporter, *New York Times*, 2013

**Sources**

Guillermo Parra-Bernal and Denise Luna, "Brazil Basks in Record Petrobras Deal," Reuters, September 24, 2010; Simon Romero. "New Fields May Propel Americas to Top of Oil Companies' Lists," *New York Times*, September 19, 2011, A1; Simon Romero, "Petrobras, Once Symbol of Brazil's Oil Hopes, Strives to Regain Lost Swagger," *New York Times*, March 27, 2013.

The rational arguments for "stimulating the economy" and "growing our way to prosperity" reinforce a visionary faith in endless economic growth (and, implicitly, decoupled material growth). From the factory to the shopping mall, from agencies to schools, from Main Street to Wall Street, the mantra resounds: just grow the economy, consume more, be more efficient, clean up the waste, and all will be well. What is striking here in the early twenty-first century is how consistent the message is, indeed, how persistent it is, especially in the face of system failure. Such is the testament of a cultural norm a century and more in the making, born of frontier expansion and cheap energy. The commercial imperative of nineteenth-century America was sustained by a vast frontier, its original inhabitants killed or exiled. That same imperative, with consumerist variations, was sustained through the twentieth century by cheap energy, overwhelmingly fossil fuels.[46] Now humanity faces a different imperative: paying the bills for those two centuries' excesses and living within our biophysical means, without a frontier, geographic or energetic. A logical starting point is challenging the cultural norm and removing the energetic agent—fossil fuels—or at least its dominance. A central conundrum, however, is that much of what drives this patently unsustainable and inequitable system occurs behind a cloak of secrecy abetted by the very ease of use and, we'll see, invisibility of fossil fuels themselves (box 3.4).

**Box 3.4**
Limits to Growth Revisited

Research scientist Graham M. Turner examined thirty years of data since the publication of the futures modeling of the Club of Rome and its much-debated report, *The Limits to Growth* (*LtG*). He found that "the data generally continues to align favourably to the *standard run* scenario," including the likelihood of a collapse in the global economy, natural resources, and human population around 2015. Adding concepts of energy return on investment (chapter 2) and peak oil (chapter 1) only supports such resource-constraint modeling. Among the key drivers of collapse, Turner finds, is the diversion of capital from productive uses (e.g., growing food) to "extracting diminishing supplies of conventional oil or difficult extraction of non-conventional oil (e.g., tar sands, deep water, coal-to-liquids, etc.)." So while the proximate cause of a global financial crisis is likely to be financial (e.g., excess debt), the ultimate cause would be resource constraint.

Turner concludes "that the scientific and public attention given to climate change, whilst important, is out of proportion with, and even deleteriously distracting from the issue of resource constraints, particularly oil. Indeed, if global collapse occurs as in this *LtG* scenario, then pollution impacts will naturally be resolved, though not in any ideal sense." What Turner did not say is there is a politics here—a politics not of managing carbon, let alone of finding efficiencies. Rather, if these trends and scenarios are at all close to the mark in the coming years, the politics would be simultaneously one of allocating resources now available (including fossil fuels; see chapter 4) to regenerative uses and of keeping fossil fuels in the ground.

**Source**

Graham M. Turner, "On the Cusp of Global Collapse? Updated Comparison of *The Limits to Growth* with Historical Data," *GAIA* 21 (2012): 116–24.

## Easy Come, Easy Go

So boosters were no mere salespeople. They had a vision, shared widely from the homestead to the boardroom, from city hall to the halls of justice. It was a vision of enterprise and empire, of commerce and grandeur, of equality and democracy, of growth and prosperity. Propelled in the nineteenth century by vast untapped resources, visible to all, now, in the twentieth and early twenty-first centuries, it is propelled by resources nobody sees. The boosters' vision, extended into these centuries, has always been above ground, in line with man's great achievements, with its shiny new things, with that which is bigger, faster, cheaper. Beneath it all, however, is energy, much of it from fossilized life, whose very invisibility helps define the politics of the fossil fuel era. Pulling back the shades and shedding light on those politics is one step in bringing that era to a prompt, life-preserving end.

Fossil fuel's historical moment occurred when forests collapsed from overharvesting and timber became increasingly scarce; new ways to power industry were needed when other countries were capturing export markets, when aggressors threatened and a military advantage was needed. But fossil fuels, probably more than any other substance, food, or technology, propelled their own developments, each unimaginable, each with highlighted benefits and shaded costs, each with a life of its own, a life deemed by nearly everyone as inevitable and perpetual, essential and life giving. By themselves, fossil fuels are mere physical embodiments of usable energy and no different from their predecessors—wood especially, but even human, animal, wind, and water power. Fossil fuels are just easier to get in bulk, easier to transport, easier to use, the primary difference—that is, the physical difference—being the ease of combustion, of transport, of chemical transformation. As a domestic fuel, for instance, coal was easier to use than wood: it didn't have to be chopped, it weighed less per energy unit, and it created a longer-lasting, steadier fire.[47] Fuel oil and natural gas are all the easier, of course, amounting to little more than flipping a switch and setting a thermostat.

But "ease" is a two-sided coin. The ease of use, that is, of the intended use, is simultaneously the ease of displacement—of hiding, of ignoring, of denying. I burn a lump of coal, and my house is heated, my bread baked, my bathwater warmed. What I do not deal with, or necessarily need to pay for, are the upstream effects on my fellow citizens, those who, in early industrial England, were rendered a "separate race" due to their physical deformities from backbreaking work and light deprivation, and those whose livelihoods in the hollows of Appalachia now are rendered expendable with every mountaintop blown off (see chapter 6). I fill my gasoline tank but need not make a connection to offshore blowouts and rig worker deaths. In fact, even if I do make the connection, I know at least intuitively that if I refrain from pumping or, for that matter, pump a little more and drive a little more, it will make no difference in the larger scheme of things.

In that scheme, the one constructed of a fluid, global energy system where no one is really in charge, the collective action problem rears its ugly head: as an individual, even as a company or, in many cases, a country, there is nothing I can do of consequence by choosing to consume or not consume the marginal unit. The fluidity of oil (and, to a lesser extent, coal and natural gas) is the fluidity of nonresponsibility, of distancing, of displacing the ethical choice.[48] As an individual actor, I must leave it to those who extract and those who regulate such extraction to do what is right. That those actors are precisely the ones who benefit most from extraction and from displacing true costs (costs to personal and ecological health, to future generations, to "downstream" vulnerable populations) is to say that "the system" is effectively built for displacement. And it is so built at all ends, from producer to consumer, from citizen to ruler. No one is responsible. It is a system of abundance in a culture of nonresponsibility, a system supremely well designed to grow, boom, extract, and move on, leaving behind tailings, poisoned water, and climate-disrupting gases. With the downside shaded, rendered invisible by design and by the very structure of supply chains of concentrated energy and concentrated power, with supplies abundant and ever growing, it is not hard for actors at all nodes to absolve themselves of responsibility, not hard to feel entitled to a constant, cheap supply and to have others worry about the downside.

The fossil fuel order is, in short, a system of excess in a culture of entitlement. Why would this be true for fossil fuels and not other energy sources? Part of the answer is the inherent invisibility of the fuels—in place and in use.

### From Invisibility to Secrecy: The Power of Powerful Fuels

Fossil fuels' invisibility starts with the fact that except for the occasional oil seep, gas vent, or hillside coal exposure, these fuels are indeed underground, out of sight. Hardly anyone actually sees coal anymore, what generates more than half of the electricity in the United States; or the oil that powers cars, trucks, and trains; or the natural gas that supplies virtually all the fertilizer. For me, it was only when I sat in the observation deck of a passenger train crossing the American plains that I saw heaps of coal in mile-long freight trains heading east. It is only when I'm careless at the gas pump that I see a few drops of gasoline. It is only when my water heater leaks gas that I smell, but don't see, natural gas (and even then it is not the gas itself I'm smelling but the mercaptan scent added to the natural gas). As a boy growing up in Walla Walla, Washington, my friends and I played at the "tar pit." I doubt anyone made the connection to the fuels that heated our homes or powered our cars.

The tar pit was just a curiosity. If wildcatters and petroleum geologists had not established that such occurrences were, or could be, signs of vast underground pools, we, like our ancestors, would go through life never knowing what stocks of energy, accumulated over millions of years, lay beneath our feet. And when the oil, coal, and gas are pulled out of the ground, it quickly goes into pipelines, tankers, and refineries. Even those who work with fuels and solvents and industrial feed stocks see only a tiny fraction of the fossil fuels that ooze and flow and tumble through industrial and residential infrastructure (box 3.5).

**Box 3.5**
Oil, Oil Everywhere, Not a Drop to See

Few of us ever come into direct contact with a fossil fuel. And yet we see, touch, smell, and hear their products daily. We just don't see the oil, natural gas, or coal that goes into them: out of sight, out of mind, out of ethical purview. Fossil fuels are both ubiquitous and invisible. A typical barrel of oil (approximately 42 US gallons) yields 19 gallons of gasoline, 11 gallons of diesel, 4 gallons of jet fuel, and 7 gallons of other products. Among those "other products" are these:

- Aircraft windows
- Police car glass
- Storm doors
- Bath and shower doors
- Lighting fixtures
- Taillight lenses
- Dials and knobs
- Bathroom plumbing materials
- Bottles
- Medical and dental parts
- Ink
- Paper coatings
- Magnetic tape
- Transparent plastics
- Plastic tiles
- Plastic bags
- Cosmetics
- Perfumes
- Flame retardant
- Resins
- Disinfectant
- Film
- Agricultural chemicals
- Latex paints
- Beverage bottles
- Hoses
- Insulation
- Bowling balls
- Bedding
- Cushions
- Car seats
- Milk bottles
- Insulation pipes
- Diaper covers
- Toys
- Detergent
- Aspirin
- Mouthwash
- Wood glue
- Tires
- Carpet backings
- Rope
- Tape
- Nonwoven fiber
- Draperies
- Curtains
- Awnings
- Paint rollers
- Speakers
- Grills

**Box 3.5**
(continued)

- Telephones
- Floor polish
- Diapers
- Ink
- Crayons
- Dishwashing liquids

- Deodorant
- Eyeglasses
- CDs and DVDs
- Ammonia
- Heart Valves

**Sources**

U.S. Energy Information Administration, "Oil, Crude, and Petroleum Products Explained," http://www.eia.gov/energyexplained/index.cfm?page=oil_use; "Petrochemicals," PDF file created by American Fuel & Petrochemical Manufacturers; U.S. Energy Information Administration, http://www.eia.gov/energyexplained/index.cfm?page=oil_refining.

Instead, what we moderns "see" is clean electricity or, more precisely, incandescent (or fluorescent or LED) light, bread toasted to perfection, beverages piping hot or icy cold. We see cars whizzing by and airplanes jetting overhead. These are services for which there is no obvious downside aside from bills to pay and tickets to buy.

At one level, all this is a great achievement. Who would want to see filthy piles of coal at one's doorstep, as was once the norm? Who would want to touch and smell the gasoline that powers our cars? Who would want natural gas to be its "natural" self—odorless, colorless, and explosive? What's more, those who provide the fuel are just providing a service. "The 'service we render' and the ongoing investments we make from our earnings are critical," stated a CEO of ExxonMobil in 2010. "Simply put, delivering energy in a safe, secure and responsible manner improves the lives and opportunities of billions of people the world over."[49]

It's all about service. And yet even in a service station, few see the gasoline. So at another level, there is a politics in such invisibility, a politics that plays out as secrecy, as backroom deals, as "what you don't see won't hurt you," as "trust us, we have everything under control," as writing the rules of the game and delivering the goods, all perfectly rational, sensible, indeed, "secure and responsible." Or so it would appear.

To see the inherent strategic element in energy invisibility and secrecy, consider preindustrial energy sources. For the great bulk of human history, the only energy source was food and then, with pastoral and settled farming practices, animals and wood. These were for all to see. Certainly a chieftain could claim the best river bottom and the best oxen, but such claims could not be hidden. And what is out front for everyone to see is what tends to be held to account. Not so when the great bulk of an energy source is in pipes and tanks and bins.

John D. Rockefeller discovered the strategic advantages, arguably the strategic necessity of secrecy, at the very beginning of the fossil fuel era, when gasoline and fuel oil were minuscule parts of an oil company's products. Rockefeller, founder of Standard Oil Company (now Exxon-Mobil), pioneered a number of practices to stabilize prices and eliminate excess capacity, all by driving out competition and getting legislators to write favorable laws. Kickbacks from refiners, predatory pricing, and communication in code all depended on secrecy. "The Standard men, moving in great secrecy," writes historian Yergin, "operated through firms that appeared to be independent to the outside world, but had in fact become part of the Standard Group."[50] When challenged, Rockefeller replied in a way that revealed the nature of what he called "The Great Game": "But I wonder what General of the Allies ever sends out a brass band in advance with orders to notify the enemy that on a certain day he will begin an attack?"[51] Today the Rockefeller tradition continues, only with much more technical and marketing sophistication. As a result of such tactics, by 1879, Standard Oil controlled 90 percent of America's refining capacity and most of its pipelines, and it dominated transportation. By 1891, it produced a fourth of America's total output of crude oil.[52] When the Supreme Court dissolved Standard Oil in 1909, the chief justice concluded that "no interested mind can survey the period in question [since 1870] without being irresistibly drawn to the conclusion that the very genius for commercial development and organization ... soon begat an intent and purpose to exclude others ... from their right to trade and thus accomplish the mastery which was the end in view."[53]

The physical invisibility of fossil fuels translates to the social secrecy of fossil fuels and, as a result, a logic of mastery and power. With Rockefeller and his competitors and their successors, it was economic power initially, then political power, and it played out internationally as well as domestically. When in the 1940s Venezuela deigned to sell oil directly onto world markets (rather than through the "Seven Sisters," the handful of big oil companies that dominated oil distribution worldwide), its president explained that "the veil of mystery over the marketing of oil—behind which the Anglo-Saxons had maintained a monopoly of rights and secrets—was removed forever."[54] Then the Organization of Petroleum Exporting Countries established its own "rights and secrets."

Although from a democratic perspective, concentrated power is problematic in its own right, the point here is that its precondition, secrecy, is inherent in a fossil-fueled political economy. If a player, private or governmental, should operate otherwise—that is, with transparency—it would be destroyed in a competitive environment widely described as ruthless. It is more than coincidental that activist groups working inside countries experiencing the "resource curse" (the decline in social well-being as oil wealth increases) have chosen transparency as a key to overcoming that curse. It is for the same reason that key players resist, vehemently.[55]

Holders of fossil fuels—whether companies, countries, or investors—also have power, however implicit, when they control an oil field, a pipeline, a sea-lane, or a line of credit. The tacit bargaining power of a holder of concentrated physical power is the power to supply and the power to deny, the power to feed and the power to bleed. That power is very much a function, as are all forms of bargaining power, of alternatives. When users have no alternative, when they are utterly dependent on a fossil fuel, the holders have power over them. Similarly, when oil is taken to be a globalized product, particular to no place, it is the producers who always have alternatives. So, for example, when the United States threatened to restrict Canadian oil imports of Alberta tar sands, ExxonMobil declared it would simply sell the oil to Asia.[56] Russia has made similar threats regarding its natural gas exports to Europe. "Throughout oil history," writes Norwegian oil historian Helge Ryggvik, "the ownership of pipelines and the choice of routes has always been very significant, both for who secured the oil rent and for what economic side benefits were to come of the industry."[57] Ownership means that others make concessions, compromise, bargain, sign skewed contracts, sweeten the pot, do favors. In short, power, especially in a globalized world, is upstream.

It is the globalized status, in fact or in rhetoric, that Steve Coll, in his extensive history of modern ExxonMobil, *Private Empire*, stresses is seen in two very different ways. Oil industry people tend to see oil, as do economists, some policymakers and pundits, and other globalization boosters, like any other commodity: one that benefits everyone when traded freely around the world. "The central reality is this: The global free market for energy provides the most effective means of achieving U.S. energy security," says an ExxonMobil CEO. "In the global market, the nationality of the resource is of little relevance. ... Energy made in America is not as important as energy simply made wherever it is most economic." Many policymakers and citizens, by contrast, see oil as "ours" or "theirs." They understand shortages not as temporary blockages in the global market but as manipulations by the industry or leaders or as premonitions of the end of oil (or, more accurately, the end of cheap energy). These ideological differences define in yet another way the culture of fossil fuels. To the extent the globalized view is official policy (as it effectively has been in the United States for several decades, since oil production peaked in the United States in 1971 and the world's greatest consumer became increasingly dependent on imports), then it is in the interest of the United States and other importers "to promote policies—free trade, open markets, low taxes, maximized oil production everywhere—that would fill the global pool with as much new oil as possible," writes Coll.[58] More than that, though, it is the globalized view that confers the most concentrated power, putting decisions in the hands of global investors and those who must respond to investor demands for "reserve replacement" and outsized returns—that is, fossil fuel producers, private and state. In this view, it is everyone else—that is, those downstream—who takes the price at the pump, cleans up the messes, and adapts to extreme weather.

Producers, private and state, can use their oil power to raise prices (to an extent), and they can use it to influence elections; depose rulers and install designates; and lobby, bribe, and threaten legislators. The history of the fossil fuel era is replete with such stories; whether they are the rule or the exception, they are clearly the baseline from which all strategic calculations are made. Finally, and maybe most consequential, these actors can use their power to create illusions like "climate-friendly" natural gas and "clean coal": "Now that nine out of every ten tons of the nation's coal vanishes into power plants," writes coal historian Freese in 2003, "many Americans can harbor the illusion that coal is no longer a major energy source or a big environmental threat, even while the nation burns more of it than ever."[59]

The invisibility of fossil fuels is thus a defining feature of a fossil fuel culture: the stuff is just there, somewhere, always available yet out of sight. But when it is out of sight of the many, it is in the hands of the few. Holders of fossil fuels have power. They can use it for both economic and political purposes. Economically, that power comes not from the so-called downstream operations—refining and distribution, let alone the efficiencies of combustion—but upstream, where there is the "ownership and control of the oil in the ground," says Yergin.[60] Writing the rules of that game for gaining access, exploring, drilling, and pumping is where the real power is. This is why, as oil historian Ryggvik puts it, many oil "companies are playing a strategic game where the goal is first and foremost to secure access to large oil fields."[61] A perusal of industry histories and case studies suggests that no industry, with the possible exception of the finance industry, which itself is deeply connected to the energy industry, exercises such power so consistently, so thoroughly, and so invisibly.

The power of fossil fuels is thus not primarily physical, or even economic; it is political. That politics starts with physical access and cost-effective exploration and then winds its way through nearly all aspects of an economy—most notably its food, manufacturing, and financial systems—and eventually to its political system. Because the political system cannot generate its own revenues, it must extract them from others and, in the process, become beholden to others. What's more, to play the political game requires skin in the game, which fossil fuel players have in spades. And it requires winning. Those who win tend to be those who write the rules of the game. And so it goes. As long as access is assured, and buyers have the money, the stuff keeps flowing in pipelines and getting loaded on railcars, the game keeps going and the power keeps its locus in the hands of the few. To confront that power, to question access and entertain the idea of keeping the stuff in the ground, is to challenge the prerogatives of those who have played the game so well and expect to continue playing it, of those who know well the benefits for them and others, those who are confident the costs are well accounted for or, with rational policies, can be. To confront that power—power, to repeat, that is upstream, not down—is not a matter of listing impacts or lining up costs against benefits, as if such a ledger can be meaningfully constructed for fossil fuels and all their effects. Rather, a starting point for that confrontation, aside from shining light on its politics, rendering them more visible, is to understand how such substances, seemingly neutral in their chemistry and energetics, skew a culture. (See table A3.1 in the appendix at the end of the chapter.)

## Cultural Skewing

No single substance like coal or oil can singlehandedly change a society, rearrange its cultural norms, shift its power distribution, or exploit its vulnerable peoples. But the historical record suggests that in combination with other patterns—industrialization, colonization, postcolonial independence, globalization, financialization—it is very difficult for a society not to be swayed, if not structurally skewed, by such largely invisible commodities and the powerful actors who command them.

In Britain, for instance, the original coal miners in Newcastle were serfs working as virtual slaves for the church, followed by immigrants from across the isles. Landless, desperately poor, and shunted into work camps, they soon became a class apart in what was already an extremely class-based society. One resident described them as "lewd persons, the Scums and dreggs of many [counties], from whence they bine driven."[62] Kept separate from resident populations, they worked long hours at grueling tasks. Deprived of sunlight, Vitamin D deficiency made rickets common, stunting their growth and deforming their bodies, further distinguishing them from other classes. Consequently, coal miners were seen not just as a lowly working class but as a separate race, social outcasts, ostracized by society as a whole. In Scotland, entire families were bonded for life to a coal mine; if a mine was sold, the families were sold with it. This internal cultural insult prefigured Britain's external insults of colonization and slave trading. All told, the cultural impact, both internal (an entirely new class, physically and economically distinguishable) and external (the brutal subjection of peoples in Africa and elsewhere), was monumental, defining Britain in its imperial heyday and, arguably, to this day.

Another example of cultural skewing with a fossil fuel is that captured by the term *resource curse* (see chapters 1 and 4). Even under the strictest of definitions, the impacts of becoming an international oil supplier are profound. In the exemplar oil-producing country, Norway, for instance, the one that took to heart the lessons of state failures in Africa, the Middle East, and elsewhere, internal critics find their country's culture has been compromised (chapter 10). According to historian Ryggvik, although Norway has been able to avoid the extreme features of the oil curse, it nevertheless failed to maintain a national mandate for a "moderate pace of extraction" and long-term oil policy, resulting in a "very oil-dependent economy."[63] That economy, writes Ryggvik, is one "whose main actors [the national oil company, Statoil] have the same underlying interests as those companies (e.g., BP, Exxon) which early Norwegian oil policy aimed to protect the country from." Rather than preserving public control over the rate and extent of extraction of a public resource, "Norway had developed an oil-industrial complex which in many ways had its own interests that were in opposition to society as a whole."[64] The oil sector, primarily through Statoil, has taken on a life of its own, attempting to extend its reach into the Norwegian Arctic, as it is now doing in Africa and Asia. The result, aside from direct economic impacts, has been to compromise "a strong self-image of Norway as an environmentally conscious nation."[65] Canada, says political scientist Thomas Homer-Dixon, has similarly skewed its economy and democracy: "Capital and talent flow to the tar sands [in Alberta] ... and anyone who questions the industry is [labelled] unpatriotic" (see box 11.1 in chapter 11).[66]

Finally, if a measure of a society's integrity, its claim to being "civilized" or "developed" or "progressive," is its treatment of children, then consider the impact of fossil fuel on children. A British parliamentary commission in the 1840s found that "in all the coal mines, in all the districts of the United Kingdom, the care of these trap doors [to keep air moving and prevent the accumulation of deadly gases] is entrusted to children of from five to seven or eight years of age, who for the most part sit, excepting at the moments when persons pass through these doors, for twelve hours consecutively in solitude, silence and darkness."[67] And then, when narrower coal seams required tunnels too low for horses or adults, children would haul the coal. "Chained, belted, harnessed like dogs in a go-cart, black, saturated with wet, and more than half naked—crawling upon their hands and feet, and drafting their heavy loads behind them—they present an appearance indescribably disgusting and unnatural."[68]

On the consumption side, death records in early industrial London revealed that the largest category was lung related and that a third of all deaths were children under four or five years old.[69] In India today, one study estimated that some 70,000 children work in some 5,000 coal mines, many so-called rathole mines with one entrance and few safety features.[70]

Maybe these cultural impacts are not unique to coal, oil, and natural gas. Maybe any sudden infusion of unearned wealth always skews the values and commitments of the beneficiary. Hitting the jackpot brings misery to the winner; shipping hoards of gold to the mother country brings down the empire. But in the past century or so, no other substances have permeated a society so thoroughly, no other substances are as useful, as strategically essential, and as readily converted to raw power as fossil fuels, especially oil. And no other substances have come from nature in such a seemingly endless supply, indeed, an endlessly increasing supply. "Black gold" is nothing of the kind. Gold may make fine jewelry and symbolize wealth and power. But oil and coal actually generate wealth and power; they stimulate, multiply, and concentrate wealth and power (figure 3.3). And they do so not as showpieces but as largely invisible undercurrents, as the lifeblood that rarely spills into view.

A society fueled by the invisible and driven by concentrated wealth and concentrated decision making is unlikely to be a democratic society in a participatory, equal opportunity sense.[71] If coal created an underclass to rival India's Untouchables and oil-cursed poor countries struggle for independence and self-determination (and maybe a few rich countries too), and children have borne an unconscionable portion of the depredation, then net beneficence is at least questionable just on cultural grounds. Add the economic (assuming one can calculate the net discounted value of future costs of present consumption), the ecological (from weather and hormone disruption to sea level rise and airplane-assisted pandemics), and the ethical (the "losers" in this game seem to grow by the day, from Newcastle to north China, from New Orleans to New York), and fossil fuel net beneficence is not just questionable but highly doubtful.

## A New Story

From a systems perspective on endless growth and costless fossil fuel consumption, the prime movers of positive, self-amplifying feedback loops of concentrated energy and concentrated power (figure 3.3) are (1) a vision of the good life propagated by boosters for more than a century and (2) the physical invisibility of the fuels with their social correlate, secrecy, and their political outcome, power. In general, centralized systems (think electric power grids and their power plants) concentrate power; decentralized systems (think solar panels on every rooftop and not connected to the grid) disperse power. The energy in well-functioning ecosystems is dispersed. Even if fossil fuels and their waste sinks were unlimited, their inherent invisibility, secrecy, and centralizing tendencies make for a brittle system, one dependent on continuous and continuously

[energy]  →  [wealth]  →  [economic power]  →

[political power]  →  writing the rules of the game

→ [energy]

**Figure 3.3**

Concentrated power. Brackets, following chemical notation, indicate a process of concentration. "Energy" here is fossil fuels. The "positive" or self-reinforcing feedback loop results in expansion: each step amplifies the next, converting the concentrated power of the fuel to the concentrated power of decision making. But no system expands forever. Collapse or planful downshift (keeping fossil fuels in the ground) are the two basic options.

growing inputs and new places to absorb the wastes. Such a system is not adaptive or resilient because, in the long term, amplifying "positive" feedback eventually leads to collapse. Positive feedback loops may be positive for many—those with more income, consumption, prestige, and power, say. But eventually "they drive growth, explosion, erosion, and collapse in systems," wrote system analyst Donella Meadows. "A system with an unchecked positive loop ultimately will destroy itself." In well-functioning systems "a negative loop [usually] kicks in sooner or later." Slowing positive feedback loops "gives the many negative loops, through technology and markets and other forms of adaptation, time to function."[72] From this systems perspective, one that notably accounts for both the physical and the political, the contemporary fossil fuel order is a bubble waiting to burst, a house of cards about to fall, an extractive economy that has run its course, an empire that rose and now must fall. It is a race car without brakes. And so it is time to slow down, if not pull off the road and park.

From a cultural perspective, one that sees no necessary separation between the physical and the social, the economic and political, the symbolic and practiced, the contemporary fossil fuel order is one whose story no longer works. All societies tell their story: they create, pass on, and enact their myths, and they practice their rituals. High-integrity, resilient societies—those perceived as legitimate by all sectors of the society and can adapt to circumstances without degrading—find congruence among their biophysical conditions, their institutions, and their distribution of power. The fossil-fueled story of bonanzas for all, the myth of endless economic growth (decoupled from material growth), and the rituals of getting car keys as an initiation rite, of invading foreign countries as guarantees against the interruption of the global flow of oil, and so on, are all suspect now. The masses are addicted (Who can get off fossil fuels even if they desperately wanted to?), the elites are bloated (they have trouble spending their money), and decision makers defer, with rare exceptions, to the commercial, the expansionist, the technological. If there is legitimacy in this cultural pattern, it is the legitimacy of a boom town, a never-ending frontier, a space race where the winner plants a flag. It is the legitimacy of lucky strikes, land rushes, and electricity too cheap to meter.

As mythologist Joseph Campbell once said, well before climate disruption and other "wicked problems" emerged, we need a new story. The new story would be, at a minimum, (1) ecological—it would situate humans as denizens, not conquerors (however happy are the "conquerors and conquered") of natural systems, as stewards, even "guests" of nature; and (2) ethical—the society's dominant norms and principles, its ethic of resource use would be grounded in ecosystem functioning, including the limits of that functioning. Many people are trying to write that story, in theory (the authors of this volume included) and in practice (those on the ground who have inspired us authors). A precondition, however, seems to be a concerted delegitimization of the old story and its fossil fuel enablers. For that, new language is needed.

### The Special: Getting the Language Right

*Housekeeper. Keeper of the pantry. Gatekeeper. Riverkeeper. Keepers of the faith.*

These terms imply a degree of care, a set of deliberate acts of kindness, of respect for that which is special or sacred. They also imply *limits*: no house, community, landscape, or temple is completely open, its resources there for the taking by the highest bidder. Keepers sustain by protecting, and that requires limits.

What is special at this historical juncture is the emerging right, indeed the necessity, of living as if, as a species, we have just one planet; living as if, as a society, life is better when we acknowledge limits, indeed embrace them, and live within them. Keepers live within their means. Regarding fossil fuels, the only safe place to keep them is fast proving to be in the ground, not in storage tanks, on railcars, and certainly not as carbon dioxide, methane, and various persistent toxic substances winding their way through critical life-support systems.

The evidence, scientific and experiential, is overwhelming: a significant number of people are living beyond their means. From a material throughput perspective, *excess* defines their lives, their economy, their politics. No substances are more implicated in, indeed more symbolic of that excess, than oil, coal, and natural gas. Or, put systemically, no human relation to a portion of its natural surroundings is more material to humans' well-being, now and into the future, than its relations with fossil fuels. Getting that relationship right will most certainly not cure all society's ills—after all, slavery, drug trafficking, torture, and war preceded the fossil fuel era—but the hyperenergized, globalized, consumerized, overpopulated world we now inhabit is certainly visiting untold depredations on vast numbers of people and places, with more to come.

Getting fossil fuels right is in part a scientific, technological, and economic task, but it is also a linguistic task. The language of fossil fuels has been one and the same with the language of boom times and boosterist dreams (box 3.6). Fossil fuels are labeled "goods," "necessities," "strategic commodities," "lifeblood" (chapter 13). They acquire an air of normalcy, of being commonplace, of being something one cannot do without. Such normalization converts what was once unimaginable to have to what is unimaginable not to have. In a commercial society the benefits of such "goods" are routinely highlighted as their costs are shaded. Goods are good, and more goods are better. Thus it is that coal, oil, and gas proceeded along a historical path well lit by "the good," while leaving behind, in the dark, well shaded, obscured, invisible "the bad." In an empty world, one with inconsequential human impact, such processes can go on indefinitely: there is always a new frontier, a new commodity, a new use for which there is "latent demand."

In a full world and over a sufficiently long time, ecologically and geologically sufficient, the detritus and poisons of past abuses eventually cycle back. The linguistic task becomes an ethical task. Ethically, those who are dumped on—the "separate race" of coal miners, the expendable rig workers, the downstream residents who don't matter, those with asthma—say enough. The British and other Europeans slowly came to realize that for all the benefits of sugar trade, the costs were far greater:

**Box 3.6**
The Language of Fossil Fuels

In cultural phenomena, language matters—not just the words and grammar, but the meanings, however explicit or implicit. Symbols, metaphors, concepts, principles, norms, and stories are all essential parts of a material's meaning, its place in society, its elicitation of behaviors.

Notice in the following list, gleaned from histories and popular accounts, how many connote chance, gamble, or fate. And many connote excitement, thrill, nerve, daring. But they also suggest that fossil fuels are not so much produced as they are extracted, that their benefits are one-time, a result of get-in-and-get-out practices. They are inherently short term, lacking in a sense of place, personal connection, or community commitment. These terms contrast, in other words, with terms like *permanence, attachment, blessing, gratitude, stewardship,* and *common good* (chapter 13).

| | |
|---|---|
| rush | gusher |
| boom | bonanza |
| frenzy | mother lode |
| fever | jackpot |
| the play | prize |
| luck | elephant (oil field) |
| lucky strike | supergiant (oil field) |
| strike it rich | Big One (oil field) |

**Box 3.6**
(continued)

| | |
|---|---|
| man camp | baron |
| wildcatter | production (extraction of oil) |
| roustabout | disposal (of extracted and |
| roughneck | processed oil by marketing and |
| landman | selling it) |
| tycoon | |

Book titles are similarly revealing of the culture of fossil fuels:

- *The Prize: The Epic Quest for Oil, Money and Power* by Daniel Yergin
- *Private Empire: ExxonMobil and American Power* by Steve Coll
- *Oil and World Power* by Peter R. Odell
- *The Race for What's Left: The Global Scramble for the World's Last Resources* by Michael T. Klare
- *Internal Combustion: How Corporations and Governments Addicted the World to Oil and Derailed the Alternatives* by Edwin Black
- *The Oil Curse: How Petroleum Wealth Shapes the Development of Nations* by Michael L. Ross
- *Oil, Dollars, Debt, and Crises: The Global Curse of Black Gold* by Mahmoud A. El-Gamal and Amy Myers Jaffe
- *Oil on the Brain* by Lisa Margonelli
- *The Paradox of Plenty: Oil Booms and Petrostates* by Terry Lynn Karl

The language of fossil fuels is, in short, the language of transience, not permanence; of moving on, not settling in; of expanding, not sustaining; of creating "power over," not "power with"; of building the powers of domination not the powers of self-determination. It is the language of consuming, not investing; of taking, not giving; of cursing, not blessing; of getting, not being.

---

slavery, among other features of the system, eroded slave and master alike, both African society and European society. Similar social learning is arguably occurring in Germany and Australia regarding coal and nuclear power, in Ecuador and Norway regarding oil, in El Salvador regarding gold (part 2) and many other places. Becoming keepers rather than masters implies a certain social sufficiency—that sense, that experience of enoughness and "too muchness," that knowledge of taking only what ecosystems can yield and waste sinks can absorb.[73]

Humans' relations with fossil fuels can be special. But as highly distanced, commoditized, placeless global substances, they are routinely treated as anything but special. They are ubiquitous and cheap, available with the flip of a switch or the click of a nozzle. All uses are legitimate—growing rice in a desert, heating driveways in mountain resorts—as long as the user can pay for it (the consumer sovereignty principle at work). Interruptions are, like interruptions to food supplies, an offense no politician wishes to encounter.

So how would fossil fuels become special? How would a language of keeping and sufficiency emerge? Like fine jewelry or a dangerous weapon, the simplest way is not to use them, or use them only on special occasions for special purposes and to keep them under lock and key at all other times. To lock away fossil fuels is to say that we will use them only when we must because no alternative exists—including the alternative of nonuse—only when we are sure, or as sure as we can be, that such limited use will not cause cascades of depredations across landscapes and through the body politic as they are now doing. Maybe it is time to resurrect the early magic in fossil fuels that made such substances special. Updated for the twenty-first century, that magic would necessarily derive from highly limited use. Limits here would not be external constraint—draconian, top-down, governmental measures, say—but self-imposed, internal restraint, which ennobles and enlivens as people figure out how to live well on that which truly regenerates— fertile soil, recharging aquifers, crops and wildlife. Fossil fuels would be reserved for functions only they can perform. The magic of fossil fuels would then derive not from their scarcity (physically they are not at all scarce) but from their abundance, an abundance experienced as having plenty because their uses are so special.

Making fossil fuels special would be a cultural act to complement the materials management acts of conventional policymaking. Thus, at the same time a politics of ending the fossil fuel era begins by delegitimizing fossil fuels, that era ends by celebrating fossil fuels too. The delegitimization is about excess rates, the celebration about the remaining modest rates. The result would be a profound cultural shift, precisely what is needed in a time of great urgency and policy deadlock.

In the end, sufficiency in fossil fuel use would not be sacrifice or doing without. It would be doing well now and into the future by using less fossil fuels, much less, than the most possible now. The energetic basis of the good life, then, would be relying mostly on regenerative sources, keeping within their regenerative capacities, and, when the occasion demands, a bit on fossil fuels. For that magic and that specialness, for the keeping and the sufficiency, societies everywhere, but especially those of the Global North, need a new fossil fuel ethic, one that supplants the old imperial and boosterist ethic of endless extraction, endless expansion, endless domination.

**Table A3.1**

Exemptions and Exclusions: Fossil Fuel's Special Status

The US oil and gas industry receives exemptions and exclusions from a number of federal laws and rules intended to protect US citizens and the environment. Thorough histories of each piece of legislation would be needed to explain exactly why this industry gets special treatment. But existing histories of the oil industry itself suggest that it has a lot to do with concentrated power—industrial and financial—as well as strategic imperative. Here is a partial list of those exemptions and exclusions.

| Law/ Regulation | Year Established/ Modified | Purpose | Fossil Fuel Exemption (emphases in bold added) |
|---|---|---|---|
| National Environmental Policy Act (NEPA) | 1978/2005 | "Title I of NEPA contains a Declaration of National Environmental Policy which requires the federal government to use all practicable means to create and maintain conditions under which man and nature can exist in productive harmony."[a] | "Prior to any significant action that affects the environment NEPA requires agency to perform an environmental assessment to determine if environmental impact will be significant and then an environmental impact statement if the initial assessment determines that an impact statement is necessary. The Energy Policy Act of 2005 created the 'rebuttable presumption' that allowed several **oil and gas related activities to be assessed by a less stringent 'categorical exclusion'** process. Given the rebuttable presumption, a full assessment through NEPA only occurs when the public proves 'extraordinary circumstances' for a proposed oil and gas activity."[b] |
| Clean Water Act (CWA) | 1972/2005 | 1972 "The objective of … (CWA) is to restore and maintain the chemical, physical, and biological integrity of the nation's waters by preventing point and nonpoint pollution sources, providing assistance to publicly owned treatment works for the improvement of wastewater treatment, and maintaining the integrity of wetlands."[c] | "The term 'pollutant' … does not mean … (B) **water, gas, or other material which is injected into a well to facilitate production of oil or gas, or water derived in association with oil or gas production** and disposed of in a well, if the well used either to facilitate production or for disposal purposes is approved by authority of the State in which the well is located, and if such State determines that such injection or disposal will not result in the degradation of ground or surface water resources."[d] Congress expanded the exemption in the 2005 Energy Policy Act, amending the CWA exemption to more broadly cover "all field activities or operations associated with exploration, production, processing, or treatment operations, or transmission facilities, including activities necessary to prepare a site for drilling and for the movement and placement of drilling equipment"[e] |

**Table A3.1**
(continued)

| Law/Regulation | Year Established/Modified | Purpose | Fossil Fuel Exemption (emphases in bold added) |
|---|---|---|---|
| Emergency Planning and Community Right-to-Know Act (EPCRA) | 1986 | "The objective of the Emergency Planning and Community Right-to-Know Act (EPCRA) is to: (1) allow state and local planning for chemical emergencies, (2) provide for notifications of emergency releases of chemicals, and (3) address communities' right-to-know about toxic and hazardous chemicals."[f] | "Industry resists disclosing chemicals used for underground injections citing trade secrets. **Oil and gas producers do not need to comply with EPCRA's annual reporting for toxic releases** if each release individually falls below emergency thresholds."[g] |
| Comprehensive Environmental Response, Compensation, and Liability Act (CERCLA) | 1980 | "Commonly referred to as Superfund … this law created a tax on the chemical and petroleum industries and provided broad Federal authority to respond directly to releases or threatened releases of hazardous substances that may endanger public health or the environment."[h] | Contamination from "federally permitted release" cannot trigger CERCLA liability. **"Federally permitted"** is defined as "any injection of fluids or other materials authorized under applicable State law (i) for the purpose of stimulating or treating wells for **the production of crude oil, natural gas**, or water, (ii) for the purpose of secondary, tertiary, or other enhanced recovery of crude oil or natural gas, or (iii) which are brought to the surface in conjunction with the production of crude oil or natural gas and which are reinjected."[i] Natural gas and petroleum are excluded from the definition of "hazardous substance." "The term '**hazardous substance**' … **does not include petroleum**, including crude oil or any fraction thereof which is not otherwise specifically listed or designated as a hazardous substance under subparagraphs (A) through (F) of this paragraph and the term does not include natural gas, natural gas liquids, liquefied natural gas, or synthetic gas useable for fuel (or mixture of natural gas and synthetic gas)."[j] |

Table A3.1
(continued)

| Law/Regulation | Year Established/Modified | Purpose | Fossil Fuel Exemption (emphases in bold added) |
|---|---|---|---|
| Resource Conservation and Recovery Act (RCRA) | 1976 | "RCRA ... set national goals for 1) protecting human health and the environment from the potential hazards of waste disposal; 2) conserving energy and natural resources; 3) reducing the amount of waste generated; 4) ensuring that wastes are managed in an environmentally sound manner"[l] | The Environmental Protection Administration says that coal ash (residuals from coal combustion) can leach mercury, cadmium, and arsenic into groundwater. These chemicals are "associated with cancer and various other serious health effects."[m] The Bevill Amendment **excludes from the act "fly ash, bottom ash, slag waste, and flue gas emission control waste generated primarily from the combustion of coal or other fossil fuels"**[n] The Bentsen Amendment excludes "drilling fluids, produced waters, and other wastes associated with the exploration, development, and production of crude oil or natural gas or geothermal energy"[o] |
| Safe Drinking Water Act (SDWA) | 1974 | SDWA "is the main federal law that ensures the quality of Americans' drinking water."[p] | Halliburton exemption: "(B) excludes (i) "underground injection of fluids or propping agents (other than diesel fuels) pursuant to hydraulic fracturing operations"[q] |
| Clean Air Act (CAA) | 1970 | "Congress designed the Clean Air Act to protect public health and welfare from different types of air pollution caused by a diverse array of pollution sources"[r] | Power plants existing prior to 1970 passage of CAA are grandfathered to ease transition into stricter standards. Amended in 1990. "Major sources" of hazardous air pollutants and smaller "area sources" in aggregate are controlled. The oil and gas industry emits 500,000 tons of volatile organic compound emissions per year. And yet "emissions from any oil or gas exploration or production well ... and emissions from any pipeline compressor or pump station shall not be aggregated with emissions from other similar units, ... and in the case of any oil or gas exploration or production well (with its associated equipment), such **emissions shall not be aggregated for any purpose under this section**."[s] |

**Table A3.1**
(continued)

a. "National Environmental Policy Act (NEPA)," Environmental Protection Agency, http://www.epa.gov/compliance/basics/nepa.html.

b. Citing Energy Policy Act of 2005 sec. 390: Renee Lewis Kosnik, "The Oil and Gas Industry's Exclusions and Exemptions to Major Environmental Statues," in *Oil and Gas Accountability Project: A Project of Earthworks* (2007), 16

c. Clean Water Act, Environmental Protection Agency, http://www.epa.gov/oecaagct/lcwa.html.

d. 33 USC sec. 1362(6), US Government Printing Office, http://www.gpo.gov/fdsys/browse/collectionUScode.action?selectedYearFrom=2011&go=Go.

e. 42 USC sec. 1362(24), US Government Printing Office, http://www.gpo.gov/fdsys/browse/collectionUScode.action?selectedYearFrom=2011&go=Go. This quote explains how the USC defines oil and gas exploration and production.

f. Environmental Protection Agency, Emergency Planning and Community Right-to-Know Act, http://www.epa.gov/agriculture/lcra.html.

g. Hannah Wiseman, "Trade Secrets Disclosure, and Dissent in a Fracturing Energy Revolution," *Columbia Law Review 111* (2011): 1, 5–6, http://papers.ssrn.com/sol3/papers.cfm?abstract_id=1743650.

h. Environmental Protection Agency, "CERCLA Overview," http://www.epa.gov/superfund/policy/cercla.htm.

i. See 42 USC sec. 9607(j).

j. 42 USC sec. 9601(10)(I), US Government Printing Office, http://www.gpo.gov/fdsys/browse/collectionUScode.action?selectedYearFrom=2011&go=Go

k. 42 USC. sec. 9601(14)(F), US Government Printing Office, http://www.gpo.gov/fdsys/browse/collectionUScode.action?selectedYearFrom=2011&go=Go.

l. Environmental Protection Agency, "History of RCRA," http://www.epa.gov/osw/laws-regs/rcrahistory.htm.

m. Quoted from Uma Outka, "Environmental Law and Fossil Fuel: Barriers to Renewable Energy," *Vanderbilt Law Review 65* (2012): 1679–1721. Original source: "Frequent Questions: Coal Combustion Residues (CCR)–Proposed Rule, U.S. Envtl. Protection Agency," http://www.epa.gov/epawaste/nonhaz/industrial/special/fossil/ccr-rule/ccrfaq.htm#4.

n. 42 USC sec. 6921(b)(3)(A)(i), US Government Printing Office, http://www.gpo.gov/fdsys/browse/collectionUScode.action?selectedYearFrom=2011&go=Go.

o. Ibid, (ii).

p. Safe Drinking Water Act, Environmental Protection Agency, http://water.epa.gov/lawsregs/rulesregs/sdwa/index.cfm.

q. 42 U.S.C. sec. 300h(d)(1)(B)(ii), U.S. Government Printing Office, http://www.gpo.gov/fdsys/browse/collectionUScode.action?selectedYearFrom=2011&go=Go.

r. Clean Air Act Requirements and History, Environmental Protection Agency, http://www.epa.gov/air/caa/requirements.html

s. 42 USC sec.7412(n)(4)(A)," U.S. Government Printing Office, http://www.gpo.gov/fdsys/browse/collectionUScode.action?selectedYearFrom=2011&go=Go.

# Notes

1. William Cronon, *Nature's Metropolis: Chicago and the Great West* (New York: Norton, 1991); Mat Patterson, *Automobile Politics: Ecology and Cultural Political Economy* (Cambridge: Cambridge University Press, 2007); Sidney W. Mintz, *Sweetness and Power: The Place of Sugar in Modern History* (New York: Penguin, 1985); Wolfgang Sachs, *For Love of the Automobile: Looking Back into the History of Our Desires* (Berkeley: University of California Press, 1992); Barbara Freese, *Coal: A Human History* (New York: Penguin, 2003); Daniel Yergin, *The Prize: The Epic Quest for Oil, Money, and Power* (New York: Simon and Schuster, 1991).

2. To use *agent* as a quality of a substance, when the term in political discourse is otherwise most useful in reference to human choice (what I call the second prong of cultural change; see below), may seem a bit odd. But I choose *agency* to connote the relational aspect of fossil fuels and, for that matter, of all energy sources. That is, what is at issue here is not calories and gigajoules, not kilowatts and BTUs, or even the density and transportability of fuel, as important as all that is (chapter 2). Rather, what is at issue in the waning years of the fossil fuel era is how humans relate to fossil fuels and, in a sense, how those fuels relate to us—how they shape us, entrance us, beguile us, mislead us. Just because the relational dynamics of previous energy sources were more obvious—especially humans and domestic animals—doesn't mean contemporary sources do not have their reciprocal qualities. The bodies of ancient organisms are serving as vehicles for ancient sunlight; that such transport would be entirely neutral, the work of objective substances, is at least as much a stretch as assigning agency to them. So here I elect a relational view on energy, what may be alien, even anathema to those who tend to specialize in energy matters, whether from physics or engineering or rationalist policymaking. But from the perspective of lived experience, what we all do, it is precisely the relations that ultimately matter. For elaboration of this perspective, see David Abram, *The Spell of the Sensuous: Perception and Language in a More-Than-Human World* (New York: Pantheon Books, 1996); Jane Bennett, *Vibrant Matter: A Political Ecology of Things* (Durham, NC: Duke University Press, 2010); Thomas Princen, *Treading Softly: Paths to Ecological* Order (Cambridge, MA: MIT Press, 2010); Sachs, *For Love of the Automobile.*

3. I thank Jim Crowfoot for bringing this point to my attention.

4. Yergin, *The Prize*, 24.

5. Ibid., 20.

6. Ibid., 84, 85.

7. Richard Holmes, *The Age of Wonder: How the Romantic Generation Discovered the Beauty and Terror of Science* (London: Harper Press, 2008), 362.

8. Freese, *Coal*, 28, 47.

9. Ibid., 57.

10. Vaclav Smil, *Energy at the Crossroads: Global Perspectives and Uncertainties* (Cambridge, MA: MIT Press, 2003), 8–9.

11. Freese, *Coal*, 66–67.

12. Ibid., 67, 68.

13. Emerson quote in ibid., 10.

14. Yergin, *The Prize*, 31.

15. Ibid., 34.

16. Sachs, *For Love of the Automobile*, 5.

17. Quoted in Freese, *Coal*, 92–93.

18. Sachs, *For Love of the Automobile*, 6.

19. Quoted in Yergin, *The Prize*, 254.

20. Cronon, *Nature's Metropolis*, 200, 202.

21. Caroline Merchant, *Ecological Revolutions: Nature, Gender, and Science in New England* (Chapel Hill: University of North Carolina Press, 1989).

22. Freese, *Coal*, 57.

23. Quoted in Yergin, *The Prize*, 223.

24. Ibid., 218.

25. Ibid., 717.

26. Tillerson, quoted in Steve Coll, *Private Empire: ExxonMobil and American Power* (New York: Penguin Press, 2012), 437.

27. Freese, *Coal*, 64.

28. Quoted in Yergin, *The Prize*, 183.

29. Yergin, *The Prize*, 178, 196, 393.

30. Both quotes in ibid., 222.

31. William Leach, *Land of Desire: Merchants, Power, and the Rise of a New American Culture* (New York: Vintage, 1993); Lizabeth Cohen, *A Consumers' Republic: The Politics of Mass Consumption in Postwar America* (New York: Knopf, 2003); J. R. McNeill, *Something New under the Sun: An Environmental History of the Twentieth-Century World* (New York: Norton, 2000).

32. Coll, *Private Empire*, 305.

33. On the failures of forecasting, see Smil, *Energy at the Crossroads*, 121–80. On the politics of progress, see Christopher Lasch, *The True and Only Heaven: Progress and Its Critics* (New York: Norton, 1991); Michael Greer, "Progress vs. Apocalypse: The Stories We Tell Ourselves," in *The Energy Reader: Overdevelopment and the Delusion of Endless Growth*, edited by Tom Butler, Daniel Lerch, and George Wuerthner (Sausalito, CA: Foundation for Deep Ecology, 2012) , 95–101.

34. Quoted in Yergin, *The Prize*, 546.

35. Ibid., 546.

36. McKinsey Global Institute, "Resource Revolution: Meeting the World's Energy, Materials, Food, and Water Needs" (November 2011), 1, 4.

37. Cronon, *Nature's Metropolis*, 34.

38. Ibid., 35.

39. Ibid., 43–44.

40. Ibid., 45.

41. Ibid., 44.

42. Ibid., 53.

43. Certainly the role of coal was not lost on nineteenth-century boosters, especially in England. As one writer there put it: "With Coal, we have light, strength, power, wealth, and civilization; without Coal we have darkness, weakness, poverty, and barbarism" (in Freese, *Coal*, 10). But given that in 1876, the United States got twice as much energy from wood as from coal (ibid., 137) and that dominance continued until the 1890s and that overall fossil fuel consumption rose fifteen-fold worldwide from 1900 to 2000 (Smil, *Energy at the Crossroads*, 4, 13), it is fair to conclude that fossil fuels supplanted other natural resources as the primary driving force in industrial and economic expansion only in the very late nineteenth century and most definitively in the twentieth.

44. Yergin (*The Prize*, 210) described service stations as "the icons of a secular religion, providing drivers with a feeling of familiarity, confidence, and security—and of belonging—as they rolled along the ever-lengthening ribbons of roads that crossed and crisscrossed America."

45. Cronon, *Nature's Metropolis*, 46.

46. Fossil fuels afford just one account of the modern cultural norm of growth—that is, as a goal, as a means, as a supreme value. Only with other accounts—of colonization, technological advance, military strategy—can a complete picture be drawn. See Marta Niepytalska and Alec Hahn, "Interview with Donald Worster I. Facing Limits: Abundance, Scarcity, and the American Way of Life," Rachel Carson Center Fellow Portraits film series, 3.04 (May 2011, Munich), http://www.environmentandsociety.org/mml/donald-worster-facing-limits-abundance-scarcity-and-american-way-life; and Worster's work in progress: "Facing Limits: Abundance, Scarcitiy, and the American Way of Life."

47. Freese, *Coal*, 143.

48. On distancing as an ecological and ethical concept, see Thomas Princen, "Distancing: Consumption and the Severing of Feedback," in *Confronting Consumption*, edited by Thomas Princen, Michael Maniates, and Ken Conca (Cambridge, MA: MIT Press, 2002) , 103–31.

49. Rex Tillerson, quoted in Coll, *Private Empire*, 599.

50. Yergin, *The Prize*, 42.

51. Ibid.

52. Ibid., 53.

53. Ibid., 109.

54. Quoted in ibid., 418.

55. Terry Lynn Karl and Ian Gary, "Bottom of the Barrel: Africa's Oil Boom and the Poor" (Baltimore, MD: Catholic Relief Services, 2003).

56. Coll, *Private Empire*, 548.

57. Helge Ryggvik, *The Norwegian Oil Experience: A Toolbox for Managing Resources?* (Oslo: Centre for Technology, Innovation and Culture, 2010), 52.

58. Coll, *Private Empire*, 239.

59. Freese, *Coal*, 166.

60. Yergin, *The Prize*, 783.

61. Ryggvik, *Norwegian Oil Experience*, 47.

62. Freese, *Coal*, 45.

63. Ryggvik, *Norwegian Oil Experience*, 88.

64. Ibid., 90.

65. Ibid., 91.

66. Thomas Homer-Dixon, "The Tar Sands Disaster," *New York Times*, April 1, 2013.

67. Quoted in Freese, *Coal*, 77.

68. Quoted in ibid., 78.

69. Freese, *Coal*, 40.

70. Gardiner Harris, "Children Work in India Mines Despite a Ban," *New York Times*, February 26, 2013, A1, A11.

71. Ivan Ilich, *Energy and Equity* (New York: Harper & Row, 1974).

72. Donella H. Meadows, "Places to Intervene in a System (in Increasing Order of Effectiveness)," *Whole Earth* 91 (Winter 1997): 78–84, 81.

73. Thomas Princen, *The Logic of Sufficiency* (Cambridge, MA: MIT Press, 2005).

# 4

# The Ethical: A Fossil Fuel Ethic

Thomas Princen

A politics of fossil fuel exit, of deliberately accelerating a society's withdrawal from oil, gas, and coal dependence ahead of a geologic imperative, ahead even of economic and financial imperatives, is ultimately an ethical act. It puts front and center the harm-to-others criterion and relegates to the wings the economic and political (as in electoral and legislative politics) criteria. What is more, a politics (as in the shaping of society's core values and steering a particular path) of ending the fossil fuel era is one of temporal extension, of taking seriously humans' past and future, including their geologically and ecologically distant past and future. Temporal extension necessitates ethical extension—from resources to ecosystems, from extraction to regeneration, from human life to nonhuman life, from us to other, from present generations to past and future generations, from material gain to societal integrity and spiritual uplift, from goods-are-good-and-more-goods-must-be-better to the "good life."

A politics of ending the fossil fuel era thus entails more than hastening the next energy transition, more than arresting climate change, more than shifting to a postindustrial world. It is a moral confrontation with a wildly successful material order, an order that has heretofore been presumed net beneficial, salutary, indeed essential and just. In this chapter I confront that order by arguing that there has long been an ethic of fossil fuels, however implicit, however submerged in the political discourse, however "natural" its constructors have made it out to be. That this moral order cannot last is as self-evident as the fact that fossil fuel dominance cannot last. What is not self-evident is that the moral order with respect to fossil fuels will shift fast enough to avert catastrophic change. In fact, I argue, the reconstruction of the fossil fuel ethic, from salutary to destructive, is necessary for an effective politics of accelerated exit. It is this reconstruction, and this shift, this urgent transition, at once physical and ethical, that is at the heart of the global environmental problematic.[1]

## The Current Fossil Fuel Ethic

As argued in chapter 1, to focus on emissions, on carbon management, is to take all that precedes emissions, all the upstream decisions, both private and public, as given. The result is a tame politics—scientific, rationalist, managerial, collaborative. The tug and pull is over data and modeling; the give-and-take is about using carbon judiciously, like a commodity; the distributing is about putting carbon in the right place; the influencing is about educating so everyone understands the science and the costs. The game is one of haggling over prices and pricing, of finding places to put the offending stuff, of rationalizing cleanup. The politics is, notably, of a wholly different sort from the politics of the fossil fuel complex itself where extractive industries, through campaign financing, lobbying, bringing together "informed influentials," largely write the rules of the game and decide who gets access. The politics of the extractive industries, from the earliest days of underground coal mining and land-based oil drilling, is indeed the politics of extraction (chapter 3). And that politics has had its own ethic.

If an emissions focus is problematic, if it fails to confront the source of the problem (both physical and political), then logically there is only one thing to do: go upstream to the source, to extraction, to where the power is. This would engender a politics that matches the politics that has created the current predicament. By saying "go upstream," I do not mean just the acts of drilling and pumping, blasting and shoveling. I mean, in the first instance, the implicit normative stance that industrial societies have adopted: fossil fuels, like other valued resources, *must* be used. Like standing timber and fish swimming freely in the sea, a resource is of "no value" until it is extracted, processed, distributed, and completely used. Anything less is a waste, inefficient.

In this, the dominant construction of resource use, the age-old notion of dominion combines with a theory of use value and a principle of efficiency to create an *ethic of total use*:

*A resource should be used, not left in place; it should be used completely, as much as markets demand and technologies allow, and, above all, none should go to waste.*

Clear-cuts ensure that all fiber is used; drift nets and bycatch marketing campaigns ensure that all available fish are caught and used; dams, canals, and irrigation ditches ensure that no water is wasted flowing to the ocean.[2]

If an ethic of total use made sense for renewables, it certainly did for nonrenewables: What good is coal or oil just sitting underground? It doesn't grow or propagate; it doesn't improve on its own. It just sits there, worthless. Got to use it, all of it, as much as people want and as much as we can get out given the geology, the technologies, the economics, and the politics.

It has been a powerful ethic. It has created previously unimaginable wealth, including material abundance for all who partake in its fruits, which is to say a very large portion of the earth's population. What is more, it has conferred great power on those who play the game well. It has created a politics where key players write the rules of the game, obtain access, and do very well for themselves and their constituents. It is a politics where conflicts can be swept aside as all boats rise. It is, in short, a politics of growth, including economic growth tightly coupled to material growth, a politics that only now analysts and policymakers are learning is dependent on cheap fossil fuels—cheap economically, energetically, and environmentally.

Elites and others are now coming to the realization that besides being a game of great wealth, it is also a game of great risk, a high-stakes global gamble. As a global society, we (especially the high-consuming Global North) are betting that total use—mining minerals and living things and filling waste sinks—can go on indefinitely. We're betting that business-as-usual commerce, the unending proliferation of "goods," the frantic search for consumers' "next big thing" can proceed and that some combination of high prices and technological cleverness will decouple the economy and handle all the bads. Modern market societies are betting that wealth today translates to wealth tomorrow, all as if depletion is immaterial. This gamble is without precedent. For those of us who don't believe that infinite material growth on a finite planet is possible, that depletion is entirely material to the human condition, that yesterday's goods are tomorrow's bads, we need a new game. In fact, a new game *will* emerge: biogeochemical trends guarantee it. What is not guaranteed is a transition that is peaceful, democratic, just, and ecologically sustainable. For that we need a new politics and a new resource ethic. Total use and global gambling won't do.

A starting point for that new politics and a new resource ethic is to posit that fossil fuels, for all their benefits, huge as they might be, are hugely destructive substances, directly and indirectly.[3] During crises such as mine collapses and oil spills, the direct, negative effects are clear. At all other times, the positive effects are front and center—rapid transport, comfortable buildings, abundant goods and services—while the costs are obscured, concentrated in mining towns and oil fields and people's bloodstream, dispersed through the atmosphere and oceans, displaced over years and decades, even centuries. But now the science tells us these are not just costs as in bills to be paid or, through exchange, the downside for which we realize an upside. Rather, these costs are, to deliberately shift metaphor from commerce to security, threats to our very survival. Now we have good scientific reason to believe that the instantaneous combustion of fossil fuels (geologically speaking) is incompatible with much of life on this planet as we know it. These are not costs in any conventional market sense; there is no calculation, no trade-off, however theoretical, that assigns a welfare-enhancing value to the deliberate destruction of life. Total use of fossil fuels, the prevailing resource ethic, is a proposition lying outside the economic realm.

A defense metaphor is more apt: How far can a patently self-destructive practice be allowed to go before people feel threatened? Or, put as a thought experiment, should our nation dismantle its defenses and hope that no one will invade? Invasion, after all, is highly uncertain; no one can prove it will happen or that its costs would exceed the benefits of diverting national resources from defense to other uses. And if an invasion begins, won't new technologies emerge to repel it? The absurdity of this thought experiment suggests the absurdity of equating economic costs and survival risks.[4] It suggests the imperative of securing the planet's life-support system and doing so first, as a foundational condition from which an economy and a democracy are then built. And for that it suggests the need to construct an ethic of fossil fuel use that would undergird the transition out of fossil fuels and toward an ecologically sustainable, socially just order.

### Toward a New Fossil Fuel Ethic

An ethic of fossil fuels can be framed in two ways. The first is as a *trade-off function*: we take pollution in exchange for all the benefits of readily available, energy-dense, easily transported fuels. This is the favorite construction of rational choice, economistic, commercially oriented analysts, and policymakers, however implicit. From a biophysical and ecological perspective, though, it is what enables a mining economy and thwarts a sustaining economy. The logic goes something like this:

*Fossil fuel use is rightfully determined by consumers and investors. Their choices, their allocation of income and their risk of capital, reveal that fossil fuels provide great benefits. Of course, fossil fuels also incur costs. Society, via people's choices, which are aggregated in markets, negotiated in legislatures, promulgated in agencies, and adjudicated in the courts, has concurred that benefits exceed costs.*

*When certain nonmarket costs (e.g., traffic fatalities, air pollution) exceed an acceptable level, part of the benefit stream is diverted for ameliora-tion. Otherwise, fossil fuels are deemed net beneficial—desirable, even essential.*

*The more that fossil fuel use shifts from desirable to essential, the more their sources (e.g., oil fields and coal beds) must be protected with public resources (i.e., security forces), and the more public resources are justified in finding new sources.*

If there is a dilemma in this ethical framing, it is in the question, How do we know that benefits actually exceed costs? Or, looking forward, how would we know that benefits continue to exceed costs when costs continue to mount (e.g., as toxics accumulate and weather extremes mount)? It is a question that can be, and is, endlessly debated. This fact alone suggests the need for an alternative framing. Other facts, like the injustice of concentrated and near-term benefits for a few, dispersed and long-term costs for the many, huge power asymmetries, and disrupted ecosystems, make an alternative ethical framing all the more compelling.

A second way to frame an ethic of fossil fuels might be called *rela-tional* or ecological. Here the relations of one species—*Homo sapiens* in the first instance, food and fiber species for humans in the second—with its larger environment, biophysical and social, are the starting point. Then those relations are considered over a long period of time where sources and sinks are not degraded, a process that generates the sustain-ability question. From this relational perspective and in light of current knowledge of dysfunctional relations (e.g., greenhouse gas (GHG) emis-sions, persistent toxics, soil loss, freshwater drawdown), the current ethical dilemma can be posed thus:

*If it is patently wrong to burn all available fossil fuels in the coming decades (wrong because human life is severely diminished), then how can it be right to burn another barrel of oil, another lump of coal today?*

This question cannot be answered through consumer and investor choice, which are inherently myopic and, for justification, resort to notions of consumer sovereignty and rational expectations. Nor can they be answered by calculating net present value with market-determined (or politically determined) discount rates for current and future uses. Once again, market trade-off and existential threat are incommensurate.

There is, though, a simple answer to this question, one that, upon a bit of reflection is ethically unacceptable: if each unit of fossil fuel is wrong to burn today, then we should stop now. To do so would, of course, wreak great havoc, the extent and cruelty of which is impossible to measure or imagine. In other words, to abruptly and completely stop using fossil fuels now, when fossil fuels comprise 85 percent of all energy consumed worldwide, would be to trade one global calamity for another, a primary difference being the timing—now versus the future.

But to stop using fossil fuels now is only a thought experiment. It won't happen, even if incontrovertible evidence arose that said such use is taking us promptly to oblivion. Fossil fuels are the ultimate path-dependent commodity.[5] They are the proverbial lifeblood of industrial society. But because continuation along this path is ultimately destructive of industrial society, indeed of livelihood and life as we know it, humanity *will* stop at some point, one way or another, planfully or not, equitably or not. The question is not when we will stop or should stop, but when will we *start stopping*. That in turn raises the practical question of how to start stopping. This framing—how we start stopping—goes to the crux of the current ethical dilemma with respect to fossil fuels and their mix of extraordinary benefits (historically speaking, their one-time existence makes them literally extra-ordinary) and extreme costs (death and destruction experienced by, among others, automobile passengers, those with asthma, miners and rig workers, and casualties of war).

## Start Stopping

The crucial question is thus not how and how much to reduce fossil fuel use (the current political dilemma) or even when to stop using fossil fuels (as if the matter can be deferred to the "right time"), but *how to start stopping now*. Ethically, each barrel of oil and each lump of coal burned is wrong. But each unit is also right because otherwise lives will be diminished and destroyed. The issue, then, is transition: how to transition out of fossil fuels and, along the way, how to assess the ethical trade-offs (here trade-offs do make sense) of declining fossil fuel use. More to the point, the issue is, given the changes now underway (bio-accumulating toxic substances, freshwater drawdown, heat waves and crop failures, coral die-offs, melting ice and tundra, rising sea levels), the lag times between cause and effect, and the likelihood of systemic "surprise" (discontinuities, synergisms, and unnoticed trends),[6] *urgent transition.*[7]

So because of the reality of extreme path dependency in industrial societies for fossil fuels and because lives will be diminished and destroyed without fossil fuels (just as they are currently being diminished and destroyed with fossil fuels), criteria are needed for using fossil fuels and transitioning out of fossil fuel use.[8] I posit three:

1. *Life saving.* This first-order ethical criterion says that another unit of fossil fuel burned is justified at present if:
   - Lives are protected—if it is the ambulance, hospital, firehouse, police station, or army base that uses that unit of fossil fuel or
   - The desperately poor can subsist another day because they use that unit or
   - The victims of an attack can repel the aggressor with that unit

2. *Transition.* This second-order ethical criterion says that if the next unit of fossil fuel to be burned at present is for the purpose of making the transition out of fossil fuels, if it enables substitution of renewables, for instance, or obviates the need for its use (e.g., more insulation obviates more heating fuel; a localized food system obviates centralized production and costly transport), then that fossil fuel use is justified. It is justified because it protects future lives that would otherwise be diminished or destroyed with continuing fossil fuel use.

3. *Livelihood.* This third-order criterion aims at ensuring people's capacity to self-provision, to associate, to thrive.[9] Notice that while this criterion may be a top development objective, it is subordinate here as a fossil fuel transition objective.

The "start stopping" framing has three notable features. First, because life saving is the ultimate criterion, now and in the future, there is no necessary trade-off between present and future generations. The goal is to transition out of fossil fuels with the least diminishment and loss of lives. It is not to maximize discounted present value of a benefit stream (because future lives are not discounted). The goal is not to maintain standards of living, keep the economy growing, or maintain the current structures of laws and markets (because all of those depend heavily on cheap energy). It is not to avoid disruption (because individual lives and entire societies *will* be disrupted as the climate, ecosystems, and reproductive capacities are disrupted). The goal is, to repeat, to protect lives and livelihood.

Second, if "start stopping" makes sense now—ethical and ecological sense—it makes sense to start stopping at all points in time along the current path. If no serious harm occurs from climate change (as is largely the case so far for most of humanity), it makes sense to start stopping; if thresholds are passed, it still makes sense; and if great calamity is experienced, it still makes sense to start stopping. It just makes more sense to start stopping sooner rather than later. This conclusion follows not just from the urgency of the matter, which is widely accepted in the scientific community, but from the logic of start stopping now.

The reasoning is contrary to that which derives from a limited, *a*behavioral biophysical perspective, what predominates in the environmental sciences, from biodiversity to atmospheric science, where one models and assesses likelihoods of passing thresholds and incurring surprises. Picking a point at which to level off or a point of no return is, behaviorally and politically speaking, to invite postponement. By analogy, it is like resource managers picking a maximum sustained yield. Such efforts rarely result in sustainable practice (think monocultures and pest outbreaks, economic decline, and the pressure to cut more). Other measures do—for example, following the principles in the common pool resource literature for small-scale, natural resource self-management.[10] The difference is indeed behavior, individual and collective. To put it differently, a faraway target can always be pushed further away: "the present is hard enough to deal with; don't ask me to take care of future generations, too!" A never-ending target that starts today and never ends, by contrast, one that, all along the path, makes sense to shoot for, is always relevant; it cannot be pushed away. It is, after all, part of a broader ethic, an ethic of the long term, of an ecologically and socially meaningful sustainability.[11]

Third, as posed, the "start stopping" choice is entirely anthropocentric. One doesn't need to invoke concern for other species or ecosystem integrity or even climate stability per se. And yet adding any of these concerns just makes the ethical argument more compelling.

### A Precondition for the Good Life

In sum, there has been and continues to be a fossil fuel ethic, however rarely it is articulated, let alone debated. Fossil fuels, since their rise and dominance as a power source, have been presumed net beneficial, right to be used, wrong to be left in the ground. Yet like so many other substances and practices through modern times where benefits are immediate and concentrated among the few and costs delayed and dispersed to the many, there comes a time when citizens and leaders must ask, Are fossil fuels compatible with the good life, with life as we know it, indeed with life itself (chapter 11)? The early abolitionists posed similar questions, and after decades of dogged research, public education, and politicking, they convinced their society that the answer with respect to slavery was no. If there was a good life under the institution of slavery, it was for the few (traders, planters, investors); everyone else, from Europe to Africa to the Americas, free and slave alike, was degraded. Now such questions are being posed with respect to nuclear weaponry and nuclear power. Along the way, international society has concluded that biological weapons and land mines, ozone-depleting substances, mile-long driftnets, and a dozen or so persistent organic pollutants are incompatible with the good life—that is, the good life for all, and for all time.

To engage an ethics of fossil fuels—past, present, and future, including the distant future—is to open a public debate heretofore unimaginable. It would be a debate that goes to the core of modern, industrial, expansionist, and consumerist society. It would challenge the prerogatives of the world's most powerful actors, state and nonstate. Not to open such a debate would be to abrogate responsibility for existing wrongs visited on select populations (e.g., polar peoples, mine and rig workers, extractive zone residents) and to avoid responsibility for coming wrongs that, if worst-case scenarios play out, would dwarf all previous human-caused calamities.

So I posit that the crucial question is not how and how much to reduce fossil fuel use (the current political dilemma), or even when to stop using fossil fuels (as if the matter can be deferred to the "right time"), but how to start stopping now. To answer these questions, three ethical criteria are life saving, transition, and livelihood. For all this, there will be a politics——a politics of urgent transition, a politics of ending the fossil fuel era.

## Notes

I thank Jack Manno, Pamela Martin, and Adele Santana for helpful comments on earlier drafts.

1. If the moral tone in this argument seems unduly negative—from salutary to destructive—then consider that this is only one part of the politics of delegitimization. In chapter 12, Adele Santana and I explore another part, positing a complementary politics of making fossil fuels "special" by drastically limiting their use.

2. For rich, detailed histories of total use as a resource ethic in the context of western US water exploitation, see Donald Worster, *Rivers of Empire: Water,*

*Aridity, and the Growth of the American West* (New York: Oxford University Press, 1985); Marc Reisner, *Cadillac Desert: The American West and its Disappearing Water* (New York: Penguin 1986/1993).

3. Consider this thought experiment to help visualize the mix of costs and benefits of fossil fuels. Imagine that mercury, a potent toxic substance, could be manipulated chemically to yield great amounts of energy, enough to power the world. But the by-products were still mercury with its qualities of persistence, bioaccumulation, and neurologic and endocrine disruption. Would we use it? Would we calculate costs and benefits? For related arguments, see Douglas A. Kysar, *Regulating from Nowhere: Environmental Law and the Search for Objectivity* (New Haven, CT: Yale University Press, 2010).

4. This thought experiment may not be that absurd, at least for minor players. Costa Rica actually conducted the experiment beginning in the 1950s by eliminating its army and has succeeded so far.

5. Jack Manno, *Privileged Goods: Commoditization and Its Impacts on Environment and Society* (Boca Raton, FL: Lewis Publishers, 2000); Manno, "Commoditization: Consumption Efficiency and an Economy of Care and Connection," in *Confronting Consumption*, eds. Thomas Princen, Michael Maniates, and Ken Conca (Cambridge, MA: MIT Press, 2002), 67–99.

6. Chris Bright, "Environmental Surprises: Planning for the Unexpected," *Futurist* (July–August 2000): 41–47.

7. For elaboration of these politics, see Princen, "The Politics of Urgent Transition: Fossil Fuel Exit and the Localization of Attention," in *U.S. Climate Change Policy and Civic Society*, ed. Yael Wolinsky (Washington, DC: CQ Press, 2014).

8. Notice that the posed ethic is an ethic of fossil fuel use, not of conservation, climate stabilization, or carbon distribution. Nor is it an ethic of the rights of nature or intergenerational equity. It is simply about using fossil fuels with the premise that we will indeed continue using them at some level, like it or not, safely or not, and we will be transitioning out of them at some rate. The question here is under what conditions their use is justified given that continued use, at current or near current levels, will be catastrophic.

9. Amartya Sen, "The Living Standard," in *Ethics of Consumption: The Good Life, Justice, and Global Stewardship*, ed. David A. Crocker and Toby Linden (Lanham, MD: Rowman & Littlefield, 1998), 287–311.

10. Elinor Ostrom, *Governing the Commons: The Evolution of Institutions for Collective Action* (New York: Cambridge University Press, 1990); Daniel W. Bromley, ed., *Making the Commons Work: Theory, Practice and Policy* (San Francisco: ICS Press, 1992).

11. See Princen, "A Sustainability Ethic," in *Handbook of Global Environmental Politics*, ed. Peter Dauvergne (Cheltenham, UK: Edward Elgar, 2012), 466–479; Princen, "An Ethic of the Long Term," manuscript in preparation.

# Part 2

## Keeping Them in the Ground

# Introduction to Part 2

We began part 1 with the great environmental quandary of our times: the metabolism of modern industrial civilization is nearly entirely powered by fossil fuels, such that we can't live without them and, yet, because the by-products of extracting, transporting, and burning fossil fuels are undermining the ecological underpinnings on which all prosperity, perhaps all life, ultimately depends, we can't live with them either. Then, after treating "the problem" of fossil fuels in biophysical, cultural, and ethical terms, we presented three cross-cutting themes regarding the prospects, indeed the imperatives, for confronting fossil fuel dominance, dependence, and depredation. These themes set the stage for the case studies in part 2.

## Theme 1: The Energetic

The fact that the net costs of fossil fuel use have begun to exceed their net benefits globally has much to do with certain biophysical inevitabilities: when the most energy-dense and easiest to-obtain-fuels are depleted, the energy returns on energy invested in fossil fuels decline, leaving dirtier fuels (think tar sands) and bigger messes. The early years and decades of fossil fuel use may have been clean, or so their boosters seemed to think, but eventually their by-products and unintended consequences make them unsustainable and unmanageable. The arguments of chapters 2 and 3 combine into a single overarching message: an ecologically sustainable and socially just world cannot develop and endure when the energetic basis of that world inherently concentrates power and externalizes costs, when the boom of energy ascendancy (ever-increasing availability of dense, low-cost energy sources) inevitably shifts to the bust of energy decline.

**Theme 2: The Moral**

Once the net benefit presumption is abandoned, the issue becomes who benefits and who bears the burdens. We find that now, as the costs are rising and benefits declining, continued expansion of the extractive economy becomes unjust for widening circles of populations across space and time. While miners, residents of extractive zones, and downwind communities have long known that they were not part of the implicit benefits-exceed-costs presumption, now other populations are shifting to the downside—for instance, those in flood-prone coastal areas, or dependent on glacial melt or frozen tundra, or in the way of extreme storms. If worst-case scenarios play out, the circles will eventually be global, exempting neither rich nor poor, North nor South. Economic calculations, even energetic ones, become irrelevant. Moral ones become paramount.

**Theme 3: The Political**

Making visible declining benefits, expanding costs, and increasing moral concerns of fossil fuel use leads to a politics of delegitimization: the deliberate undermining of the popular assumption that more such fuel is better than less. One result is the political possibility of choosing to leave fossil fuels in the ground, even when economic, technological, and financial boosters call for more extraction. This, we argue, is the turning point, where it become possible, even inspiring, for individuals and communities, governments and businesses, to decide to leave it in the ground, to be among the first to start stopping the fossil fuel juggernaut.

These three themes point to a two-part politics of ending fossil fuel dominance. Delegitimization focuses on the unmanageability of fossil fuel impacts while minimizing the polarizing tendencies of interest group, distributional, and rights-based approaches. "Start stopping" operationalizes delegitimization by offering criteria for current fossil fuel use, criteria that prioritize saving lives at the same time its use is for fossil fuel exit.

Drawing on these themes, we made the case for why, despite continued dependence on fossil fuels, the time has come to take the first steps, rather than waiting for global agreement, a swap-in of renewables, a miracle technology, a carbon market, or lifestyle changes. To wait for these is to skirt the politics.

The themes should be seen as propositional, grounded in the bio-physical and social to be sure, but future oriented and, given the social objective, normative and thus not absolute, not definitive, not extrapola-tive. In fact, their usefulness can only be assessed in context, in the lived experiences of those on the ground, those resisting the "slow violence" of fossil fuel extraction and combustion, those trying to exit the indus-try, those cocreating the "good life" for the post–fossil fuel era.[1] So as we developed these themes and explored these choices, we turned to researchers from around the world to provide examples of choosing to leave fossil fuels or other valuable minerals in the ground.

We offer here in part 2 six case studies where individuals, groups, com-munities, and, in one case, an entire country (Ecuador) are experimenting with a preposterous idea: not using an otherwise valuable substance. We ask readers to consider the cases not as "successes" (or failures; see below) but as first steps in what we expect will be a long process of social change. As first steps, they are imperfect and incomplete, as they neces-sarily must be at this point in human history, a point of unprecedented transition—energetic, moral, and political.

Common to nearly all six chapters are elements of a politics that suggests these cases are not unique but part of a larger movement. One element is normative underpinnings where, for example, clean water is tied to life support and a sensibility for the natural world. This is most pointed in Salvadorans' fight against gold mining and Germans' against brown coal. These communities, offered employment and other entice-ments by the extractive industries, have chosen a politics of creation based on longstanding agricultural practices and deep ties to the land—what they often call Mother Earth.

A second common element of the politics is the key roles played by national and international actors institutionalizing norms and financing initiatives. In Ecuador, it was a president who supported the growing calls of civil society for an oil moratorium. In El Salvador, it was the two major political parties that supported a ban on gold mining, only to be challenged by a transnational mining company in an international tribunal. In Germany it was a policy of phasing out coal that conflicted with plans to lessen Germany's dependence on nuclear energy.

A third political element is violence, from the dramatic deaths of oil rig workers in Norway to the slow violence of exposure to toxic substances in American Appalachia and the Australian outback and Ecuadoran Amazon. Indigenous peoples in particular experience such violence in part because, for some, voluntary isolation puts them out of sight and out of mind of the decision makers. While such violence may not be at the end of a gun, end-of-pipe solutions cannot solve the slow and devastating loss of communities, biodiversity, clean air and water, and fertile soil. A fourth political element is a growing wave of calls for something like "the good life" and "the good mind" and a reconceptualization of sustainable development based on ecosystem functioning, rather than on short-term extractive potential. This vision applies to the local and the global, adaptable to different cultures and ecologies.

## Case Studies

For those well ensconced in mainstream policymaking and policy analysis, effectively ignoring the biophysical or implicitly presuming that an endlessly growing material provisioning system (an "economy") is possible, keep it in the ground (KIIG) is indeed a preposterous idea. Delegitimizing the lifeblood of modern industrial society would be suicidal, this view would have it. Yes, there are problems, but leaders have a multitude of other issues—like economies that won't respond to stimulus packages. For the fossil fuel issue, it is better to manage emissions, promote solar and wind, implement a tax (or a cap), and grow the economy so we can afford all these correctives.

When we coeditors began this project some five years ago, we thought ourselves lucky to have one case to illustrate this preposterous idea. We thought the best we could do empirically is show that in one place, Ecuador (chapter 5), a coalition of civil society actors and government leaders could entertain the idea of deliberately not using a known oil reserve. We saw Ecuador's Yasuní project as bold and creative, but completely anomalous, not to mention trivial in the larger scheme of things (about 850 million barrels of oil, some ten days of worldwide oil consumption). And we worried that for all its promise, the Yasuní project might actually fail, which, as of this writing, it seems to have done—for now.

So we kept looking. And lo and behold, when we stopped looking under the street lamp where all the attention was on climate change and occasionally biodiversity and freshwater, we found more. In the shadows, small groups—some environmentalists but often as not local or indigenous peoples and select businesspeople and government officials—were nibbling at cherished beliefs such as total use, economic growth, market solutions, technological salvation, and ecological modernization (chapters 2 and 3). There, in Appalachia, El Salvador, Norway, Australia, and

Germany, along with Ecuador, we saw more boldness, indeed courage, and more creativity. We saw resistance to extraction and to the presumption of net benefit to locals and society as a whole. We saw the rejection of conventional development with its limited monetary measures. We saw attempts to cocreate a just and sustainable world where, in the first instance, their "world" was where they lived, in place, and from which they drew much of their sustenance and, in the second, their world included other peoples, even the entire planet. In other words, we saw in action visions of an idea that was not preposterous at all, not for those who understood the downsides of fossil fuel use (and gold and uranium use).

What we did not see were clear-cut successes—that is, permanent and complete cessation of fossil fuel use at a significant scale. Should that happen, we believe, it will necessarily be much further down the road of transition out of fossil fuels and endless material expansion. Instead, what we saw are small steps in the direction of KIIG.

We offer these cases, then, for what they are and no more: a sampling of efforts to enact something like KIIG. Their significance and their "success" is, however, highly limited. We introduce each with a note intended to establish that significance, to connect the case to themes of parts 1 and 3, and to preview new themes particular to the case.

For the methodologically concerned, we turn to social science questions of theorizing and case selection. Other readers will likely prefer to go straight to the cases, looking for elements of the three themes in practice as well as new themes and lessons, which we highlight in the concluding chapter, 13.

## Methodological Commitments

Case studies for a normative and prospective study such as this are inherently problematic. One reason is that if, in the coming years and decades, a widespread pattern of fossil fuel resistance and exit emerges, accelerates, and then normalizes (resistance is no longer isolated and sporadic, industry no longer finds exit irrational), and all this is accompanied by the cocreation of the good life without fossil fuels, then the defining characteristics of that pattern will be evident—but only after the fact. A premise of this study is that a driving force behind the end of the fossil fuel era (not to be confused, we emphasize, with the complete and sudden cessation of all fossil fuel use) is indeed resistance, a deliberate and active politics of delegitimizing fossil fuels occasioned by a broad public awareness of the net detriment of continued rates of fossil fuel extraction, conversion, and combustion. This politics includes an acceptance by powerful actors that, as Richard Heinberg put it some ten years ago, "the party's over." The one-time binge, the energy and material extravaganza, the boom as if there will be no bust is ending; it's time to exit. In the case studies, we see some pioneering efforts at such resistance at the same time we see examples of cocreation.

This view of the future—resistance, exit, cocreation, and societal acceptance—is only one scenario, however. Other plausible scenarios emphasize geologic and energetic imperatives (society effectively "runs out" of fossil fuels), technological salvation (all imaginable efficiencies are squeezed out of current operations and new technologies swap in as fossil fuels swap out), financial collapse and restructuring (various bubbles, including the "carbon bubble," burst and the capital no longer exists to extract), diplomatic achievement (a post-Kyoto process of international negotiations emerges that imposes real limits on carbon emissions). With a scenario-based approach, we are not predicting the success of or asserting the superiority of a resistance, exit, and creation politics. Nor are we dismissing other scenarios. We are merely taking seriously what others have ignored or dismissed. We have chosen one among several plausible scenarios to explore the possibilities of deliberately leaving fossil fuels in the ground or, to be precise, deliberately using fossil fuels (and other destructive substances and extractive processes) at rates dramatically less than those of the twentieth century fossil fuel era (that is, from roughly 1890 to 2005) and at rates compatible with regenerative capacities of biophysical systems and thus fit for the twenty-first century.

In this study, we are not making predictions, let alone offering hard-and-fast solutions. If pressed, we would predict some combination of scenarios, the emphasis of which would vary from one society and bioregion to another. And we would hypothesize that the societies (national or subnational) that begin the transition early and with relatively abundant resources and waste sinks to work with will be most likely to avoid catastrophic outcomes; they may even prosper. The case studies offer prospects for such societies, but only prospects; there are too many variables at play to be predictive.

At this historical juncture, a transition from energy glut to energy decline is unlike all previous ones (chapter 1) given the one-time nature of fossil fuels, overpopulation, and overconsumption. Those doing retrospective studies can often select from vast data sets and be confident in their generalizations. In a prospective study like this, we must work with the empirical base we have: cases that are necessarily nascent, incomplete, full of contradictions. Their significance is not so much in the emissions avoided or toxic substances forestalled. Their significance, rather, is in their audaciousness, their attempt to challenge core tenets of an extractive, expansionist, and externalizing global order. Perhaps most pertinent, these cases exist while being fully embedded in that order. Again, they are not "successes" in any definitive or measurable KIIG sense. Rather, they are social experiments —examples of early movers, political pioneers, moral entrepreneurs taking on the hegemon; small Davids slinging pebbles at a Goliath that still runs roughshod over anyone defying modernist norms. At any point in its development, a given case may be declared a success or a failure. In this prospective and normative study, we can draw no such firm conclusions. To do so would be like drawing conclusions about a revolution based on early skirmishes. Instead we offer provisional hypotheses and lessons, and we welcome the development of many more such cases.

The biggest claim we make regarding the import of any case is that a politics of resistance, cocreation, and early exit is possible, possible that is before geologic or social imperative makes fossil fuel exit unavoidable. We might go a step further and say that if such efforts have existed under (1) "energy ascent" (more cheap, dense energy is available year after year, decade after decade), (2) seemingly endless waste sink deposition (there is always an "away" to throw offending substances), and (3) booming economies with untold wealth generation, then under current conditions—energy leveling to be followed by energy descent (chapter 2), toxic accumulation, climate change, freshwater drawdown, soil depletion, and economies that just won't respond to more stimulus—such cases are precursors, harbingers, prefamiliarizing instances of a widespread shift away from fossil fuels.

The second reason case studies in a prospective, normative study are problematic is that actors' motives are highly variable, only one of which is a desire to end fossil fuel extraction and combustion. In Ecuador, we will see, motives include the preservation of biodiversity, protection of Indigenous peoples, promotion of democracy, and national self-determination, steering society toward a constitutional mandate for "the good life," and offering the world an alternative model of development and energy independence. In Australia, motives include fear of a world full of nuclear weapons, economic concerns for labor, and impacts on Indigenous peoples.

We see little problem with multiple motives. Peace treaties are signed for reasons quite separate from saving lives, economic agreements separate from mere economic gain, human rights measures separate from enhancing human dignity. If, in practice, fossil fuels (and, we will see, uranium and gold) are deliberately kept in the ground, it need not be because leaders and publics saw the light environmentally. Rather, multiple motives suggest multiple strategies and multiple outcomes are possible, only one outcome of which is a step toward a post–fossil fuel world. In short, a politics of ending the fossil fuel era need not be that of a singular substantive focus, as so many other environmental measures tend to be (all about "carbon," persistent organic pollutants, oil spills, endangered species, and so forth). In fact, as we argued in chapter 1 and will argue further in chapter 13, KIIG is about much more than carbon, oil, gold, or even extraction. The politics of KIIG, we will see in different ways in the coming cases, is about a shift away from a mining economy, away from an exclusive and narrow focus on a single substance (oil, coal, gold, or uranium), away from a contest between industry and environmentalists (or "locals" or place-based peoples). It is about a shift to an inclusive political economy where no cost–benefit analysis can justify sacrifice zones in far-away places, let alone in the neighborhood next door or the planet as a whole.

Given the nascent and evolving nature of these cases, here in part 2 we take a "weak case" approach. A strong case approach would offer case studies that are complete and unambiguous in their success or failure.[2] Regarding the end of the fossil fuel era, extraction would be stopped for good; the postcarbon era would have begun. But we have weak cases because it is the state of things in this growth-manic world that extraction trumps conservation, goods are good so more goods must be better, the highest bidder gets access, and so forth. Any effort to slow the extractive onslaught, let alone stop it in its tracks, is almost doomed to quashing by the powers that be; whether liberal or conservative, democrat and autocrat, East or West, North or South, dominant actors thrive on endless extraction and convenient disposal.

So at this historical juncture, the case studies are not definitive, and they cannot be. Rather, they are suggestive; they are candidates for transition, instances of early movers, of social pioneers, and of moral entrepreneurs. The chapter authors' conclusions are necessarily tentative, their lessons provisional. However "weak" these cases are, they do signal movement in many cultures to leave otherwise valuable resources in the ground. These are instances of collective action that have gotten purchase, gone two steps forward and one back (or, at the time of report-ing, two or three steps back but still climbing), and quite possibly (only time will tell) blazed a trail for others. The best we can offer are cases that have moved people to think and act in new ways, including with moral and spiritual commitment. These are instances that entertain the idea that some things humans just cannot handle and are best left untouched.

Eventually fossil fuels (and uranium and a lot of gold, we assume) will stay in the ground, one way or another. To accelerate that "keeping" it is precisely the limited and localized efforts that show the way, that keep the possibility alive, that offer lessons and construct a language of living well by living well within our means, all of which, we presume, will eventually trump the dominant ethic of total extraction (chapter 4). So the cases of part 2 and the minicases sprinkled throughout the book are small and of limited success. Empirically, we can only presume that as resource constraints tighten, they will become a focus of attention, helping set the stage for when people everywhere, not just the marginal-ized, say enough. It is our job in this project, as we see it, to document and show how these cases, however "weak," however small and limited, help point the way to a post–fossil fuel world.

In this book, we aspire to make a difference; we make no pretense of value neutrality. We believe, as international relations theorist Robert Keohane reminded one of us, that there are few social scientists who come to their craft without a driving social concern, whether it be issues of war and peace, democracy and human rights, prosperity and develop-ment, biodiversity and ecological integrity. The scholar's responsibility is not to project objectivity but to be explicit about one's normative com-mitments and the empirical bases of one's analysis. It is not just to look back and describe what is, but look forward and imagine what can be. On those counts, we try throughout this work to articulate our com-mitments and bases, but especially here in this introduction and then, in different ways, in the concluding chapter. Otherwise, we leave it to the chapter authors to make clear their commitments and bases and do so in their own way.

We can only hope that readers find our approach an honest effort to combine responsible scholarship with responsible citizenship. We hope they will agree that both roles, scholar and citizen, are embedded in insti-tutional environments that, on balance, tend to be very much conven-tional, very much perpetuating of the status quo, and very much ill-suited to twenty-first century challenges. We hope readers can agree with—or at least work with—a central premise of this work: that the world's environmental problems (and so many others) cannot be "solved," not one at a time, and not piecemeal or incrementally, such that, once solved, we can all "get back to normal." We hope readers share in some way our deep belief that the contemporary fossil fuel–driven, growth-dependent, consumer-serving, debt-laden, cost-displacing order cannot last, that it is at best flawed in the short term and illegitimate in the long term. We hope readers will see that on biophysical, moral, and political grounds, it is possible to deliberately and rationally end the fossil fuel era.

## Notes

1. Rob Nixon, *Slow Violence and the Environmentalism of the Poor* (Cambridge, MA: Harvard University Press, 2011).

2. For exemplary "strong case" social science and such a retrospective approach, see Elinor Ostrom, *Governing the Commons: The Evolution of Institutions for Collective Action* (New York: Cambridge University Press, 1990).

# 5

## Leaving Oil under the Amazon: The Yasuní-ITT Initiative as a Postpetroleum Model?

Pamela L. Martin

In this opening chapter of part 2, Pamela Martin analyzes one of the world's leading examples of the politics of keeping oil in the ground: the Yasuní-ITT Initiative of Ecuador, a nationwide effort to leave untouched a portion of known oil reserves in the Amazon. The significance of this nascent and evolving case lies largely in the fact that an entire country, indeed an oil-producing country, has entertained the idea of keep it in the ground and acted on it. What's more, the politics are more than resistance politics; they come from a national movement to live harmoniously with nature, to seek the good life, and to plan for a postpetroleum order. With rights of nature enshrined in its constitution, Ecuador is effectively practicing what we call in this book "twenty-first-century realism": after a five-decade history in an oil-dominated economy, it recognizes the reality of the destruction of human, plant, and animal life and declining fossil fuel reserves. Because the initiative would direct funds to alternative energy sources and sustainable practice, the initiative is not just about starting to stop; it is about transitioning to a post–fossil fuel economy.

The best evidence for questionable net benefit from oil extraction, this chapter reveals, is the great poverty in the Amazon, the otherwise richest area of Ecuador. So ultimately Ecuador's keep-it-in-the ground (KIIG) politics involves a process of delegitimization of extractive industries. The moral condition of sustaining the planet for future generations follows from the indigenous concept of *sumak kawsay* (good life). This position can be seen as an implicit adoption of Manno and Balogh's well-being return on investment (WROI) where well-being applies not just to the extractors and their immediate beneficiaries, but to all Ecuadorians, including those in the Amazon. But it is also a politics of fossil fuel exit where riches are viewed in terms of saving lives and livelihoods—those of Indigenous and non-Indigenous peoples and their communities and of untold species in one of the world's most pristine, biodiverse places.

As we will see at the end of this chapter, keeping oil in the soil in Yasuní is no easy task in today's fossil-fueled world, where payment for avoided extractions is an alien concept and fast-growing countries with great capital reserves (like China) are willing to fill host governments' coffers in exchange for oil supplies. In this case, Ecuadorians are attempting to counter concentrated power at the international and transnational levels with power of a different sort at the national and subnational levels.

Leaving oil underground was, and could possibly be again, a national policy in the least likely of places: in Ecuador, a country that currently depends on oil extraction for over 35 percent of its national budget. This chapter analyzes the national and global environmental governance institutions of the Yasuní-ITT Initiative of Ecuador (ITT) stands for Ishpingo Tambococha Tiputini, the three original exploration sites named by Petroecuador in 1992) and the struggle to maintain this policy as a clear example of leaving fossil fuels underground, showing that not only are such policies possible, but they have already happened: the oil is in the ground. Yet keeping it there is challenging in a world economy driven by fossil fuel extraction. What's more, this initiative has possibilities for replication by other countries that meet certain criteria. This initiative changes the global systemic dynamic of top-down, elite-driven global management, end-of-pipe solutions, and adaption and mitigation (greenhouse gas emissions), on the one hand, toward a transformative norm of no emissions, on the other. It does so by leaving 20 percent of Ecuador's oil reserves untapped in the Amazonian Yasuní National Park, a UNESCO Man and Biosphere Reserve, denoted by scientists as one of the most biodiverse places on the planet.[1] Despite President Rafael Correa's decree to rescind the initiative in August 2013, Ecuadorians, as of this writing, are organizing to continue it and to extend it to other areas of the Amazon.

The impetus for this initiative lies not only in its emission-saving graces, but also in the normative underpinnings of a plan that deliberately challenges conventional notions of development while proposing alternative visions of sustainable development based on the indigenous concept of *sumak kawsay* (*buen vivir* in Spanish and *the good life* in English). *Sumak kawsay* calls for humans to live responsibly within nature and with nature for the state of the planet now and for future generations. It is in this light that the Yasuní-ITT initiative also emphasizes the rights of two (and possibly more) Indigenous groups that live in voluntary isolation: the Taromenane and the Tagaeri. The goal of this initiative is to transition Ecuador toward a postpetroleum society by investing the Yasuní-ITT funds into renewable energy projects for the country. Ecuadorian leaders recognize the inevitable decline in petroleum production and the increasingly dangerous impacts of fossil fuel emissions on the planet. Establishing new development pathways beyond fossil fuel–based economies and societies is not without conflict and negotiation, as this chapter shows. Ultimately the Yasuní-ITT Initiative illustrates the complex process of shifting toward *sumak kawsay,* institutionalizing norms and policies at local, national, and global levels. As one of the founders of the Yasuní-ITT proposal and former minister of energy and mines, Alberto Acosta, explains regarding climate change and carbon management,

We, the ones that planned this [Yasuní-ITT Initiative], are questioning the system [of carbon trading]. We are talking about a postextractivist economy, postpetroleum. We are looking for a different kind of organization of the society.
I agree, it [the Yasuní-ITT proposal] is part of post-Kyoto but it does not fit into the logic of Kyoto. It is not like saying, "Kyoto is over, we have to make a second Kyoto," like part 2. But we are thinking it is from "Kyoto to Quito." That is our idea: make a leap. The ITT opens up a different vision. It is not the carbon market. It is not carbon bonds. It is not the traditional logic of mercantilizing the ecology.[2]

What, then, is it a matter of? I start with the proposal and its history.

## The Amazon-Sized Proposal

The Yasuní-ITT proposal calls for coresponsibility with the rest of the world ("common, but differentiated" in United Nations parlance) in avoiding emissions that the nearly 900 million barrels of oil in the ITT block could produce. The world, including states, nongovernmental organizations (NGOs), corporations, and individual citizens, would pay for avoided carbon emissions in order to protect one of the most biodiverse plots on Earth. The $350 million per year that Ecuador sought each year for ten years would be placed in a UN Development Programme (UNDP) trust fund with a board of directors that included Ecuadorians as well as members of the global community. If the trust fund, which started building in 2010, were successful, it would have been one of the largest global environmental trust funds of its kind. Its fate hangs in the balance as President Correa (as of August 2013) announced the closure of the trust fund while Ecuadorians protest in the streets and mobilize nationally and internationally, including the collection of over 756,000 signatures of the nearly 600,000 necessary for a national referendum to keep the oil of the ITT block in the ground. In May 2014, the National Electoral Commission announced only 359,000 signatures were valid, thus supporting the President's and the National Assembly's approval to extract the oil from the ITT block. The closing of the trust fund and the rejection to hear an appeal of the National Electoral Commission's decision, have only further mobilized mass global and national support for keeping oil in the ground, an indicator of the possibility of starting to stop (chapter 1).

The funds were to be directed to two funding windows: a capital fund and a revenue fund. The funding of projects would be based on Ecuador's National Development Plan for the Good Life.[3] The five main project areas of funding were deforestation prevention and ecosystem conservation, particularly in protected areas; reforestation and appropriate management of 1 million hectares of forest; renewable energy and increase in national energy efficiency; promotion of social development in the initiative's zones of influence; and support of research, science, technology, and innovation to enhance bioknowledge, river basin management, and changes in the energy matrix.[4] The ultimate goal is to transition Ecuador from an economy and society dependent on fossil fuels for its development to a postpetroleum society that focuses on sustainable development and living in harmony with nature—the politics of the good life.

**The Yasuní-ITT Trust Fund**

The UNDP Yasuní-ITT Trust Fund is an example of the very Ecuadorian flavor of this initiative where local and national governance of natural resources and funds over the international management of such issues are favored over international mandates. After a failure to sign the UNDP Trust Fund in Copenhagen in 2009 due to issues with sovereign control over the trust fund governing board, Ecuador formally institutionalized the initiative on August 3, 2010. Due to pressure from President Correa, the revised global governance structures were altered to emphasize Ecuadorian sovereignty over the initiative, as table 5.1 indicates. These changes included an additional Ecuadorian government representative (the secretary of state for the initiative, the director of the National Development Plan Secretariat, and the minister of cultural and natural patrimony) and two representatives from contributing countries to the Yasuní-ITT UNDP Trust Fund steering committee (the

**Table 5.1**

Changes to the Yasuní Fund Steering Committee

| Original Trust Fund Configuration (Pre-January 2010) | Yasuní Fund UNDP Terms, August 3, 2010 |
|---|---|
| 2 government representatives | 3 government representatives |
| 1 civil society representative | 1 civil society representative |
| 1 contributor representative | 2 contributor representatives |
| 1 technical secretary with no vote | 1 technical secretary with no vote |

*Source:* Author.

ambassadors of Italy and Spain as of January 2013). The civil society representative remained in place (an Indigenous Huaorani representative), but the voting process changed. While majority voting on issues within the Yasuní Fund UNDP steering committee was originally accepted, the August 3, 2010, memorandum of agreement between Ecuador and the UNDP outlined consensus voting if there was not a majority vote. President Correa has stated that Ecuador's government reserves the right to make all trust fund final decisions. In addition, a national-level government directorate board, which included government members from the UNDP steering committee and the ministers of renewable energy, the environment, and foreign affairs, was to meet monthly to review the initiative, its funding, and its projects.[5] Such national control over these global governance structures would prove attractive to other countries looking to replicate the plan as well.

Despite recent (as of August 2013) calls to drill in less than one-tenth of 1 percent of Yasuní National Park by President Correa, a review of the UNDP Trust Fund structure outlines the governance-sharing structures that make keeping oil underground possible—in Ecuador and elsewhere. According to the UNDP agreement with the Ecuadorian government, the Yasuní Fund steering committee would be responsible for the oversight and direction of the fund. It would review and make allocation decisions from both the capital and revenue funds, as recommended by the Government Coordinating Entity and the Technical Secretariat. It would authorize the release of funds to recipient and implementing organizations and review and prepare an annual strategic plan for the fund. In addition to third-party auditing oversight of the fund, as commissioned by the steering committee, it would also be responsible for monitoring and evaluation of the Yasuní Fund through an external "lessons learned" report. The steering committee has set the minimum threshold for reimbursement in the event that the government should drill for oil in the ITT block at $50,000 as guaranteed by the emission of certificates of guarantee Yasuní (CGYs).[6] All proposals for funding from recipient organizations out of the revenue fund or from implementing entities from the capital fund (for renewable energy projects only) would first be vetted by the government coordinating entity in consultation with the Technical Secretariat (an independent body with no vote) before going on to the approval stage by the steering committee. This structure favors Ecuadorian government control over the funding, while also ensuring global governance mechanisms and oversight. An important caveat to this structure is that only national entities may receive funding from the Yasuní Fund for projects. This may have to be amended, if there are future iterations, as technical consulting and other services may be required of international entities, such as the case of renewable energy projects.[7]

The first country to commit funds to the Yasuní-ITT Trust Fund was China, with a $1 million contribution and a $20,000 contribution from the members of its Ecuadorian embassy. Chile has committed $100,000. Spain has committed $1.3 million and the possibility of a multiyear commitment for other funds. Italy contributed funds via debt cancellation of $38 million of its $58 million debt with the country. One innovative feature of the Yasuní Trust Fund is that any person or organization may contribute funds. During the Cancun Climate Change talks of the UN Framework Convention on Climate Change in December 2010, the regional government of Wallonia in Belgium was the first European regional government to make a contribution to the trust fund: $398,000.[8] By October 2012, the UNDP Trust Fund totaled nearly $5 million in the capital fund account and $200 million total in commitments.[9] The former secretary of state for the Yasuní-ITT Initiative, Ivonne Baki, said that she expected contributions to reach $290 million by the end of 2012. These contributions, while indicating support for the initiative, were far from the goal of $3.5 billion over ten years and less than the goal of $350 million that President Correa requested to maintain the proposal active. Yet Baki and others are calling to keep the oil underground in the ITT block regardless of international contributions. This is what former Constituent Assembly and presidential candidate Alberto Acosta and Indian scholar/activist and Yasuní supporter Vandana Shiva call "Plan C."[10]

Contribution levels aside, Ecuadorians largely support the initiative, including polls that suggest nearly 80 percent wish to leave oil underground.[11] In July 2012, one group marched from the Amazonian city of Coca (near Yasuní National Park) to Quito in support of leaving oil underground in the ITT block. The march took place on the twenty-fifth anniversary of the spearing of Alejandro Labaka and Ines Arango, missionaries who tried to contact the Taromenane and Tagaeri Indigenous groups who live in voluntary isolation in Yasuní National Park. One participant in the march explained, "The forest is not for sale; we love it and we defend it. We do not want investment (in oil extraction) if it means destruction of the forest. You don't sell Yasuní; you defend it."[12]

### The History behind the Proposal

Although Ecuador has been known as a "petrostate," that status is quite recent in Ecuadorian history. Large-scale exploration and extraction began only in 1971 when Texaco Oil Company of the United States entered the northern Amazonian Lago Agrio region. In this same year, Ecuador's military government passed its first hydrocarbon law, creating legal-institutional structures to support natural resource extraction. In 1974, the government formed the state-owned national oil company, Corporación Estatal Petrolera Ecuatoriana (CEPE, now called Petroecuador), as a consortium with US-based Texaco-Gulf.[13] Today, Ecuador claims the third largest oil reserves in South America and is eighth overall on the list of crude oil countries from which the United States imports. It produces approximately 500,000 barrels per day, which it primarily extracts from the Amazon.[14] Still, as resource curse analysts have documented around the world, those who live in the extraction zone gain little and suffer the most.[15] In Ecuador after over forty years of oil production, the Amazonian region remains the poorest of the country with the highest levels of cancer and violence in the country (see table 5.2).

General Rodriguez Lara, Ecuador's military head of state in 1972, gleamed when the first barrel of oil rolled down the street of Manta, Ecuador. Among international dignitaries and Texaco Oil Company officials, a new "decade of oil development" was occurring. But as the figures in table 5.2 reveal, such "development" never reached the area from which it was extracted.[16] By contrast, the Yasuni-ITT Initiative is based on a new model of development under the norm of *sumak kawsay*, which includes the protection of communities and their right to a healthy environment. It also recognizes the limits to oil production and the absolute need for Ecuador to transition to a new energy source. This is abundantly clear in the Ministry of Nonrenewable Resources acknowledgment in its

**Table 5.2**
Poverty Percentages by City and Region, 1995–2006

| City or Region | 1995 | 1998 | 1999 | 2006 |
| --- | --- | --- | --- | --- |
| Quito | 27.3 | 19.9 | 29.1 | 20.9 |
| Guayaquil | 34.6 | 40.2 | 47.9 | 36.0 |
| Coast | 51.6 | 58.4 | 62.8 | 52.4 |
| Highland | 52.4 | 53.0 | 59.3 | 43.6 |
| Amazon | 71.5 | 63.2 | ----- | 66.8 |
| Rural | 76.5 | 77.9 | 81.6 | 72.7 |

*Sources:* Carlos Larrea, Ana Isabel Larrea, and Ana Lucía Bravo, "Petróleo, sustentabilidad y desarrollo en la Amazonía ecuatoriana: Dilemas para una transición hacia una sociedad post-petrolera," unpublished manuscript (2008). See also Instituto Nacional de Estadísticas y Censos (INEC), Encuesta de Condiciones de Vida, 1995, 1998, 1999, 2006 (www.inec.gob.ec).

new concessions for twenty-one oil blocs in the southern Amazon that states Ecuador's oil supply will last only twenty years.[17]

**Institutionalizing Yasuní and Learning from Mistakes of the Past**

The Yasuní-ITT Initiative was born not only of Indigenous norms, but also of the local Indigenous and non-Indigenous people's experiences with Texaco. Extraction began in 1972 and culminated in an internationally well-known lawsuit filed against Chevron Texaco (formerly Texaco) that began in 1993 in New York City and was transferred to a small courtroom in Lago Agrio, Ecuador (which, ironically, means "Sour Lake" in English) in 2008. The case was deemed the "David and Goliath" of its time and was filed on behalf of over thirty thousand Indigenous and non-Indigenous peoples who claim that the company destroyed the ecology of their northern Amazonian province and left behind grave social and health problems, including high cancer rates.[18] Chevron Texaco was found guilty of dumping billions of barrels of wastewater and toxic substances into the water supply in the northern Amazon and fined $19 billion. As of October 2012, Chevron Texaco's appeal to the US Supreme Court against this ruling in Ecuador was rejected and Ecuadorian plaintiffs were seeking to freeze Chevron Texaco assets in Argentina to pay $2 billion of the $19 billion fine.

International and local Ecuadorian NGOs, including Oilwatch, Amazon Watch, Rainforest Action Network (RAN), Pachamama Alliance, Acción

Ecológica, and Frente de la Defensa de la Amazonía, have been protesting the inadequacy of Texaco's alleged cleanup and calling for compensation. Writing in her book *Amazon Crude,* Judith Kimerling claims that the oil spill is larger than that of the *Exxon Valdez* in Alaska.[19] But the "Yasuní Park Depends on You" (the original name of the campaign sponsored by Acción Ecológica with other international nongovernmental organizations) international campaign is different, as Esperanza Martínez of Acción Ecológica argues, because the world community now knows of the destruction of the rain forest and the devastating impacts that oil extraction and industry, as illustrated in figure 5.1, can bring to not just plant and animal life but to the daily lives of Indigenous peoples.[20]

Pablo Fajardo, lawyer for the local peoples in the Chevron Texaco campaign, environmental activist, and Goldman Environmental Prize

**Figure 5.1**

The memory of Texaco in Ecuador and demonstration against oil development in the Amazonian rain forest. Public demonstrations involving Indigenous peoples, activists, and everyday citizens have been key to the initiation and continuation of Ecuador's attempt to leave oil in the ground. Notably, such bottom-up organizing has been matched (with exceptions) by top-down governmental policymaking, both national and international. *Source:* Author.

winner of 2008 from the northern Ecuadorian Amazon, explains the importance of the Yasuní-ITT Initiative:

[Yasuní] decreases emissions; we have to lower emissions. I know that many business sectors see this as a threat to their economic growth, but growth for what? To where do we grow? Accumulating more money for what? If what one wants is to live, and the majority of humanity does, then let us live. And think today if one sector has accumulated a lot of money, but we are destroying the future for the next generations, not thinking even about my children, but about the seventh or eighth or tenth generations. What will we leave then? ... For me, Yasuní is a fantastic place that we have to conserve because it is a heritage site. I know this area. It is an amazing heritage site; there are humans, uncontacted peoples, whom we are killing. I would not want for them what happened to the Indigenous peoples who lived where Texaco operated. I would not want this same story for them. ... The people who live there [in Yasuní], we do not understand their language or hear what they have to say, but what they want is to live in peace, and we need to respect that.[21]

Learning from the previous campaign surrounding the Chevron Texaco case and others in the southern Amazon, such as Sarayacu, activists, researchers, and scholars began calling for a moratorium on oil drilling in this region in the mid-1990s. As NGO activist, founder, and leader of Ecuadorian NGO Acción Ecológica Esperanza Martínez explains:

[The Yasuní-ITT Initiative] is the vindication of the fact that there are local peoples who do not want petroleum and have a right to a distinct model. And finally, it is the request after various years of proposing a moratorium on oil extraction, which Ecuador had in the southern Amazon. [Yasuní-ITT] is the result of three things: a postpetroleum Ecuador, a moratorium, and the protection of this zone.[22]

In 2000, Alberto Acosta and Acción Ecológica published a book entitled, *El Ecuador Post Petrolero*, which called for a moratorium on oil extraction in the Amazon and a move toward alternative energy sources for the country.[23] This laid the groundwork for a larger plan that included opposition to global climate change, support for those portions of the developing world not included in the Kyoto Protocol, and protection for the rain forest and uncontacted peoples living in voluntary isolation.

In October 2006, an international NGO, Oilwatch, and Acción Ecológica jointly sponsored the Forum on Human Rights, Oil, and Reparation in the Amazonian town of Coca. The forum hosted leaders and people from all over the world in the rain forest to take a tour of its devastation and exchange their problems, strategies, and hopes for a better future. It also celebrated ten years of Oilwatch's international mobilization on behalf of rainforests in the Southern Hemisphere. The

**Figure 5.2**

Pipelines and place. The power disparities between oil companies, independent and national, with their associated governmental and financial agencies, on the one hand, and resident peoples trying to survive, indeed thrive, on the other, is best seen on the ground, where these peoples live. Many people, across Ecuador and around the world, have come to the conclusion that living in place, indeed, living the good life (*buen vivir*) is possible only when oil stays in the ground. *Source:* Author.

forum began with a tour of contaminated sites in the Amazon as Figure 5.2 demonstrates, followed by a march for human rights and dignity of the peoples who live in affected areas. It culminated in an exchange of experiences from peoples all over the global South. This event was a key element in discussing alternatives to petroleum extraction and fostered information sharing among leaders in similar situations around the globe.

## Gaining Momentum and Political Opportunity

The forum roughly coincided with the presidential election of Rafael Correa, who began his term in January 2007. His election is significant, as the newly formed cabinet and supporters were members of the anti–oil extraction community. Furthermore, his first minister of foreign affairs, María Fernanda Espinosa Garcés, was formerly the International Union for Conservation of Nature regional director for South America and senior advisor on biodiversity and Indigenous peoples. When President Correa announced his official support of leaving oil underground in the ITT block in June 2007, he designated the study of the proposal to the Ministry of Foreign Affairs, headed by Espinosa.

According to Alberto Acosta, former minister of energy and mines, the first phase of strategizing about an official proposal began in January 2007 after Correa's election. Before becoming minister, Acosta had worked with Acción Ecológica leader Esperanza Martínez on a plan for leaving oil underground. Once he was appointed minister, he worked with Martínez to refine the proposal to present to President Correa. Acosta remembers the period between January and June 2007 as one of "tension" between the ministry and the executive president of Petroecuador, Carlos Pareja Yannuzzelli, an illustration of the push and pull between governmental institutions involved in such a process. While Acosta supported the proposal to leave oil underground, Pareja sought outside contracts from la Industria Petrolera de Venezuela, Sinopec of China, Petrobras of Brazil, and Enap of Chile in an attempt to convince the president to drill for oil. However, during the Petroecuador board meeting on March 30, 2007, Correa accepted the proposal to leave oil underground.[24]

In 2008, Acosta left his role of minister of energy and mines to lead the constituent assembly that wrote the 2008 Constitution, which enshrined rights of nature. The constitutional process consisted of a constituent assembly that was democratically voted into power, as well as civil society organizations that supported the process and gave feedback. Ecuadorians accepted the constitution in September 2008 by a democratic referendum vote. Within this document, natural resource extraction is prohibited in national parks, and Indigenous peoples living in isolation have the right to continue to do so.[25] Thus, the Yasuní-ITT proposal incorporates norms and laws from Ecuadorian society.

Supporting these norms toward the well-being, or the good life, of all Ecuadorian citizens, Acosta sums up the ethical responsibility of the Yasuní-ITT proposal:

It is 900,000 hectares that will be protected; very little really. It's not a great quantity of land, but it has enormous significance. It is consciously saying we are going to stop extracting petroleum, not for costs, not because we do not have technological capacity, but rather consciously because this represents respect for biodiversity, the great biodiversity of this zone. The quantity of native trees is superior to most other places in the world; the quantity of animal species; and the existence of two or three uncontacted indigenous groups, the Tagaeri, the Taromenane, and surely the Oñamanene are also there. There are human groups there that could lose their lives if there is petroleum activity. Thus, it is an ethical question that can have legal connotations because a president who enters there can be tried for genocide at the international level and should be tried at the international level for this.[26]

At the global level, the proposal supports human rights for Indigenous peoples living in voluntary isolation, following the Interamerican Commission on Human Rights ruling of precautionary measures in 2006 and Human Rights Council guidelines for Indigenous peoples living in voluntary isolation.[27] Furthermore, it attempts to advance global norms on climate change in four areas: solutions from the developing world, particularly those from fossil fuel– dependent, megadiverse countries; a shift in thinking from carbon sequestration and limits to avoided emissions; the social consequences of climate change; and a replicable plan to protect biodiversity in one of the most biodiverse places on the planet. On a more controversial level, the Yasuní-ITT proposal represents alternative economic solutions for postpetroleum societies. It provides one pathway to start stopping, based on a version of well-being indicators, in this case *sumak kawsay* (chapters 1 and 2).

### Challenges Facing the Yasuní-ITT Initiative and a Postpetroleum Society

Institutionalizing a postpetroleum state involves not only the interaction of state institutions and NGOs, but also the market and its actors, multinational corporations. In this case, oil companies are part of the Ecuadorian government's plan B: to drill for oil in the ITT block if funding to keep it underground is not found. Plan B places actors in favor of keeping oil underground in competition with oil companies and the governmental institutions that support oil extraction, such as the Ministry of Energy and Nonrenewable Resources and the state-run oil company, Petroecuador. Thus, the Yasuní-ITT campaign is illustrative of the complex dance between developing countries' traditional government institutions that seek revenues from natural resources and institutions, such as the Ministry of the Environment and environmental NGOs, that seek new forms of energy and environmental governance. As we will later examine, even post-Kyoto policies, such as the United Nations Programme on Reducing Emissions from Deforestation and Forest Degradation (which aligns with the Ecuadorian program Socio Bosque to combat deforestation called) and the Yasuní-ITT Initiative, compete for funding and authority within the Ministry of the Environment, and derail global funding sources from postpetroleum initiatives. President Correa has also used them to argue that oil can be extracted in part of the block and funding maintained by these sources to preserve the park, an option not supported by the national Indigenous organization, CONAIE, and environmental activists.

While the Ministry of Foreign Affairs and President Correa have called for an "ecological revolution" and the Yasuní-ITT proposal was one of the pillars of their foreign policy as part of the national and global plan to reshape sustainable development toward the good life on national and global levels, plan B (drilling in the ITT block) still loomed large, even as the Yasuní-ITT proposal was still active. In late October 2009, according to the newspaper *El Universo*, President Correa met with his Russian counterpart, Dimitri Medvedev, during which time both countries signed an agreement of cooperation to allow the Russian companies Oao Zarubehzneft and Oao Transgaz with Petroecuador "to explore and exploit Block 31," the adjoining block to ITT, also within Yasuní National Park and the Huaorani Ethnic Reserve.[28] Recently, *National Geographic* photographers and scientists, through aerial photos taken in 2012 when the trust fund was still open, discovered roads being built in Block 31, just next to the ITT block, that extend only 2 kilometers from Yasuní National Park. Such tactics have outraged members of civil society who have been told any work in this sensitive area would be done without roads—given that roads are known to be the leading causes of tropical deforestation and devastation to communities living in voluntary isolation.[29] Since the August 2013 closure of the initiative and the news of road building, mobilization to leave oil underground in the southern Amazon has increased.

In March 2010, President Correa announced the process of applying for licensure for drilling in the ITT block, which raised another red flag regarding his commitment to the Yasuní-ITT proposal and hurt the national and international campaign for its funding. Moreover, some question the protection of the ITT block when renegotiation of China's oil block 14 (Petro Oriente in Ecuador) expanded it to border the ITT block. Furthermore, the former secretary of state for the Yasuní-ITT Initiative, Ivonne Baki, sought funding from Iran for the initiative, a country with which Ecuador shares a spot on the list of countries that the Financial Action Task Force designates as having deficiencies in its financial regulations regarding money laundering. Such associations hurt the proposal and its tenuous future funding.[30] The policy disjuncture between the leadership of the initiative and the president, and presumably the Ministry of Nonrenewable Resources, outlines the complications of moving toward a postpetroleum society and state while still functioning within the current, fossil fuel–driven economy and system, even with the best of intentions and a national development plan that outlines *sumak kawsay* and includes the Yasuní-ITT plan.

Put in a global perspective, these developments suggest that the geopolitics of the fossil fuel era no longer involve only the Middle East. The financial crisis that began in 2008 has put a crunch on Ecuador's international sources of finance. This tightening of credit has put emerging market economies like China in the position of being able to tie their financial support and loans to extractive industries in places as diverse as Sudan, South Sudan, Russia, and Ecuador. While such financial strings threaten the Yasuní-ITT Initiative to leave oil in the ground, it also strengthens the Ministry of Nonrenewable Resources in Ecuador and the equivalent in other developing world countries as new sources of funding for energy enter the government coffers. In 2011, 11 percent of Ecuador's GDP was owed to Chinese development banks, totaling about $7 billion. Repayment for some of the loans is in the form of crude oil to China, including one report that identified an increase of 72,000 barrels per day.[31] China is also one of the primary investors with Venezuela in a $13 million oil refinery on Ecuador's coast, which expects to be refining 300,000 barrels per day by 2015.[32] These global pressures for fossil fuel energy and ease of funds to governments show how the concentrated power of oil interests coincides with that of financial interests, all to make efforts like the Yasuní-ITT Initiative to leave crude oil under the ground exceedingly difficult.

## From the Global Challenges to the Local Peoples

Including local leaders and Indigenous peoples in the initiative has varied. According to a survey of residents in the Francisco de Orellana province, where the park is located, only 27.93 percent of those living in the park's region knew that it was a national park. While 85.91 percent of those surveyed from twelve cantons within the province claimed that they were interested in supporting conservation measures for the park, only 11 percent were aware of the Yasuní-ITT Initiative and proposal. Yet when local peoples were asked if they would be interested in preserving the park and maintaining it for reasons of biodiversity and natural resources, over 92 percent responded favorably. Thus, local support for the goals of the proposal exists. At the national level, 78.6 percent of Ecuadorians support the initiative. In fact, the government raised $3 million from citizens in a one-day telethon, demonstrating national buy-in for the initiative.[33] Still, my informal conversations with Ecuadorians throughout the country reveal their trepidation about leaving this oil in the ground if the goals of the plan—$350 million per year for ten years—are not met, again highlighting the difficulties of implementing transformative change within a fossil-fueled economy. If there were a referendum on the issue, as many in Ecuadorian civil society called for, some expect that the leave-it-in-the-ground initiative would be pitted against poverty in the country, that is, that oil extraction would be offered as a way to decrease poverty. Such is the insidious culture of fossil fuels (chapter 3).

### Challenges to Funding the Initiative

Compared to funding an initiative like Yasuní-ITT to leave oil underground, funding adaptation and mitigation initiatives is almost easy. The UNFCCC COP 16 talks in Cancún concluded with a promise of $100 billion of joint funding per year by 2020 for mitigation and adaptation purposes through a Green Climate Fund with the World Bank as its trustee.[34] Having rejected World Bank financing for other projects (although it reestablished its ties in 2013), Ecuador might have jeopardized funding for the Yasuní-ITT Trust Fund. In addition to the Green Climate Fund, the World Bank, with other multilateral banks, has established the Clean Technology Fund, a $40 million fund that supports technology and programming for lower carbon emissions. There is also the South-South Trust Fund through the World Bank, in which developing countries share technology and ideas on sustainable development initiatives from the South. Finally, World Bank Climate Investment Funds have totaled over $6.14 billion in donations from the industrialized world to support climate programs in the developing world, most specifically in the area of pilot projects for mitigation and adaptation, for offsets and trades, none of which actually prevents fossil fuels from coming out of the ground.[35] The Yasuní-ITT Initiative is thus fundamentally different. It effectively challenges the global status quo institutions to go beyond end-of-pipe solutions.

The financing of the United Nations Programme for Reducing Emissions from Deforestation and Forest Degradation (UN REDD ) is another issue with a direct impact on the Yasuní-ITT Trust Fund and its ultimate success. REDD+ is a leading competitor for funds at the global level. Rather than combating climate change directly from the source, fossil fuels, it attempts to mitigate carbon emissions through forest protection and absorption of carbon. The UN Environment Programme and the Food and Agriculture Organization also manage funding for REDD+ through a Multi-Partner Development Fund through the UNDP. While preserving the world's forests is a key part of moving toward reduced atmospheric carbon, UN REDD+ does not confront the issue of subsoil rights, which are generally owned by the governments in the developing world; national sovereignty trumps UN REDD authority regarding land rights. Additionally, Indigenous communities in Ecuador and around the globe have argued that participating in the program in some countries can threaten their land rights. Thus, UN REDD may protect the forests but not the soil below them, leaving room for owners to be paid for forest protection *and* for oil extraction below it. The Yasuní-ITT Initiative directly confronts leaving oil unemitted and in the ground.

The challenge for the Yasuní-ITT Initiative and for others who may want to replicate it is to convince leading industrialized nations to go beyond above-ground measures, whether protecting forests or erecting windmills, and fund leaving fossil fuels in the ground. UN REDD provides an easy alternative for industrialized nations to claim to act on climate change initiatives without having to reduce their own fossil fuel production and consumption. They are effectively hoping that the remaining forests on the planet in the developing world will offset emissions from the global polluters. Germany was one of the initiative's initial supporters and had promised Yasuní-ITT Trust Fund support. Recognizing the above-below ground gap, the German parliament unanimously supported the Yasuní-ITT proposal in 2009 and has given 300,000 euros toward researching the trust fund and its global governance mechanisms.[36] Yet it backed out of its commitment in 2010 in favor of funding Socio Bosque, REDD programs in Ecuador that protect only the land above the ground, thus leaving open the possibility of oil extraction. Why such a sudden change in course?

While one answer to the problem of financing could be President Correa's suggestion to license the ITT block (or renegotiate oil contracts), another answer could be at the global level. Many European donor countries have diverted their deforestation and climate funds either directly or indirectly to REDD funding in an effort to consolidate their deforestation initiatives. Norway, Denmark, and Spain are currently giving the majority of funds to the multidonor trust fund. However, Germany has also stated its support of REDD activities through the German International Cooperation Fund (in German, Deutsche Gesellschaft für Internationale Zusammenarbeit).[37] Other European countries that have vocalized support for Yasuní-ITT, including France and Italy, are also donating funds to REDD activities. The US Agency for International Development (which was asked to leave Ecuador by President Correa and did so in July 2014), contributed $5.7 million to REDD activities in the country. The Long-Term Cooperation Action Ad Hoc Working Group proposals focused on new mechanisms beyond Kyoto may have the unintended effect of creating competition among new initiatives like REDD and Yasuní-ITT. The real solution is to protect the forests above and below the soil.

The danger of failure of the Yasuní-ITT Trust Fund is not just one of empty coffers, but a lost opportunity to institutionalize the norms around which the initiative was built. This includes a national dialogue and policy of achieving its goals of operating in harmony with nature, moving toward the good life, protecting Indigenous peoples contacted and uncontacted, and transitioning to a postpetroleum economy, all with constitutionally protected rights to nature. Unlike REDD programs that attempt to protect forests (only above ground), the Yasuní-ITT Initiative represents a much more ambitious national effort of transforming key social norms including those surrounding subsoil rights.[38] In Ecuador, the constitution does grant subsoil rights to the state. Thus, a change in policy course and funding toward REDD activities would not guarantee avoided emissions from natural resource extraction (in this case, petroleum), let alone the associated consequences to forest ecosystems and their peoples. The loss of such an initiative would also be in its applicability to other megadiverse countries of the developing world and the protection of rights to their peoples and forests. Thus, competing sovereignties exist above and below ground and must be aligned to ensure a postpetroleum future for REDD-protected forests.

### Hopes for Replication beyond Yasuní

Concerns persist about replication. Some analysts fear that wealthy oil-exporting countries like Saudi Arabia may try to join such initiatives if accepted as part of a post-Kyoto accord. Among other issues, they fear such an initiative could tighten global supply and thus increase oil prices, creating a windfall for oil exporters. However, Ecuadorian officials who crafted the plan argue that their approach avoids these problems by limiting the initiative's applicability to:

• Developing countries. One of this mechanism's main attractions to the international community for buy-in is that it simultaneously fulfills three objectives: it combats climate change, maintains biodiversity, and reduces poverty and inequality. The initiative also promotes sustainable development.

• Megadiverse countries located between the Tropics of Cancer and Capricorn, where tropical forests are concentrated. These countries host most of the planet's biodiversity.

• Countries that have significant fossil fuel reserves in highly biologically and culturally sensitive areas.

Among the countries satisfying all of these conditions are Brazil, Colombia, Costa Rica, the Democratic Republic of Congo, Ecuador, India, Indonesia, Madagascar, Malaysia, Papua New Guinea, Peru, Bolivia, the Philippines, and Venezuela. Therefore, it would exclude countries like those mentioned and would avoid an excessive number of similar projects.[39]

The Amazon Cooperation Treaty Organization (OTCA), founded in 1978 by the eight Amazonian countries (Brazil, Guyana, Surinam Ecuador, Peru, Colombia, Venezuela, and Bolivia) with a governing headquarters in Brasilia, is a regional organization that has the potential to unite the region in support of the Yasuní-ITT Fund. In November 2009, the organization hosted a conference in Brazil on Indigenous peoples and their territories with the goal of outlining common policies to protect them.[40] The Ecuadorian Ministry of Foreign Affairs is working with OTCA member states to support rights of nature in Amazonian countries, a basis of the Yasuní-ITT initiative.[41] However, the organization is weak and lacks authority within the member nations. Supporters of the Yasuní-ITT Initiative have called on the Amazonian region to unite behind the proposal, but natural resource exploitation and the resources it brings to state budgets have hindered any real coordination among state members. As box 5.1 illustrates, there are glimmers of hope in the Latin American region for replication, or at least a plan to start stopping.

## Institutionalization and Internalization of Norms

At the national and local levels, Ecuador's new constitution and the support of the initiative from the former minister of energy and mines,

**Box 5.1**
Oil Moratorium in Costa Rica

Ecuador isn't the only country in Latin America trying to limit fossil fuel extraction. With known oil reserves offshore, Costa Rica, a small Caribbean country with high biodiversity, a strong tourism sector, and no standing army, enacted a moratorium in 2002 on oil extraction, citing its ecological and social damages. In his 2002 inaugural address, President Abel Pacheco declared, "Costa Rica will become an environmental leader and not an oil or mining enclave." He went on to say, "Costa Rica's real oil and real gold are its waters and the oxygen produced by its forests."[a] Despite a brief encounter with the oil industry in the 1980s and recent considerations of natural gas exploration, it has maintained its stance against this industry in favor of ecotourism and alternative energy sources, and has achieved high human development indicators, as evidenced by the United Nations Development Programme reports. In June 2013 Costa Rica's antioil stance was challenged when a visit of China's president, Xi Jinping, led to a proposal by China of a $1.5 billion investment to upgrade an oil refinery in Costa Rica, a refinery capable of processing 65,000 barrels of oil a day.[b] Costa Rica's minister of environment and energy commented that there "is no other viable solution to the country's transport problems."[c] However, the project has been halted in the face of widespread criticism from both national experts and international authorities and the accusation that Costa Rican law has been broken. The general controller of Costa Rica stated that a company that had a relationship with the Chinese petroleum company conducted the feasibility study for the project, a violation of Costa Rican law.[d] And so, as global drivers push extractive processes to new depths and places, Costa Rica will be a case to watch to see if they continue to keep it in the ground.

**Sources**

a. "Ecological Debt and Oil Moratorium in Costa Rica," Oilwatch International, August 2005, http://www.oilwatch.org/doc/campana/deuda_ecologica/deuda_costarica_ing.pdf.

b. Guy Edwards, "Chinese Loan for Oil Refinery Clashes with Costa Rica's Climate Policies," *Intercambio Climatico*, June 20, 2013, http://www.intercambioclimatico.com/en/2013/06/20/chinese-loan-for-oil-refinery-clashes-with-costa-ricas-climate-policies.

c. Gervase Poulden, "Chinese-Funded Oil Project Halted amidst Flood of Criticism," *China Dialogue*, June 27, 2013, accessed October 1, 2013, http://www.chinadialogue.net/blog/6151-Chinese-funded-oil-project-in-Costa-Rica-halted-amidst-flood-of-criticism/en.

d. Jarmon Noguera González, Andrés Bermúdez Aguilar, and César Blanco Fajardo, "Contraloría frena nueva refinería," *La Prensa Libre*, June 21, 2013, http://temwww.prensalibre.cr/nacional/85029-contraloria-frena-nueva-refineria.html.

Alberto Acosta, and President Correa have moved the initiative through national politics and international agendas. Furthermore, Ecuadorian NGOs, such as Acción Ecológica, no longer seek approval for the alternative norms and how they frame the initiative; rather, they collaborate with their global colleagues to adapt such concepts to international levels.

Regarding the possibility of institutionalizing the initiative elsewhere, it is worth pointing out that it began as an idea from civil society and, over a two-year period, became a measure within a new political institution, the Commission for the Administration and Direction of the Yasuní-ITT Fund (CAD), represented by the Ministries of the Environment, Foreign Affairs, and Natural and Cultural Patrimony. A secretary of state, an executive-level position within the Correa regime, was its national and international spokesperson.

Can the norms expressed through the Yasuní-ITT Initiative be internalized? In the current order, when, for example, geopolitics still presumes abundant and cheap energy, it is hard to imagine. Nevertheless, some NGOs have expressed a need for a postpetroleum politics, as well as economic and political policies toward harmony with nature and the protection of human rights. In June 2012, the Ministry of Foreign Affairs held its first conference on the good life, Buen Vivir. It was discussed as a form of national and global sustainable development, an alternative to the dominant understanding of development where links to ecosystems are ignored. The Yasuní-ITT Initiative is Ecuador's shining example of such development. For all the problems with financing and threats of further oil extraction, the initiative can represent Ecuador's commitment to transformation, to enacting a policy of *buen vivir* and according constitutional rights to nature. Even with such governmental support, it is not clear if Ecuadorian society as a whole has bought in. Such a test will come when and if the state chooses to extract oil from the Amazon not only in the ITT block, but beyond it into the southern Amazonian region. As he did for the ITT extraction proposal, the president will have to present such a decision for a vote, according to the new constitution. This vote will demonstrate the strength of the alternative norms, which may be determined sooner versus later given the announcement to extract in part of the ITT block and the support of the National Assembly to do so as an issue of national interest.

**Downriver: Kyoto and the Future of the Amazon and Our Planet**

With only vague commitment on post-2012 climate change agreements from the United States and China, among other leading industrialized nations, the developing world needs to seek solutions now to climate change problems that have an impact on their environments, societies, and future generations. Brazil ranks fourth on the list of the largest carbon emitters in the world.[42] This is due to its extreme level of deforestation, which increased in 2008 and is about 20 percent of its Amazonian rain forest—about the size of Greece.[43] In fact, Brazil is responsible for 48 percent of all deforestation on the planet.[44] Approximately 30 percent of Ecuador's Amazonian region has been deforested.[45] Ecuador's greatest areas of deforestation follow oil access roads in the Amazon with a deforestation rate of 98,842 acres per year.[46] In both countries, deforestation of the Amazon is caused by natural resource extraction, infrastructure projects, and agricultural development, in addition to legal and illegal logging.[47]

In the GEO Ecuador 2008 report, the Ministry of the Environment emphasizes the destruction of biodiversity in the country's rain forest caused by deforestation and reliance on natural resource extraction—namely, oil and mining industries.[48] Researchers directly correlated the rise and fall of Ecuador's gross domestic product with the price of oil from 1980 to 2006.[49] During this time, poverty levels in the Amazon increased, and environmental threats and degradation rose as well. In conjunction with the Ministry of the Environment and the UNEP, researchers called for a moratorium on oil extraction in the Amazon, which suffers not only from forest loss but threats to biodiversity and human loss as well. In addition, they urged Ecuador to "internalize" international agreements such as the Convention on Biodiversity, Agenda 21, and the Millennium Development Goals.[50] Given the grim losses in the Amazon, compounded by the violence and threat of loss of Indigenous groups, action in this region is necessary for Amazonian countries and citizens of the world alike.

While researchers have found the western Amazon to be the richest in terms of biodiversity, they have also found this region the most threatened from oil concessions. One hundred eighty oil concessions spot the map in this area of the rain forest. Peru in 2008 added sixteen blocks to its forty-eight oil blocks, totaling sixty-four oil concessions in its rain forest. According to ecologist Matt Finer,

Peru is the most troubling country in the region in terms of oil and gas activities. The situations in Ecuador and Bolivia are certainly cause for alarm, but the scope and pace of the recent proliferation of oil and gas concessions in Peru is unprecedented. The vast majority of the Colombian Amazon, on the other hand, is not currently threatened by hydrocarbon activities, presumably because of the dangers posed by the FARC [Fuerzas Armadas Revolucionarias de Colombia].

Given the large amount of oil and gas known and suspected to lie under the Amazon, and the subsequent build up of hydrocarbon concessions, Ecuador's Yasuní-ITT Initiative is really one of the last hopes for a lot of areas. If it works, it may serve as precedent for similar conflict zones, like Bolivia's Madidi National Park. If it doesn't work, there is not much else to hold governments back from drilling.[51]

The Yasuní-ITT proposal may call on the world to protect one small plot of rain forest, but its ramifications are much larger. It is an innovative and imaginative guide to a future without oil dominance. It offers one way to protect areas that mean so much to both its local peoples and the global citizenry. While the economic and environmental benefits of the initiative may be clear, both in avoiding emissions and pursuing energy sources beyond oil, the mystery in the Amazon is what is yet to be discovered. Given the competing sovereignties in Ecuador (chapter 11), if a country tries to start stopping, this may be one method of doing so; it creates a global precedent with an established framework. At the very least, it suggests a pathway for local, national, and global actors to implement such a plan. Ecuador has made this option thinkable. It has normalized a dialogue and many in civil society are working to make it happen. It has dared to counter dominant norms to create a political will to "keep it in the ground." The question is whether it will continue the struggle to change or opt for the boosterist economies of the past without regard for a future based on the good life.

## Notes

1. Margot S. Bass, Matt Finer, Clinton N. Jenkins, Holger Kreft, Diego F. Cisneros-Heredia, et al., "Global Conservation Significance of Ecuador's Yasuní National Park," *PLoSONE* 5, no. 1 (2010)1): e8767, doi:10.1371/journal.pone.0008767.

2. Alberto Acosta, interview by author, Quito, Ecuador, February 27, 2009. Translation by author.

3. "Firmas Recolectadas por Yasunidos No Alcanzan para la Consulta Popular," *El Comercio*, May 6, 2014, http://www.elcomercio.com/actualidad/politica/firmas-recolectadas-yasunidos-no-alcanzan.html ; National Secretary for Planning and Development (Senplades), Ecuador, *Plan Nacional de Desarrollo para el Buen Vivir 2009–2013*, http://www.planificacion.gob.ec/plan-nacional-para-el-buen-vivir-2009-2013.

4. Multi-Partner Trust Fund Office, United Nations Development Group, *Ecuador Yasuní-ITT Trust Fund: Terms of Reference*, July 28, 2010, 5–6, http://mdtf.undp.org/yasuni.

5. Daniel Ortega, director of environment and climate change, Ministry of Foreign Affairs, interview by author, Quito, Ecuador, December 13, 2012.

6. Daniel Ortega, "Yasuní-ITT: The Initiative to Change History" (presentation at the Conference on Alternative Sustainable Development and Buen Vivir, Ministry of Foreign Affairs, Quito, Ecuador, June 8, 2012). As of August 2013, the Ecuadorian government is working with the United Nations Trust Fund Agency to review contributions over $50,000 and their remittance. All contributions have either been returned or placed in a fund for the Yasuní National Park.

7. Multi-Partner Trust Fund Office, United Nations Development Group, Yasuní-ITT Trust Fund.

8. Ministry of Natural and Cultural Patrimony, Republic of Ecuador, *Iniciativa Yasuní-ITT, Sección Noticias*, accessed December 15, 2010, http://yasuni-itt.gob.ec/blog/seccion/noticias.

9. Multi-Partner Trust Fund, *Yasuní-ITT Initiative*, July 23, 2012, http://mptf.undp.org/factsheet/fund/3EYR0; Ortega, "Yasuní-ITT: The Initiative to Change History. The UN REDD Multi-Partner Trust Fund through the UNDP has $173 million in commitments comparatively, http://mptf.undp.org/factsheet/fund/CCF00.

10. "Un Plan C Se Propone para el Yasuní ITT," *El Comercio*, August 29, 2012, , http://www.elcomercio.com/negocios/Plan-propone-Yasuni-ITT_0_763723784.html.

11. Encuesta de Perfiles de Opinion Sobre Yasuní ITT, *Amazonia por la Vida*, accessed August 25, 2013, http://www.amazoniaporlavida.org/es/Noticias/encuesta-de-perfiles-de-opinion-sobre-la-iniciativa-yasuni-itt.html.

12. Jose Olmos, "Homenaje a religiosos se volvió una denuncia contra Yasuní-ITT," *El Universo*, July 22, 2012, http://www.eluniverso.com/2012/07/22/1/1447/homenaje-religiosos-volvio-denuncia-contra-yasuni-itt.html.

13. Alberto Acosta, *La Maldición de la Abundancia* (Quito: Abya Yala, 2009), 37–40.

14. International Energy Agency, *Crude Oil and Total Petroleum Imports*, accessed September 17, 2013, ftp://ftp.eia.doe.gov/pub/oil_gas/petroleum/data_publications/company_level_imports/current/import.html.

15. Michael L. Ross, *The Oil Curse: How Petroleum Wealth Shapes the Development of Nations* (Princeton, NJ: Princeton University Press, 2013).

16. Constituent Assembly of the Republic of Ecuador, *First Barrel of Oil*, (Quito, Ecuador, July 26, 1972), video, http://www.youtube.com/watch?v=D9DsiXxxLQI.

17. Ministry of Nonrenewable Resources, Republic of Ecuador, "Gobierno ecuatoriano alista nueva ronda de licitación petrolera," July 23, 2012, http://rionapocem.com.ec/comunicacion/noticias/237-gobierno-ecuatoriano-alista-nueva-ronda-de-licitacion-petrolera-.html.

18. William Langewiesche, "Jungle Law," *Vanity Fair*, May 4, 2007, http://www.vanityfair.com/politics/features/2007/05/texaco200705.

19. Judith Kimerling, *Amazon Crude,* (New York: Natural Resources Defense Council, 1991).

20. Esperanza Martínez, "De Kyoto a Quito," *Llacta! Acción Ecológica,* May 9, 2007, http://www.llacta.org/organiz/coms/2007/com0096.htm.

21. Pablo Fajardo, interview by author, Quito, Ecuador, February 6, 2009. Translation by author.

22. Esperanza Martínez, interview by author, Quito, Ecuador, January 20, 2009. Translation by author.

23. Alberto Acosta, "El petróleo en el Ecuador: Una evaluación critica del pasado cuarto del siglo," in *El Ecuador Post Petrolero,* eds. Alberto Acosta and Esperanza Martínez. (Quito: *Acción Ecológica,* Idlis, and Oilwatch, 2000).

24. Esperanza Martíne, "Dejar El Crudo en Tierra en el Yasuní—Un Reto a la Coherencia," *Revista Tendencia* 9 (April 2009): 1–13.

25. 2008 Constitution of the Republic of Ecuador, http://pdba.georgetown.edu/Constitutions/Ecuador/english08.html.

26. Acosta interview.

27. Interamerican Commission on Human Rights, "Precautionary Measures for Tagaeri and Taromenane Peoples of Ecuador," accessed September 17, 2013, http://www.oas.org/en/iachr/indigenous/protection/precautionary.asp. Various Ecuadorian environmental and indigenous NGOs called for the IACHR investigation that resulted in this finding.

28. "Correa firmó ayer amplio pacto económico con Rusia," *El Universo,* October 30, 2009, http://www.eluniverso.com/2009/10/30/1/1355/correa-firmo-ayer-amplio-pacto-economico-rusia.html.

29. Nongovernmental Organizations interviews after Yasuní-ITT Commission Meeting, interview by author via telephone, Washington, DC, November 2009; Matt Finer, Varsha Vijay, Salvatore Eugenio Pappalardo, and Massimo De Marchi "Exclusive: Stunning Aerial Photos Reveal Ecuador Building Roads Deeper into Richest Rainforest on Earth (Yasuní National Park), http://news.mongabay.com/2013/1112-yasuni-secret-oil-road-finer.html.

30. Financial Action Task Force, public statement, July 12, 2012, http://www.fatf-gafi.org/topics/high-riskandnon-cooperativejurisdictions/documents/fatfpublicstatement-22june2012.html.

31. Katerine Erazo "La 'revolución' trae 'boom' económico de China en el país," *El Universo,* June 3, 2012, http://www.eluniverso.com/2012/06/03/1/1356/revolucion-trae-boom-economico-china-pais.html.

32. "Ecuador Minister: Met with China's ICBC on Strategic Projects," *Wall Street Journal,* July 20, 2012, http://online.wsj.com/article/BT-CO-20120720-700049.html.

33. Grupo Faro, "Encuesta de Percepción Ciudadana Francisco de Orellana," e-mail message to author; Ortega, "Yasuní-ITT: The Initiative to Change History."

34. Draft decision/CP.16 Outcome of the work of the Ad Hoc Working Group on long-term Cooperative Action under the Convention, 2010, accessed February 10, 2011, http://unfccc.int/files/meetings/cop_16/application/pdf/cop16_lca.pdf.

35. "World Bank Climate Investment Funds," accessed February 10, 2011, http://www.climateinvestmentfunds.org/cif.

36. Republic of Ecuador, Decreto Ejecutivo 1572, 2; budget information per interview with Natalia Greene, Quito, Ecuador, February 12, 2009.

37. Multi-Partner Trust Fund, United Nations Development Group, UN REDD Programme, accessed September 17, 2013, http://mptf.undp.org/factsheet/fund/CCF00.

38. Kathleen Lawlor, Erika Weinthal, and Lydia Olander, "Institutions and Policies to Protect Rural Livelihoods in REDD+ Regimes," *Global Environmental Politics* 10, no. 4 (2010): 1–11.

39. Carlos Larrea, Natalia Greene, Laura Rival, Elisa Sevilla, and Lavinia Warnars, "Yasuní-ITT Initiative a Big Idea from a Small Country," October 2009, mdtf.undp.org/document/download/4545. Joseph Henry Vogel's critique of the Yasuní-ITT Initiative is that it needs to include carbon-rich and economically poor countries rather than the geographical delineations suggested by the Ecuadorian government's technical team.

40. Organización de Tratado de Cooperación Amazónica, "Declaración de los Jefes de Estado sobre la Organización," November 27, 2009, http://www.itamaraty.gov.br/sala-de-imprensa/notas-a-imprensa/2009/11/27/declaracion-de-los-jefes-de-estado-sobre-la.

41. Ortega interview.

42. Al Gore, *Our Choice: A Plan to Solve the Climate Crisis* (Emmaus, PA: Rodale, 2009), 173.

43. Rhett Butler, "Deforestation in the Amazon." *Mongabay,* May 20, 2012, http://www.mongabay.com/brazil.html.

44. Gore, *Our Choices*, 174.

45. Jefferson Mecham, "Causes and Consequences of Deforestation in Ecuador" *Centro de Investigación de los Bosques Tropicales—CIBT* (May 2001), http://www.rainforestinfo.org.au/projects/jefferson.htm.

46. Matt Finer, Matt, Clinton N. Jenkins, Stuart L. Pimm, Brian Keane, and Carl Ross. "Oil and Gas Projects in the Western Amazon: Threats to Wilderness, Biodiversity, and Indigenous Peoples," *PLoS ONE* 3, no. 8:e293 (2008): 2, doi:10.1371/journal.pone.0002932.

47. Gore, *Our Choice*, 173–75; Butler, "Deforestation in the Amazon."

48. Guillame Fontaine, Iván Narváez, and Paúl Cisneros, eds., *GEO Ecuador 2008: Informe sobre el Estado del Medio Ambiente* (Quito: UNDP and FLACSO, 2008), 16.

49. Ibid., 26.

50. Ibid., 148–154.

51. Matt Finer, e-mail message to author, July 10, 2009.

# 6

## Appalachia Coal: The Campaign to End Mountaintop Removal Mining

Laura A. Bozzi

In this case of mountaintop removal for coal in the Appalachian Mountains of the United States, Laura Bozzi explores the delicate insider-outsider tension of keep-it-in-the ground (KIIG) politics. Mountaintop removal activists recognize both the deep sense of place, history, and culture of the peoples of Appalachia and the impacts of mountaintop removal and coal on local and global ecosystems. This chapter shows how the quick violence of destroying mountains, streams, and rivers creates a slow violence of lung cancer and other diseases, along with diminished educational, employment, and retirement opportunities. Appalachian peoples are effectively pursuing a KIIG politics based on the reality of decreasing coal reserves, ever-increasing mechanization, and declining market share on the one hand, and a dire need for a solution that marries well-being and livelihood on the other. Delegitimization thus combines the economic, the ecological, and the ethical with an eye to the long term.

This chapter also explores the uneasy politics of transition when local peoples have few alternatives and fear losing their way of life. The coal industry's manipulation of public opinion and lack of transparency have long thwarted such a transition despite the industry's decline. But now local groups are engaging both in what we call a politics of resistance and a politics of creation. Such groups are effectively saying that the good life cannot be had in a region deeply entrenched in a fossil-fueled, boosterist economy and with such extreme power imbalances. They are implicitly asking for investments of time and money that yield returns to their own well-being rather than to just company shareholder well-being. Bozzi tells how her own experience in an anti-mining protest makes clear the great difficulties of KIIG politics and at the same time shows the need for urgent transition.

In the keynote address to Power Shift 2011, a major youth climate conference, activist Tim DeChristopher called on the young leaders to come to West Virginia and take action to end a massive and destructive form of surface coal mining called mountaintop removal.[1] Looking out at the 10,000 people in the audience, he laid out a plan in which, "with these people, just right here," they could shut down a mountaintop removal mine for a year: thirty people each day, every day, occupying the mine site and so forcing the operators to halt their coal extraction. Power Shift had collected these young activists in Washington, D.C., but rather than spur them to rally Congress to pass legislation curbing carbon emissions, DeChristopher directed them to where the climate change problem physically begins: fossil fuel extraction. In his conceptualization, mountaintop removal and West Virginia are the front line for addressing climate change, and direct action is the most expedient way to stop the extraction.

Galvanized by DeChristopher's challenge, the nonviolent direct action group RAMPS (Radical Action for Mountain People's Survival) made an open call for people to come to southern West Virginia in July 2012 and participate in the movement to end mountaintop removal. The group planned a "mountain mobilization" where they would use direct action as a tactic to shut down a surface mine for a day. In their public call, RAMPS justified the escalated tactic in these terms:

To win our struggles against the extraction industries, we will have to band together. … If we want strip mining to end and restoration work to begin; if we want a post-coal future that is more than devastated landscapes, rampant fracking, and deepening poverty; if we want a healthy and whole Appalachia, we must escalate our resistance.[2]

Ultimately, approximately fifty people walked onto the largest mountaintop removal mine in West Virginia, Patriot Coal's Hobet mine on the Lincoln/Boone county border, stopping its operations for about four hours. Some locked themselves onto heavy machinery, one onto a tree, and others unfurled banners reading, "Coal Leaves, Cancer Stays" and "Restore Our Mountains, Re-Employ Our Miners." Twenty were arrested.

What does the RAMPS Mountain Mobilization say about how a movement to keep fossil fuels in the ground might operate, grounded in a place so historically tied to that extraction process as central Appalachia? I had traveled down to West Virginia to participate in the action in order to better understand that question. Setting the mobilization within a historical trajectory of activism against mountaintop removal more widely, my experience revealed the tensions and the difficult strategic and ethical choices participants confront. A group like RAMPS chooses its relationship with local and national participants, a choice that plays out within a larger contestation over who is a legitimate stakeholder in the conflict, an insider-versus-outsider divide. A group also chooses where on a regulate-to-ban continuum to select its policy goals on mountaintop removal and, relatedly, whether to take a stance against mountaintop removal but accept other coal mining or instead call for the end of all coal extraction. These kinds of decisions are the stuff of politics; they define the incredibly powerful scope of conflict. They are also particularly relevant to KIIG movements, since those contestations sit at the nexus between the local (the political economy at the site of extraction) and the global (the threat of the climate and other environmental crises).

The chapter begins with a description of mountaintop removal and a brief review of its consequences. Because the structural power of the coal mining sector is a primary force pushing back against ending mountaintop removal, I offer a basic description of the region's economic relationship with coal. Then by chronicling shifts in the movement over time, I open a window onto the dynamics of anti–mountaintop removal advocacy's choices. Circling back to the present day, I review the RAMPS mobilization with a discussion of the action in light of the historical tensions. The chapter concludes with a commentary on what, based on this case study, a KIIG movement could look like and the debates it will undoubtedly encounter. There is no necessarily clear or correct solution to the choices and tensions that groups confront, though I suggest that in the end, choices about who participates and what the policy goals should be will shape how the movement is perceived at the source and so may affect the ultimate effectiveness.

## Mountaintop Removal and Its Consequences

Technically, mountaintop removal mining has been a practice applied in the Appalachian region (also called the coalfields) since before Congress passed the federal surface mining act in 1977. In fact, it is a specific mining technique that is sanctioned in that federal act, specified as when the whole mountaintop or ridge is removed, exposing the full seam of coal. Operators are allowed to leave the area as a flat plateau, a more economical means to mine the coal as long as they make plans for specific economic development activities on that land. These early mountaintop removal mines, however, were small and much less common than the regular contour mines, which strip mined along the sides of mountains.

Starting in the mid-1980s, coal companies, under the pressure of low coal prices and heightened competition with western mines, innovated so as to increase the economic efficiency of mining the thin seams of low-sulfur coal within the steep slopes of central Appalachian mountains. Whether or not the mines fell under the specific "mountaintop removal" clause of the surface mining act, these enormous mining operations flattened mountains and lowered their height by at times 500 feet, dumping the rubble into the adjacent valleys. Such large-scale mountaintop removal is centered in eastern Kentucky and southern West Virginia, as well as western Virginia and northeastern Tennessee.[3] Current statistics on the area affected by mountaintop removal are difficult to find. The Environmental Protection Agency (EPA) estimated that between 1985 and 2001 in this region, mountaintop removal deforested nearly 400,000 acres of biodiverse forest, buried 724 miles of streams, and adversely affected an additional 1,200 miles of streams.[4] In 2002, mountaintop removal permits covered an area of 630 square miles, with a projection to double over these 2002 levels by 2012, amounting to an area the size of Rhode Island. In sum, one geologic study identified coal mining in this region as the greatest contributor to earth-moving activity in the United States.[5]

A defining characteristic of mountaintop removal is that it results in a large amount of rock and earth—what the industry calls "overburden"— which the miners then deposit in adjacent valleys, creating valley fills and burying streams. Burying the headwaters harms the entire stream length, as it destroys habitat for important macroinvertebrates that are key elements to the stream ecology, as well as reduces the flow of nutrients necessary for downstream health.[6] Furthermore, as runoff filters through the valley fills, it picks up metals, salts, and other compounds toxic to the biological life in the streams, carrying these to downstream users.[7]

While much of the early science has focused on the ecological effects of mountaintop removal, important new published research gives credence to local residents' longstanding concerns about how the mining affects their health. Chemicals and toxins are found in the drinking water in areas near the mining sites, as well as in hazardous airborne dust. Rates of mortality; lung cancer; and chronic heart, lung, and kidney disease are all elevated as a function of county-level coal production.[8] New research has found that birth defects are significantly higher in mountaintop mining counties compared to other counties in the region, controlling for risks associated with socioeconomic disadvantage, such as mother's health and education, prenatal care, and race.[9] Removal of vegetation, compaction of soil, and other impacts at the mined sites cause greater storm runoff and increased frequency and magnitude of downstream flooding.[10] The end product, coal, however it is mined, contributes to global climate change, acid rain, and mercury contamination, among other pollution effects, on combustion.[11] All of this adds up to a slow violence that, by itself, is out of sight and out of mind for decision makers and the public alike.

Aside from the slow violence of the physical, ecological, and health effects, however, this mining practice raises a unique moral question for society: Is it right to permanently remove a mountaintop? Is it right to intervene into a landscape in a way that is irreversible on a geologic timescale? And if, with such practices, people are knowingly causing irreversible damage to humans and the planet, is it right to continue such activity?

**The Coal Economy?**

The coal sector's supporters emphasize mining's role as a primary economic engine and source of jobs in the central Appalachian region. Indeed, the coal industry contributes significantly to regional economies. This is most pronounced in West Virginia, where, following personal income and consumer sales taxes, severance taxes on coal in West Virginia provided the third largest source of income for the state's general fund (about 10 percent of total general revenue in 2011).[12] In all of central Appalachia, surface coal mining employs about 13,500 people and underground coal mining about 24,000 people. While this figure leads to only a small percentage of mine workers at the state level, it is more significant in "coal counties," where it can represent 10 percent of the workforce, compounded by the indirect, economic multiplier impacts.

From a historical perspective, these employment figures are at the bottom of a steep decline. As coal mining in the region shifted from underground to the surface and became increasingly mechanized, the number of jobs the industry provided has declined even while productivity (tons per worker) has stayed high. In 1973, coal mines employed 124,000 workers in the Appalachian region; in 2003 there were 46,507 miners; meanwhile production during that period stayed nearly constant (about 380,000 short tons coal).[13] According to sociologists Shannon Bell and Richard York, through cultural manipulation, the coal industry has effectively masked this drop in employment in order to maintain its powerful public influence. Bell and York point to ways in which the coal industry promotes itself and "seeks to convince coalfield citizens that the industry is central to the region's economy, identity, and way of life": media campaigns; coal education programs in public schools; sponsoring of sports events, scholarships, or cultural events; and the creation of front groups like Friends of Coal.[14]

The critique of the coal economy can be taken to a further, structural level. Especially in the 1960s and 1970s, Appalachian scholars put forward a culture-of-poverty model to explain the persistently distressed economic conditions in the region.[15] The model viewed Appalachian people as ignorant and lazy, a condition that the model explained mountain culture reinforced; in other words, individuals are faulted for their own poverty. The policy prescription resulting from this problem definition was to create social and economic programs to bring Appalachia into the dominant culture and economy. The Appalachian Regional Commission, for instance, focused on road building, ostensibly to connect the region to the rest of the country. The practical effect, however, was to facilitate the trucking out of raw materials.[16]

In the 1970s, other scholars put forward the internal colonialism model, which argued that the region's integration into, rather than isolation from, the larger market system was a structural explanation for poverty.[17] The model posits that outside industrialists exploited the region for its natural resource wealth without reinvesting the profits in economic development and diversification. To the extent these structural factors exist, they highlight the challenges an antiextraction campaign in the Appalachian coalfields faces. With a local economy built up for a century around coal mining, the region lacks a diversified set of economic drivers, often the underlying conditions to shepherd in new alternatives when they are proposed. It may be that there is also a cultural lag in the perception of the mining industry, which economist T. M. Power coins the "rearview mirror' problem.[18] He asserts that extractive industries are historically entrenched and create a shared vision about the economic livelihoods of a community. As that economic pattern changes (in this case, mining employment numbers decline), the vision is slow to adjust. Power concludes that the "conventional wisdom about the local economy is the view through the rearview mirror, focused on the past rather than the present and dismissing all economic alternatives as unreliable or inferior."[19]

Anti–mountaintop removal groups are aware of these two explanations of poverty. The groups often tell a story of how, following the culture-of-poverty model, there has been a history of outsiders who come into Appalachia to "fix" things, then leave once they realize the task is harder than it seems. The groups are careful to frame their involvement as one of solidarity with and deference to the local communities. The internal colonialism model also offers a starting point for their critique of the coal industry and absentee landowners.

Meanwhile, geologic projections suggest that the coal reserves in central Appalachia are running out. More specifically, the remaining reserves are those that are more costly to mine due to higher stripping ratios (the ratio of coal to overburden) caused by thin seams buried beneath hundreds of feet of mountain.[20] For instance, a consensus report by researchers at West Virginia University found that "the depletion of low-cost reserves in the southern part of the state leads to increased mining costs that can make the [sic] southern West Virginia too expensive for the market."[21] Similarly, the US Energy Information Administration, projecting coal production until 2035, reports substantial expected declines from current levels, "as coal produced from the extensively mined, higher cost reserves of Central Appalachia is supplanted by lower cost coal from other supply regions."[22] Diminishing energy returns on energy (and capital) investments are now being felt (chapter 2).

In fact, analysts have been sounding a warning about diminishing returns on and depletion of the central Appalachian reserves for a long time. There has similarly been recognition of the need to diversify the economy. That neither of these calls has been well heeded by the political authorities makes the call to keep it in the ground seem all the more an abrupt and extreme transition for which to advocate. Those who call for the end of coal mining become the locus for blame, rather than the long history of repeated decisions and nondecisions that further entrench the region in the coal economy. In response, environmental groups seek to shift the public's framing of the issue toward the companies. They argue that the companies do little more than exploit the region, taking the resources but leaving little wealth. Kentuckians for the Commonwealth (KFTC), a grassroots social justice organization, states in its 2007 position paper on coal,

the coal industry has not and will not bring prosperity to coalfield communities. They provide an ever dwindling number of jobs and a big economic windfall to a few, well-placed political figures. ... Coal has been mined in eastern Kentucky for over one hundred years. If the coal industry was going to produce prosperity for us, shouldn't they have done it by now? ... Government on every level, federal, state and local has failed the people of the coal producing region for generations and has been complicit in allowing the extraction of billions of dollars worth of coal while not compelling the industry to contribute towards building a high quality of life.[23]

In short, concentrated energy (coal) tends to result in concentrated wealth and power (chapter 3).

## History of Activism

Over the course of activism around surface mining—and mountaintop removal in particular—groups continually encounter a series of choices with respect to their goals and the most appropriate pathways through which to achieve those goals. Among these choices, which together delineate the mountaintop removal movement, this chapter probes two: whether the goal is to better regulate mining or ban it altogether and whether to be against coal no matter the extraction technique or to be against a specific type of mining like mountaintop removal (and then perhaps support underground coal mining). In both of these, the group has to decide who will be invited to participate in the advocacy. Will they be from coalfield communities only or part of the wider public? If participants are from the outside, how will they justify their participation, and how will they relate to local concerns? Groups have made different choices about these questions over time. To give a sense of this trajectory and the tension across these themes, I provide a brief history of mining activism in the region, highlighting in particular KFTC, before returning to RAMPS and the Mountain Mobilization.

In the years following the passage of the federal surface mining act in 1977, surface mining opposition generally fell in two categories. One was a technical and professional one, in which national groups like the National Wildlife Federation and the Environmental Policy Institute mounted strategic and programmatic litigation to ensure that the implementing regulations retained the stringency Congress intended, fighting in particular against the Reagan administration's efforts to weaken the rules. The other category was localized and in response to specific threats, with neighbors coming together to protest particular impacts of mines in their communities. These local groups used a variety of tactics, including meetings with the state regulatory agency, attendance at permit hearings, and occasionally rallies and other public awareness events. In both cases, the advocates by and large did not oppose the mining altogether; rather, they wanted it to be done more responsibly. Rather than call for its prohibition, they sought to improve the enforcement of the existing laws and, at times, create new laws, in order to reduce the negative impacts of surface mining.

Meanwhile, mountaintop removal expanded rapidly in the 1980s, unbeknown to many people even in the surrounding communities. How could these massive mines have gone unnoticed? Reflecting back on those early years, people often say that the mines were hidden behind a row of trees, away from public roads and, of course, up a mountain. Word got out only slowly about the practice. In 1987, for instance, the *Washington Post* published an exposé in the Sunday edition with an oversized photograph of a denuded, flattened mountaintop.[24] In the coalfields, the news tended to spread by word-of-mouth. A coalfield organizer recollected, "Stories drifted through the hollows, about this terrible thing that was happening in the next county over, and that they had to pay attention to. That's how people would hear about mountaintop removal."[25]

### Kentuckians for the Commonwealth and the Regulate-to-Ban Continuum

Kentuckians for the Commonwealth is a grassroots citizens' organization actively involved in the campaign to end mountaintop removal mining. The organization has undergone a key shift in its analysis of the problem of coal mining, moving from calls for a more responsible mining sector to a systemic critique that marries the stance of community groups and traditional environmental groups. That is, KFTC has traveled along the regulate-to-ban continuum toward a qualified conclusion that coal should remain in the ground.

Through much of the 1990s, KFTC opposed one coal mine at a time in response to community member requests for help to address concerns like blasting or dust from living close to the mine sites. Due to this bottom-up issue selection, KFTC's position on coal was one of regulation rather than abolition. It limited its campaigns to calling for mining companies to obey laws rather than questioning whether coal benefited Kentuckians in the first place. By 2002, the organization's leadership realized that while they had made significant strides and were "winning many battles," their approach meant they were still "losing the war."[26] Consequently, following support from its membership across Kentucky, the organization shifted its critique from destructive mining practices to coal extraction in general. The Canary Project, adopted by the membership in 2003, gave structure to this broader focus. The project goals are (1) enforcement of existing laws for coal mines; (2) adoption of new mining-related laws where existing laws are inadequate to protect homes and communities; (3) creation of a sustainable economy with good jobs, and (4) promotion of "survivable" energy sources. On this last goal, KFTC advocates for renewable energy and energy-efficiency projects in the state.

To achieve the Canary Project's goals, KFTC set out a series of objectives including immediately halting mountaintop removal and "other forms of radical strip mining that are eliminating the mountains of Eastern Kentucky," as well as the associated valley fills. Another objective is to "accelerate the inevitable transition back to underground mining." Support for underground mining is a position that many take in the anti–mountaintop removal movement. It allows for continued support of the coal mining jobs—and, in fact, advocates argue that there are more jobs in deep mining than surface mining—to dampen the transition effects for the region. Nevertheless, KFTC also emphasizes that government funds should be diverted from supporting the coal toward investment "in locally generated, sustainable economic development for the coalfields and clean renewable energy sources for the country."[27]

As a state-based group, KFTC can take an insider's stance: "We are Kentuckians, which means we are coal miners, the families and friends of coal miners, and the descendants of coal miners." This position pushes back against the critique that only outsiders oppose mountaintop removal, a main rhetorical tactic by the coal industry and its supporters to delegitimize the advocates.

Thus, KFTC as an organization has shifted along the regulate-to-ban continuum, an evolution based on its many years of advocacy across communities affected by coal mining. Its attention to economic diversification for the region reflects how the group is tied to the region's well-being rather than to a specific environmental goal. In 2007, KFTC adopted a position statement on coal that formalized the Canary Project. Depletion of the state's coal reserves provided a starting point for the platform: "Coal is here today and tomorrow—but for how much longer?" But it is not only coal's inevitable depletion that drives the organization's call for an end to coal mining and a transition to a sustainable alternative economy. KFTC also takes a position on the debate over whether whatever economic benefits the sector provides justify the mining's negative impacts. The position paper states explicitly, "We believe if a block of coal cannot be mined without causing the physical, emotional, spiritual, and cultural destruction that we experience so often today, that block of coal should be left in the ground."[28] That is, given the destruction wrought by mountaintop removal, there is no room for an intermediate, compromise position. The mining cannot be carried out "better": the government cannot improve enforcement and avoid destruction. Mountaintop removal is inherently destructive. The only choice is to keep it in the ground.

**Nationalizing Awareness and Action on Mountaintop Removal**

In spite of strong local activism, the issue of mountaintop removal has remained unnoticed by the national public for much of its history. By the early to mid-2000s, local and regional groups increasingly employed strategies to raise awareness across the country, all with little support from the large national environmental organizations. For instance, Appalachian Voices was founded by an Appalachian State University professor in 1997 to help support grassroots groups in their campaigns against mountaintop removal (and other issues), providing analysis, communications, and other functions that local groups often lack the capacity to fully address. With online tools like the "My Connection Tool" where people around the country can type in their postal codes and learn whether their electricity provider uses mountaintop removal coal, Appalachian Voices helps make tangible the link between the broad public and this geographically defined issue. Other organizations seek to move people from individual (and often online) activism and toward collective actions. Modeled loosely on the Mississippi and Redwood Summers, Mountain Justice (previously Mountain Justice Summer) began in 2005 as a way to bring volunteers down to the coalfields and train them to join the anti-mountaintop removal movement.

**Mountaintop Removal and Climate Activism: The Shift from Emissions to Extraction**

From the middle to the end of the first decade of the twenty-first century, with much of mainstream environmental advocacy centered on passing federal climate legislation, anti–mountaintop removal advocacy groups had to decide whether to frame their issue in terms of climate change. To do so might offer a wider audience, as well as support from national groups, but it made untenable the compromise position of supporting deep mining as a bridge alternative to mountaintop removal. To oppose coal completely, these local groups would risk losing support from a segment of the Appalachian public who disliked mountaintop removal but supported the coal economy generally. Other groups expressed concern common to coalition politics—a fear that their issue of mountaintop removal could be used as a bargaining chip in gaining support for climate policy. This fear was particularly pronounced given the possibility that the technology of carbon capture and storage could, some thought, allow power plants to continue burning coal while reducing their greenhouse gas emissions (see chapter 1).

The tenor of national climate advocacy now has changed following Congress's failure to adopt federal climate legislation. Many advocates have shifted away from Washington, DC, and away from regulating emissions, moving instead toward place-based action like opposing individual coal-fired power plants. This may work to the benefit of issues like mountaintop removal. While messaging on climate change brings with it the challenge of making the impacts feel tangible, that is not the case with MTR; mountaintop removal has direct and immediate impacts and unavoidably raises major questions on social injustice.[29]

National organizations' move away from emissions reductions and toward place-based action is well demonstrated by how the RAMPS Mountain Mobilization coincided with a number of other actions against fossil fuel extraction undertaken in the summer of 2012. Coal Export Action, a week-long sit-in at the Montana capitol, protested a large coal mining permit that activists saw as the start of a spike in western coal mining aimed for the export market. In another event, thousands marched in Washington, D.C., for "Stop the Frack Attack" in opposition to the hydraulic fracturing boom taking place across the country. Also that summer, activists in Texas created a human blockade to protest construction of the southern segment of the Keystone XL Pipeline, which would pump Alberta's tar sands oil to a Texas refinery.

These and other events became collectively known as part of the Summer of Solidarity, a name that raises two important points. First, it represents the emerging conceptual and organizational link across the sites of local resistance. These links help avoid falling into not-in-my-backyard strategic positions, since the individual campaigns reference and support each other and avoid taking policy stances that trade their struggles for another's. Many of the actions were supported officially or in more informal ways by climate activism organizations like 350. org and Rising Tide North America. In addition, *solidarity* suggests that those outside the frontline communities have a legitimate and appropriate role in taking action in support of those on the inside. Within the Summer of Solidarity, it seems that groups can unite and be linked to a concern about climate change, but without an explicit statement as such. At its best, it allows people from outside to lend support to a particular action, strengthening that action and widening the concern beyond the local issue. The struggles in Montana, for example, are part of the larger project of stopping fossil fuel use at its source, and yet it is Montanans who maintain the leadership and authority in setting out the action and its frame.

## Radical Action for Appalachian People's Survival and the Mountain Mobilization

In late July 2012, I drove from my urban university campus down to southern West Virginia to join others from around the country at the RAMPS Mountain Mobilization action.[30] I arrived at the remote training camp in early evening, rolling onto a grassy field that had become a makeshift parking lot. I scanned the license plates: Missouri, Mississippi, Vermont, and other long drives away like I had made. Others had come from across Appalachia and its mountaintop removal landscape: West Virginia, Kentucky, Tennessee, and Virginia. In the back corner of the field, an old, hand-painted green school bus parked for a short stay along what I later learned was its rambling trip east from Oregon, having picked up wandering activists along the way. A number of the participants also came after learning about the action at the Earth First! Rendez-Vous in Pennsylvania a few weeks prior (and where they conducted a direct action to protest hydraulic fracturing for natural gas, or fracking).

For the next few days, I joined trainings where RAMPS organizers prepared the volunteers and facilitated planning of the action itself. The trainings made clear that this situation was far different from the ritualistic protests and arrests that have become a popular tactic in the environmental activist's toolbox, like those in front of the White House where the arrested are often processed and released after just a few hours, never even entering a jail cell. Occupying a West Virginian mine site, however, could result in serious consequences, perhaps days in jail, police brutality, civil suits by the coal company, and a criminal record that would preclude future choices like certain jobs or adopting a child.

In another set of trainings on nonviolent direct action and deescalation, the RAMPS organizers told us emphatically that all participants in the action must respect the RAMPS mission statement. It is critical, they explained, that this action not hurt the local organizing efforts, as people within the coalfields have been working for far longer on the campaign than RAMPS had. There will be a public reaction locally to the action, an organizer stated; most of the people at the camp will not be around to deal with the repercussions, but the local resistance will be. Participants were called on to be careful in their messaging and strategic in the action and to act with dignity, particularly in front of the media, and to abide by the tenet of nonviolence. Those who could not follow this code of conduct were advised to leave.

This code caught some of the new participants by surprise. Some disagreed with the stance of nonviolence or on taking responsibility for the action (rather than doing something undetected). For others, the RAMPS commitment to community partners and the local movement felt like an infringement on their own autonomy. Deference to the wishes of local activists, however, is common in anti–mountaintop removal actions, in part because of the historical legacy of the culture of poverty.

Organizers had already set out the action's messaging, framing it so as to "encompass the whole rather than feed a 'environmentalist' versus 'miner' divide."[31] The list of messages included:

Restore Our Mountains, Re-Employ Our Miners | We Want Healthy Communities | MTR Kills Communities | MTR Poisons Our Water | Mountaintop Removal Destroys Our Health | Coal Leaves; Cancer Stays | Keep It Underground

On the Saturday morning, fifty protesters walked onto Patriot Coal's Hobet mine, shutting down operations for about four hours. Ten people used locks to attach themselves to a massive dump truck, dropping a banner reading, "Coal Leaves, Cancer Stays." Another protester climbed a tree and attached himself to it, unfurling a banner along the tree trunk reading, "Stop Strip Mines." Ultimately the police arrested twenty of the protesters. Many of the other thirty who were not arrested then had to walk for four hours, down off the mine site and along nearby roads, before they could meet the shuttles waiting for them. Along the way, they encountered counterprotesters, who harassed them. Ten of the arrested protesters stayed in jail for eleven days, nine for six days. One arrested protester said he was beaten by police and denied medical treatment.

That morning as well, RAMPS hosted a training and media event in the Kanawha State Forest near Charleston. I joined this, choosing to take the role as a "peacekeeper" along with a few others from the camp. The idea of the event was that people who had not come to the training camp earlier in the week were to attend a shortened version at the forest. That is not, however, what happened. For whatever reason, hardly any additional anti–mountaintop removal activists arrived at the training. Nevertheless, anticipating conflict between the activists and counterprotesters, about twenty state troopers and other police had stationed themselves at the training. Then a crowd of counterprotesters swelled to about seventy over the course of the morning. They held signs reading, "Friends of Coal," or, "Coal Feeds Us," and many were dressed in navy blue mining uniforms with orange reflective strips.

For a few hours, my group of peacekeepers and a few other activists shared with the counterprotesters a small strip of grassy area in the park between a fishing lake and the small road. The police then required us to stand in opposing lines, one side of this combustible conflict facing the other. The tight quarters prompted interaction between the two groups. The counterprotesters asked the activists questions: "If the coal industry is shut down, what will replace it?" They asked personal questions too: "Where are you from? Why are you here? Who paid you to come here?" To most of these questions, the activists were largely silent. (As peacekeepers, we were told not to engage because it would distract us from monitoring the overall scene.) There were, however, a couple who took leadership to speak in response to the questions and taunts. In fact, a lengthy dialogue ensued between an activist and the collection of antiprotesters, with a level of courtesy and restraint on both sides not common in such confrontations.

Over the course of the morning, this civil dialogue was punctuated with anger and threats from some counterprotesters. One of the RAMPS-affiliated organizers later reflected on the scene:

> I witnessed an incredibly well organized group of people I would call a mob, mainly dressed in mining stripes, some sadly dressed up with coal on their faces, lining up and doing everything in their power to seem menacing, cruel and mean by singling people out from the group and picking apart their identity. ... Most of it was non-sequitur, just any old comment meant to be threatening, which helped show the hopelessness these folks feel about the situation—the mines going bankrupt, the water polluted, jobs scarce, etc., but hate speech all the same.[32]

The counterprotesters conveyed the general message that the activists were unwelcome outsiders and that the coal economy fed their families and sustained their communities. They expressed anger about activists' coming down to West Virginia every summer and causing trouble. They viewed the activists not just as outsiders but also as ignorant of the real conditions of the coalfields and to its history. On the popular *Charleston Gazette* blog, Coal Tattoo, a commenter explained what might reflect well the sentiment of those miners and their families:

How more radical can you be by protesting something they really have no idea about? They are mostly outsiders who spend a few weeks running around here claiming to be here to save West Virginia, gather 15 minutes of fame by breaking state law and are gone again until next year's 15 minutes. Really, if they are not mostly out of state, paid protestors going to college, how can they come down here and spend their summer "organizing"? Most of us have to maintain a job 12 months out of the year. They have no solutions to bring to the table. They can't answer the hard questions about their cause as we [have] seen Saturday morning in a face to face in Kanawha State Forest. Really, they are just paid protestors doing someone else's dirty work.[33]

Some of what angered the counterprotesters was the very nature of what the activists planned: people coming into the area for just a few days and undertaking the confrontational action of occupying a mine site. It may have been this tactic that kept local anti–mountaintop removal supporters from participating, feeling uncomfortable with its risks and public nature. Yet by telling the activists to go home, the counterprotesters also were trying to circumscribe where mountaintop removal politics can take place and who can have a voice in the decision. Furthermore, that local activists might have been afraid to participate is also not a neutral decision but one that is influenced by coal's continued cultural power within Appalachian communities.

A common goal of actions like the Mountain Mobilization is to attract media attention. Articles appeared in local newspapers including the *Charleston Gazette* and the *Williamson Daily News*, though without splashy front-page photographs. Other than mention in progressive sites like *Democracy Now* and the *Huffington Post*, however, national outlets gave the event little coverage. In fact, the environmental news website, grist.org published an article, "A Weekend of Protests Barely Makes the Papers," on how the RAMPS action and other antiextraction events that weekend also passed by largely unnoticed.[34]

## Conclusion

What does it mean to be a movement grounded at the point of extraction? Does this phrase mean truly grassroots or local, or does it mean going to the source? Whose campaign is it, anyway? Political scientist E.

E. Schattschneider's classic work, *The Semisovereign People*, argues that such divides are constitutive of an issue's politics: "The most important strategy of politics is concerned with the scope of conflict."[35] That is, this very question of who is an insider or an insider, or who is a "legitimate" participant, is a strategy of political conflict in and of itself. Nationalizing an issue, or more generally expanding the bounds of an issue, offers the chance to "break up old local power monopolies," which suggests why those seeking change would want to expand the conflict while those privileged by the status quo would want to keep it restrictive.[36] During the US civil rights movement, for instance, the controversy was not only about the rights of southern blacks to protest but also about the rights of outsiders to intervene.[37] From this perspective, counterprotesters at the RAMPS rally were trying to restrict the scope of conflict by saying that a national public does not have a place in the decision about mountaintop removal's legality or appropriateness. Meanwhile, RAMPS's efforts to invite participation from across the country and gain wide media attention with its direct action on the mine site were strategies to expand the scope of conflict and break apart the Appalachian coal industry's local power monopoly.

Nevertheless, to see that debate as only one of strategic positioning ignores the actual localized repercussions that are being raised. During the rally, a man came up to me and asked, "Say the mines are shut down tomorrow. Then what do I do?" I fumbled with a response, saying that people should not have to choose between the home they love and a paycheck. Others talked about economic alternatives. Our answers felt vague and theoretical. He said we had no real response, and that in fact no one does.

When climate advocacy groups shift from their downstream emissions-based approach (see chapter 1) to specific places of extraction such as an Appalachian coal community, the dynamics necessarily change. Such a shift moves the political conflict from the global (or national) scope to the local and, as a result, the specific, the particular, the grounded. This is not to say that the current carbon management approach does not play out at the local level (with assessments of the localized costs and benefits of greenhouse gas regulation), but going to the source makes those costs and benefits more concrete and imaginable for a particular place and for those who engage that place, however much they come from afar. For instance, gaining public support to keep coal in the ground in central Appalachia would likely require compensating that same region with benefits (e.g., funds to kick-start renewable energy production). This is a very different scenario from a national greenhouse gas policy that might lead instead to reduced coal production in central Appalachia but investment in renewable energy in the Midwest.

A homegrown resistance is likely to be more cognizant of these concerns and the need for a just transition away from extraction and toward sustainable economic alternatives. Indeed, Appalachian groups that oppose mountaintop removal tend to be very conscious of coal's cultural significance and its contribution (though declining) to regional employment and tax revenue. Like KFTC's Canary Project, their campaigns to end mountaintop removal often include at least one of two transition elements. The first is to call for an end to mountaintop removal while supporting a return, whether temporary or permanent, to underground mining. Underground mining is more labor intensive than mountaintop removal, among other reasons for groups to support it as an alternative. Of course, such a proposal does not keep the coal in the ground. The second route is to counteract the region's extractive resource dependence by helping to build up a diversified economy. KFTC, for instance, supports projects to train workers in energy-efficiency trades, and it is part of a coalition advocating for state policy to incentivize renewable energy in Kentucky. The RAMPS action had this theme as well, for instance, dropping a banner reading, "Restore our mountains, reemploy our miners."

This book calls for an inquiry into the exit strategies of the fossil fuel industry itself (chapter 12). Unlike modern coal companies that are usually diversified energy conglomerates and so are quite resilient to market shifts, the communities themselves remain vulnerable because they often lack alternative economic opportunities.[38] The advocacy efforts reviewed in this chapter have recognized, with varying degrees of commitment, that coalfield regions also need an exit strategy. To achieve this would not only lessen the impact Appalachian residents feel from the inevitable end of the coal era, but it might also soften resistance to the early exit necessary to address the climate crisis.

These two place-specific elements can combine with the ethical imperative and cultural shift developed in chapters 2 and 3, which scratch at coal's unquestioned position in the dominant fossil fuel paradigm, to collectively contribute toward delegitimizing coal in the region. For this to be effective, these ethical and cultural alternatives must have a local resonance and so also must be grounded in the place of extraction. There are many troubling issues to point to: mining's declining employment, the region's chronic poverty amid the coal companies' wealth, the cancer clusters in communities surrounding mine sites or the streams running rust orange from mine toxics, the irreversible destruction of mountains and Appalachian heritage. If this is posed alongside an honest hope for a just transition—for the good life—a politics of creation may open the space for a viable movement to keep coal in the ground in Appalachia.

## Notes

1. Available at http://www.youtube.com/watch?v=81EZUkYzrxU.

2. Available at http://rampscampaign.org/mountain-mobilization.

3. There is disagreement over the term used to name these mines. Industry and government tend to use *mountaintop mining*, while environmental activists use *mountaintop removal.* Since this chapter centers on activists' strategies, I use their nomenclature to reference all large-scale surface mining in the central Appalachian coalfields.

4. US Environmental Protection Agency, "Mountaintop Mining/VF Final Programmatic Environmental Impact Statement" (Philadelphia,: EPA Region 3, 2005).

5. Roger. L. Hooke, "Spatial Distribution of Human Geomorphic Activity in the United States: Comparison with Rivers." *Earth Surface Processes and Landforms* 24 (1999): 687–92.

6. Margaret A. Palmer, Emily S. Bernhardt, William H. Schlesinger, K. N. Eshleman, Efi Foufoula-Georgiou, Michael S. Hendryx, A. Dennis Lemly et al., "Mountaintop Mining Consequences." *Science* 327, no. 5962 (2010): 148–49; US Environmental Protection Agency, "Mountaintop Mining/VF Final Programmatic Environmental Impact Statement."

7. Kyle Hartman, Michael Kaller, John Howell, and John Sweka, "How Much Do Valley Fills Influence Headwater Streams?" *Hydrobiologia* (2004): 91–102; "Mountain Mining Damages Streams: Study Shows That Stripping Mountains for Coal Has a Much Greater Impact Than Urban Growth," *Nature* August 12, 2010. 806.

8. Michael S. Hendryx and Melissa Ahern," Mortality in Appalachian Coal Mining Regions: The Value of Statistical Life Lost," *Public Health Reports* 124 (2009): 541–50.

9. Melissa Ahern, Michael S. Hendryx, Jamison Conley, Evan Fedorko, Alan Ducatman, and Keith J. Zullig, "The Association between Mountaintop Mining and Birth Defects among Live Births in Central Appalachia, 1996–2003," *Environmental Research* 111, no. 6 (2011): 838–46.

10.  Palmer et al., "Mountaintop Mining Consequences."

11. Norbert Berkowitz, *The Chemistry of Coal* (New York: Elsevier Science, 1985).

12. *West Virginia Executive Budget Fiscal Year 2012: Volume I Budget Report* (Charleston: State Budget Office of West Virginia, 2012).

13. Energy Information Administration, *Coal Production in the United States—An Historical Overview* (Washington, DC: Energy Information Administration, 2006).

14. Shannon E. Bell and Robert York, "Community Economic Identity: The Coal Industry and Ideology Construction in West Virginia," *Rural Sociology* 75, no. 1 (2010): 111–43; Shannon E. Bell, "Review: *Removing Mountains: Extracting Nature and Identity in the Appalachian Coalfields* by Rebecca R. Scott," *American Journal of Sociology* 116, no. 6 (2011): 2034–2036.

15. Dwight B. Billings and Kathleen M. Blee, *The Road to Poverty: The Makings of Wealth and Hardship in Appalachia* (Cambridge: Cambridge University Press, 2011); Robert L. Lewis and Dwight B. Billings, "Appalachian Culture and Economic Development," *Journal of Appalachian Studies* 3, no. 1 (1997): 43–69.

16. Ron D. Eller, *Uneven Ground: Appalachia since 1945* (Lexington: University Press of Kentucky, 2008).

17. Billings and Blee, *The Road to Poverty*.

18. Thomas M. Power, *Lost Landscapes and Failed Economies: The Search for a Value of Place* (Washington, DC: Island Press, 1996).

19. Ibid.

20. Rory McImoil and Evan Hansen, *The Decline of Central Appalachian Coal and the Need for Economic Diversification* (Morgantown, WV: Downstream Strategies, 2010).

21. Randall A. Childs and George W. Hammond, *Consensus Coal Production Forecast for West Virginia, 2009–2030* (Morgantown: Bureau of Business and Economic Research, West Virginia University, 2009).

22. Energy Information Administration, *Annual Energy Outlook 2011 with Projections to 2035* (Washington, DC: Energy Information Administration, 2011).

23. KFTC, "The Canary Project: For a Better Future Beyond Coal," position paper (2007).

24. Cass Peterson, "Bulldozers Driving through Holes in 1977 Strip Mining Law," *Washington Post*, May 30, 1987.

25. Interview by the author with former West Virginia citizens' organization staff member, April 16, 2011.

26. Ibid., May 25, 2011.

27. KFTC, "The Canary Project."

28. Ibid.

29. Interview by the author with national environmental law organization staff member, September 28, 2011.

30. The following account is based on the author's participant observation in the RAMPS Mountain Mobilization, July 26–28, 2012.

31. Available at http://rampscampaign.org/our-eco-chaplains-take-on-the-mountain-mobilization/#more-2820.

32. Ibid.

33. Available at http://blogs.wvgazette.com/coaltattoo/2012/07/31/coal-protest -was-this-really-so-radical.

34. Philip Bump, "A Weekend of Protests Barely Makes the Papers " grist.org. July 30, 2012, http://grist.org/news/a-weekend-of-protests-barely-makes-the-papers.

35. E. E. Schattschneider, *The Semisovereign People: A Realist's View of Democracy in America* (New York: Holt, 1960).

36. Ibid.

37. Ibid.

38. The energy companies' portfolios often include mines across different regions across the country (e.g., low-sulfur, low-cost Powder River Basin; high-sulfur, underground mines of the Illinois Basin). Arch Coal, for instance, is the most diversified American coal company, with 15 percent of the country's coal supply at mining complexes in Wyoming, Utah, Colorado, Illinois, West Virginia, Kentucky, Virginia, and Maryland (http://www.archcoal.com/aboutus). Others, like CONSOL Energy, produce both coal and natural gas. Admittedly, these are varied forms of fossil fuel extraction, but it still underlines the point that these companies are buffered against risk of change in particular regions. William R. Freudenburg and Lisa J. Wilson, "Mining the Data: Analyzing the Economic Implications of Mining for Nonmetropolitan Regions," *Sociological Inquiry* 72 no. 4 (2002): 549–75; William R. Freudenburg and Robert Gramling, "Natural Resources and Rural Poverty: A Closer Look," *Society & Natural Resources: An International Journal* 7 no. 1 (1994): 5–22.

# 7

## El Salvador Gold: Toward a Mining Ban

Robin Broad and John Cavanagh

In this chapter, Robin Broad and John Cavanagh show how one small country, El Salvador, long at the mercy of foreign powers and powerful transnational corporations, has achieved near consensus that gold mining, and the destruction of water, land, and livelihood that go with it, must stop. Although the substance is not fossil fuel, this case acts as a bridge to other movements that resist destructive extraction. It challenges conventional notions of development, implicitly saying that a return on well-being, not on capital, is the proper return on investment. It shows how citizens, especially those whose livelihood depends on clean water and land, are in a struggle with the slow violence of short-term extraction and long-term toxicity.

Power here resides largely with the transnational corporation, which can circumvent national sovereignty by reaching beyond El Salvador's borders, in this case to the International Centre for Settlement of Investment Disputes, housed at the World Bank. Much like power in the fossil fuel industry, here it is opaque, capable even of challenging the sovereign authority of the Salvadoran state to implement its own moratorium on gold mining.

At the same time, that distant power is forcing Salvadoran activists to create their own politics of local resistance as well as of creation. Theirs is an attempt to transition out of a mining economy and into a "solidarity economy" where rivers are free of cyanide and arsenic, opponents of corporate-driven development are not threatened and killed, and residents can pursue livelihoods free of environmental and political threats. For them, the transition is indeed urgent, as it is for the rest of the planet. Here, keeping gold in the ground is one example of what marginalized communities face as they seek a decent life at the same time they confront what in this book we call twenty-first-century realism.

Gold and other precious metals have been mined for centuries in different parts of the world. Over the past decade, there has been a rapid expansion of mining activities as the prices of most minerals have increased dramatically. Gold, for example, sold for under $300 an ounce as recently as 2000, yet its price passed $1,700 an ounce in 2012.[1] In this bonanza of skyrocketing prices, mining firms have looked back to many geographic areas where they abandoned mining over the past century (including in the United States) to sign new concessions and begin mining again.

With this rapid expansion of mining, social movements in many countries have gathered forces to oppose the mining, from South Africa to the Philippines to Guatemala to Peru. They have raised many factors in their opposition. Paramount is the environmental havoc that large-scale mining firms wreak on communities and nations. Most commercial mining requires large quantities of water. In addition, most large-scale gold mining separates the gold from the rest of the rock with cyanide, which is highly toxic and can contaminate both soil and water. Frequently the cyanide separation process also releases toxic arsenic from the surrounding rock. A number of health problems often result. Unlike the struggle against fossil fuels, the mining struggles are not just or necessarily about climate change. Rather, the environmental challenges from mining are more local and national, although they too can be transboundary.

There is also social and political fallout from mining activities. Mining firms often bring money into communities where they want to mine as a way to win over politicians and the public. That money, and the promise of more money once the mining starts, frequently creates conflict between the pro-mining and anti-mining forces. In some countries, people have been killed as the tensions rise. Some mining companies have donated funds to politicians, which increases conflict as well.

El Salvador is an instructive case study of such conflicts. The northern half of El Salvador is dominated by old volcanic mountains that contain a great deal of gold and other minerals. Foreign firms started mining this gold over a century ago, but much of the mining stopped in the 1930s when global mineral prices plunged.

In the mining boom of the 2000s, many Canadian and other mining firms expressed interest in dozens of potential mining sites. A number of these firms signed contracts with the Salvadoran government to explore for minerals with the hope that they would discover lucrative enough deposits to move on to the exploitation stage. Local groups began to oppose the mining after 2005, and by 2007, one poll showed that close to two-thirds of Salvadorans opposed mining. The opposition was so strong that both major political parties in the 2009 elections said they would stop mining if elected.

The winner of that election, Mauricio Funes of the progressive FMLN Party, kept his promise and did not issue any new mining permits during his five-year term. Hence, El Salvador has become the first mineral-rich country to "keep the gold in the ground," at least from 2009 to 2014.[2] Funes's successor, Salvador Sánchez Cerén, elected in March 2014, has pledged to continue this de facto moratorium on metals mining.

We visited El Salvador in 2011, 2012, 2013, and 2014 to research the anti-mining movement and the government response. This case study of the struggle against metals mining in El Salvador reveals a great deal about the dynamics of keep it in the ground (KIIG) struggles at a local, national, and global level and provides a dynamic study to compare and contrast with case studies of fossil fuels.

### Water for Life

Over thirty years ago, several thousand civilians in the northern Salvadoran community of Santa Marta in the province of Cabañas quickly gathered a few belongings and fled the US-funded Salvadoran military as it burned their houses and fields in an early stage of the country's twelve-year civil war. Dozens were killed as they crossed the Lempa River into refugee camps in Honduras.[3]

Today, residents of Santa Marta, some born in those Honduran refugee camps, are among those leading the fight against US, Canadian, and Australian mining companies eager to extract the rich veins of gold buried near the Lempa River, the water source for more than half of El Salvador's 6.2 million people.[4] Once again, civilians have been killed or are receiving death threats.

The goal of groups opposing mining, who have come together to form a national coalition, La Mesa Nacional Frente a al Mineria Metálica (National Roundtable on Mining), is to make El Salvador the first nation to ban gold mining permanently. We traveled to El Salvador in April and May 2011, in July and August 2012, and again in May 2013 to find out if this struggle to keep gold in the ground could be won. Our investigation led us from rural communities in the country's gold belt to ministries of the progressive government in San Salvador and ultimately to free trade agreements and a tribunal tucked away inside the World Bank in Washington, D.C.[5]

In 2011, we were greeted at the airport by Cabañas native Miguel Rivera, a quiet man in his early thirties whose face is dominated by dark, sad eyes. Miguel is the brother of anti-mining community leader Marcelo Rivera, who was "disappeared"—tortured and assassinated—in June 2009 in Cabañas in a manner reminiscent of the death squads of the 1980s civil war.[6] We had first met Miguel in October 2009, when he and four others active in El Salvador's National Roundtable on Mining traveled to Washington to receive the Institute for Policy Studies' Letelier-Moffitt Human Rights Award, a prize that brought international recognition to this struggle.

As we drove on the mountainous roads that lead to Santa Marta and other towns in the northern department (the equivalent of a US state) of Cabañas, we commented on the starkly eroded parched hills that look like landslides waiting to happen. "We are the second most environmentally degraded country in the Americas after Haiti," Miguel explained through an interpreter.[7] "How did you come to oppose mining?" we asked. Miguel pointed to our water bottle and said simply: "Just like you, water is our priority." Over the next days and again in 2012, 2013, and 2014 we would hear testimonies from dozens of people in Cabañas, many of whom are risking their lives in the struggle against mining. Almost all started or ended their stories with some variation of Miguel's answer: "Water for life," for drinking, for fishing, for farming—not just for Cabañas but for the whole country.

Miguel drove us to the office of his employer, the Social and Economic Development Association (known as ADES), where local people talked with us late into the night about how they had come to oppose mining. As ADES organizer Vidalina Morales acknowledged, most did not begin as anti-mining activists: "Initially, we thought mining was good and it was going to help us out of poverty ... through jobs and development."

The mining corporation that had come to Cabañas was Vancouver-based Pacific Rim Mining Corporation, one of several dozen companies interested in obtaining mining exploitation permits in the Lempa River watershed. In 2002, Pacific Rim acquired a firm that already had an exploration license for a Cabañas site bearing the promising name El Dorado.[8] That license gave Pacific Rim the right to use techniques such as sinking exploratory wells to determine just how lucrative the site would be.

Francisco Pineda, a corn farmer and organizer with the Environmental Committee of Cabañas, invited us to spend an afternoon with eighteen of his fellow committee members, some of whom had walked or been driven a long way to join us. One after another, each stood up to tell his or her story. Francisco, who received the 2011 Goldman Environmental Award (which some call the Environmental Nobel Prize), kicked off what became a five-hour session.[9] He talked about watching the river near his farm dry up: "This was very strange, as it had never done this before. So we walked up the river to see why. … And then I found a pump from Pacific Rim that was pumping water for exploratory wells. All of us began to wonder, if they are using this much water in the exploration stage, how much will they use if they actually start mining?"

Francisco, Marcelo, Miguel, Vidalina, and others set out to learn everything they could about gold mining. From experience, they already knew that Cabañas was prone to earthquakes, some potentially strong enough to crack open the containers holding the cyanide-laced water used in the mining process. Community members traveled to mining communities in neighboring Honduras, Costa Rica, and Guatemala, returning home with stories about the contamination of rivers and lands by cyanide and other toxic chemicals. They learned that arsenic, another highly toxic substance, was often released into the ground or water as a result of the use of cyanide. They turned to water experts, university researchers, and international groups like Oxfam. A number of people attended seminars on mining in San Salvador and in the process became water experts on top of their previous profession. Miguel, for instance, was an accountant; many were farmers.

They also discovered that just a tiny share of Pacific Rim's profits would stay in the country. In addition, the El Dorado mine was projected to have an operational life of only about six years,[10] and many of the promised jobs required skills that few local people had. And as a study by the International Union for the Conservation of Nature pointed out, people in Cabañas "living near mining exploration activities began to notice environmental impacts from the mining exploration—reduced access to water, polluted waters, impacts to agriculture, and health issues."[11]

In community meetings, Pacific Rim officials claimed they would leave the water cleaner than they found it (the Pacific Rim website is filled with promises about "social and environmental responsibility").[12] But many local people, given their new knowledge and expertise, were wary of the company's intentions and honesty. Three people recounted how a Pacific Rim official boasted that cyanide was so safe that he was willing to drink a glass of a favorite local beverage laced with the chemical. The official, we were told, backed down when community members insisted on authentication of the cyanide. "The company thought we're just ignorant farmers with big hats who don't know what we're doing," Miguel said. "But they're the ones who are lying."

## Gold and Conflict

As the anti-mining coalition strengthened with support from leaders in the Catholic Church and the general public (a 2007 national poll showed that 62.4 percent opposed mining), tensions within Cabañas grew.[13] These emerged in the context of other challenges, including the increasing use of Cabañas as an international drug transshipment route, with the attendant problems of corruption and violence. While questions remain, many mining opponents believe that pro-mining forces, including local politicians who stood to benefit if Pacific Rim started mining, are ultimately responsible for the 2009 murder of Miguel's brother, Marcelo Rivera. Marcelo, a cultural worker and popular educator from the Cabañas town of San Isidro, was an early and vibrant public face of the anti-mining movement.[14]

In San Isidro, Rina Navarrete, director of the Friends of San Isidro Association (ASIC), whose founders included Marcelo, stressed that his work lives on through local groups focusing on cultural work and youth leadership development. Members of another citizens' group, MUFRAS-32, led us on a walking tour of this small farming town. At the renamed Marcelo Rivera Community Center, a yellow and red mural with Marcelo's face above a line of dancing children covers the front wall.

Four other murals painted by young people, on the outside walls of houses owned by sympathetic residents, make it impossible to forget Marcelo's mission or his assassination. One, for example, offers a dramatic contrast between two alternative paths of development: on one side of the mural, dark and gloomy "monster" projects, including gold mines, dump waste into a river that bisects the wall. On the other side of the mural's river, sunlight bathes healthy agricultural land and trees (see figure 7.1).

ASIC, MUFRAS-32, and other groups continued to organize theater and artistic festivals. Jaime Sánchez, a former theater student of Marcelo now in his mid-twenties, told us more: "We use theater, songs, murals and other cultural forms to show resistance. We use laughter." Jaime described ADES's creation of a radio station, Radio Victoria, that teaches young people to become deejays and production engineers or to fill the

**Figure 7.1**

A mural in San Isidro, El Salvador, expressing resistance to mining.

other roles of running a station. These young people also took courses on mining and spread what they learned over the airwaves.

In the months after Marcelo's assassination, the conflict intensified. Over six days in late 2009, two more local activists were killed—one a woman who was eight months pregnant; the two-year-old in her arms was wounded.[15] ADES's Nelson Ventura barely escaped an attack. Hector Berrios and Zenayda Serrano, lawyers and leaders of MUFRAS-32, had their home broken into while they and their daughter slept, and documents related to their work were stolen. As Hector lamented, "Clandestine organizations still operate with impunity in this country."

One person told us (in front of others) that he had turned down an offer of thirty dollars a week to meet with representatives of Pacific Rim to inform on anti-mining activists. Mourned another, "Now in our communities, you don't trust people you've trusted your entire life. That's one of the things the mining companies have done."

Many of the people we interviewed, including youths at Radio Victoria, have received death threats. In May 2011, just after we returned home from our 2011 visit, the death threats against individual youths at Radio Victoria escalated, with such ominous untraceable text messages as: "look oscar [sic] we aren't kidding shut up this radio or you also die you dog."[16] And in June 2011, nearly two years after Marcelo Rivera's murder, the body of Juan Francisco Duran, a student volunteer with the Environmental Committee of Cabañas, was found; he had been assassinated execution style, two bullets in his head. "The last time he was seen by fellow environmental activists was … distributing fliers against metallic mining in [Cabañas] in preparation for a public consultation about the mining sector taking place nearby," emphasized the press release of the anti-mining coalition in El Salvador. The press release's title implored, "Not another mine, not another death."[17]

## "A Pact with the Devil"

In 2012, we traveled from Cabañas to the community of San Sebastian in the province of La Union in the northeast corner of El Salvador. "The water's bright orange!" we exclaimed with revulsion while balancing ourselves precariously on rocks alongside a spring. Above us stood a mountain with a prominent slash where firms from the United States and elsewhere mined gold for over a century. The mountain is a key watershed for this area.

"I've seen this water cranberry red and also bright yellow," our companion said before reminding us not to touch the water. "Last time I was here, I slipped and ended up with rashes all over my leg and stomach where I got wet." Experts from the Salvadoran government's Ministry of Environment and Natural Resources were here in July 2012 and found levels of cyanide and iron that were through the roof.[18]

What at first seemed odd to us was that there had not been commercial gold mining in this area for years—not since Commerce Group, a US company based in Milwaukee, was kicked out by the government.[19]

But as we learned in San Sebastian, a decade or two can be a blink of an eye for the environmental havoc wreaked by gold mining. These ancient mountains contain not only gold and many other minerals but also sulfide. Once mining excavations expose sulfide to the air and rain, it converts to sulfuric acid. With each new rain, the acid unleashes new toxic substances that flow down the mountain and into the springs and streams.

The now-orange spring water flows into a now-lifeless stream that flows into the San Sebastian River, which in turn flows into the Santa

Rosa River. Along the way, many communities use it before it enters the Gulf of Fonseca far to the south and continues its journey.

The water is not the only thing affected. "This land is heavily contaminated," sighed Father Lorenzo, the priest from the nearby city of Santa Rosa de Lima when we met him in 2012. In the words of a local man whose father worked for Commerce Group, large-scale mining involved "making a pact with the devil."

The technical term for the environmental nightmare that is unfolding in front of us is *acid mine drainage* or *acid rock drainage*. Acid mine drainage has plagued mine sites from Pennsylvania to El Salvador for centuries. Indeed, as the late mining expert Robert Goodland stressed, some communities near ancient Roman mines in England and Spain continue to suffer the effects of acid mine drainage over two thousand years after the mines were closed.[20] And remediation—or cleanup—is technically and financially challenging. As a local college student who has become an expert on the San Sebastian situation explained, even if the funding were available, "we wouldn't be able to clean this up even in one hundred years."[21] The mining thus creates not only the brutal violence of conflicts but also continuous destruction—what postcolonial literary scholar Rob Nixon calls "slow violence" of people and their ecosystems for very, very long periods of time.[22]

With Father Lorenzo, we hiked up the mountain to the site where Commerce Group had mined. On the way, we walked by homes built on land covered by finely crushed rocks that are the tailings, or what remains after the mining companies separate the gold from the surrounding rock. Often this separation process also releases naturally occurring arsenic.[23] We walked on roads and paths built from such tailings. As is the case with many other gold mining operations, Commerce Group used cyanide to extract the gold. As a result, the cyanide is left as part of the waste mine tailings.

We found the mines near the summit: jagged holes carved into the side of the mountain, with piles of rock below. Here and there a glint suggested the ore that lies within.

It was not hard to figure out where Father Lorenzo stood in terms of mining: a dramatic mural outside his church depicted a denuded landscape darkened by machinery and tailings. We asked him how he and the community first came to understand that the mountain we were standing on was a toxic nightmare. He began his answer by explaining that he was not always against mining; indeed, his father was a miner in a nearby area. But then, more than a decade ago, he was assigned to this parish, and he and others noted San Sebastian's high incidence of kidney failure, cancer, skin problems, and nervous system disorders. "My first clue," he told us, "was that I would visit farms in this community, and I would wash my hands and notice that there were no suds from the soap. The water was too acidic to make suds. This is how we made the link between the mining and the water."

The environmental nightmare actually goes back to the original miner here, US metallurgist Charles Butters, who commercialized the use of cyanide in gold extraction over a century ago. In 1927, Butters was kicked out of nearby Nicaragua by the revolutionary leader Augusto Sandino, but he had already found a much more compliant government in El Salvador. Butters purchased the San Sebastian mine in the early years of the 1900s for $100,000 and within a year had mined $1 million worth of gold—24 carat we were told. We could not help but admire Butter's gold-sourcing skills and his workers' stamina as we envision him searching for the vein and them hauling ore out of these steep mountains before any roads were built.[24]

And thus began the sad saga.

The most recent company to extract gold from this mountain was Wisconsin-based Commerce Group, which acquired the San Sebastian gold mine in 1968. Commerce Group made money off the gold for many years, stopping temporarily during El Salvador's civil war and ceasing mining activity by 1999 primarily because of the low price of gold. Yet as gold prices skyrocketed soon thereafter, Commerce Group filed for and was awarded a new mining license from the government. Then, in 2006, a study by the research group CEICOM found that the San Sebastian River was 100,000 times more acidic than uncontaminated bodies of water in the same region. This was so toxic that the Environmental Ministry of El Salvador's then right-wing government revoked the company's permit for the San Sebastian mine in 2006.[25]

### Democratic Spaces

We traveled from mining country to San Salvador, visiting the sprawling Cuscatlán Park. Along one wall is the Salvadoran version of the US Vietnam Veterans Memorial, in this case etched with the names of about 30,000 of the roughly 75,000 killed in the civil war. Thousands of them, including the dozens killed in the Lempa River massacre of 1981, were victims of massacres perpetrated by the US-backed, and often US-trained, government forces and the death squads associated with them.[26]

Peace accords were signed in 1992, and successive elections delivered the presidency to the conservative and pro–free trade ARENA Party until 2009, when the progressive Farabundo Martí Liberation Front (FMLN) won the largest bloc in the Congress and, two months later, the presidency.[27] (In the 2012 legislative elections, ARENA regained the largest number of seats.[28]) Anti-mining sentiment was already so strong in 2009 that both the reigning ARENA president and the successful FMLN candidate, Mauricio Funes, came out publicly against mining during the campaign.[29]

Much of the credit for this anti-mining sentiment goes to the La Mesa Nacional Frente a la Minería Metálica—the National Roundtable on Mining, often referred to as La Mesa or The Roundtable. The Roundtable was formed in 2005 as leaders in Cabañas began meeting with groups from other departments where mining companies were seeking permits, as well as with research, development, legal aid, and human rights groups in San Salvador. Roundtable facilitator Rodolfo Calles enumerated the goals its members collectively agreed on after arduous deliberations: to help resistance at the community level; to win a national law banning metals mining; to link with anti-mining struggles in Honduras and Guatemala, since the Lempa River also winds through those two countries; and to take on the international tribunal in which Pacific Rim and Commerce Group are suing El Salvador for not allowing them to mine. Part of what moved the Roundtable to the "complete ban" position, Francisco Pineda explained, "was the realization that the government lacked the ability to regulate the mining activities of giant global firms."

We were eager to understand how the FMLN-led government was deciding whether to ban metals mining. Roundtable members told us that, in 2009, the Funes government had announced it would grant no new permits during his five-year term and that it was considering a permanent ban. They also told us the government had initiated a major strategic environmental review to help set longer-term policy on mining.

We visited the Ministry of the Economy, which, along with the Environment Ministry, led the 2011–2012 review. In 2011, the man overseeing it, an engineer named Carlos Duarte, explained that the goal was to do a scientific analysis, with the help of a Spanish consulting firm (with Spanish funding). We pushed further, trying to understand how a technical analysis could decide a matter with such high stakes. On the one hand, we posed to Duarte, gold's price has skyrocketed from less than $300 an ounce at the turn of this century (and for much of the 1980s and 1990s) to more than $1,500 an ounce then,[30] increasing the temptation in a nation of deep poverty to consider mining. We quoted former Salvadoran finance minister and Pacific Rim economic adviser Manuel Hinds, who said, "Renouncing gold mining would be unjustifiable and globally unprecedented."[31] We also quoted the head of the human rights group and Roundtable member FESPAD, Maria Silvia Guillen: "El Salvador is a small beach with a big river that runs through it. If the river dies, the entire country dies."

In 2011–2012, the government-hired Spanish consulting firm, backed by four technical experts from other countries, carried out a lengthy study of the issues and consulted with stakeholders affected by mining, ranging from mining companies to the Roundtable groups. While Duarte hoped this process would produce a consensus, he admitted it was more likely that the government and the consulting firm would have to lay out "the interests of the majority," after which the two ministries would make their policy recommendation. (In 2011, Roundtable members told us that the first group consultation, about ten days earlier in San Salvador, had turned into a pitched debate between them and representatives of the mining companies.) "If new laws are necessary," Duarte informed us, "then it will go to the legislature."

We proceeded to the national legislature, its hallways a cacophony of red posters bearing the photos of FMLN leaders (and the ever-present martyr, Archbishop Oscar Arnulfo Romero, assassinated in 1980 by the right) competing with offices adorned with posters of the leading opposition party, ARENA. We came to meet FMLN members of the legislature's environment and climate change committee, including Lourdes Palacios, a three-term member from San Salvador. In 2011, Palacios explained that they were ready with a bill to ban metals mining, but at the request of the executive branch, they were waiting for the outcome of the strategic review before introducing it.

A representative from the department of Chalatenango, just west of Cabañas and an FMLN stronghold, expressed impatience at how long the review was taking and his conviction that "economic and political powers" were "putting pressure on non-FMLN legislators." For the FMLN legislators, he stressed, "the pressure is the will of the people, and we are convinced that the majority of the people don't want mining." The FMLN did not have a majority in the legislature; still, those present expressed confidence that the ban could pass if the executive branch recommended it. One legislator suggested that El Salvador might have an easier time saying no to gold mining than countries already dependent on revenues from gold exports.

At the time of our visit in 2012, the strategic environmental review was completed, and the executive branch submitted its proposed legislation for revising the country's mining laws to Palacios's committee. As of the end of the Funes administration, the legislative branch had not yet passed new mining legislation—be it La Mesa's proposed country-wide ban or the Funes administration's 2012 post–strategic review draft legislation to define environmental and social no-go zones for mining.

Given the human rights situation in Cabañas, we also interviewed the government's human rights ombudsman, a post created after the 1992 peace accords, to be selected by and report directly to the legislature. Through mid-2013, the ombudsman was Oscar Luna, a former law professor and fierce defender of human rights—for which he too received death threats. We asked Luna if he agreed with allegations that the killings in Cabañas were "assassinations organized and protected by economic and social powers." Luna replied with his own phrasing: "There is still a climate of impunity in this country that we are trying to end." In 2011, he was pressing El Salvador's attorney general to conduct investigations into the "intellectual authors" (or masterminds) of the killings. Several people were arrested in connection with Marcelo Rivera's assassination, but the attorney general's office appeared to be dragging its feet in digging deeper into who ordered and paid for the killings. Three of the six who were convicted were already released by July 2012. Critics told us that the attorney general, appointed by the legislature as a compromise candidate between ARENA and the FMLN, had failed to investigate aggressively a number of sensitive cases involving politicians, corruption, and organized crime.

Our interactions in Cabañas and San Salvador left us appreciative of the new democratic space that strong citizen movements and a progressive presidential victory have opened up, yet aware of the fragility and complexities that abound. The government faces an epic decision about mining, amid deep divisions and with institutions of democracy that are still quite young. As Vidalina Morales reminded us when we parted after our first stay in Cabañas, the complications were even greater than what we found in Cabañas or in San Salvador, because even if the ban's proponents eventually win, "these decisions could still get trumped in Washington."

## A Tribunal That Can Trump Democracy

Protesters around the globe know the sprawling structures that house the World Bank in Washington, yet few are aware that behind these doors sits a little-known tribunal that will be central to the Salvadoran gold story. The Salvadoran government never approved Pacific Rim's environmental impact study, and thus never gave its permission for Pacific Rim to begin actual mining.[32] In retaliation, the firm sued the government under the 2005 Central American Free Trade Agreement and El Salvador's investment law.[33] Like other trade agreements, CAFTA allows foreign investors to file claims against governments over actions—including health, safety, and environmental measures and regulations—that reduce the value of their investment. The affected farmers and communities are not part of the "investor rights" calculus. The most frequently used tribunal for such "investor-state" cases is the International Centre for Settlement of Investment Disputes (ICSID), housed at the World Bank (but there are others, such as UNCITRAL).[34]

In the words of lawyer Marcos Orellana of the Center for International Environmental Law, who assisted the Roundtable in drafting an amicus brief for the tribunal in the hopes that ICSID would dismiss the Pacific Rim case on "jurisdictional grounds," Pacific Rim "is trying to dictate El Salvador's environment and social policy using CAFTA's arbitration mechanism." Pacific Rim's "claim amounts to an abuse of process." The 2011 amicus brief methodically lays out how Canada-headquartered Pacific Rim first incorporated in the Cayman Islands to escape taxes, then lobbied Salvadoran officials to shape policies to benefit the firm. Only after that failed, in 2007, did Pacific Rim reincorporate one of its subsidiaries in the United States to try to use CAFTA in addition to the investment law to sue El Salvador.[35]

Commerce Group, the US company that had operated in San Sebastian until the Salvadoran government revoked its license on environmental grounds, brought a $100 million case against El Salvador in ICSID. ICSID tribunalists actually ruled against the Commerce Group in 2011—but on grounds that did not include the substance of whether El Salvador had the right to limit "investor rights" based on environmental and other concerns.[36] Commerce Group appealed the dismissal; the appeal too was dismissed by ICSID in 2013 (this time, based on the fact that Commerce Group was behind in payments to ICSID). While it is a major victory that the Salvador government did not lose in a $100 million case, it is tempered by the fact that ICSID still ruled that the Salvadoran government had to pay $1.4 million for its own legal fees and half of the arbitration costs from the nonappeal stage fees. And the Commerce Group case set no precedent in terms of a country's right to protect its people and its environment, or a mining company's responsibilities in that regard. Rather, the case reinforced the perception that an investor-rights institution such as ICSID is not likely to condone a government's decision to stop granting mining licenses on these grounds. And the ICSID ruling had no impact whatsoever on the ground in San Sebastian, where the Commerce Group–catalyzed environmental nightmare continues.

Meanwhile, by 2013, Pacific Rim had increased it original demand for compensation by the Salvadoran government (from at least $77 million) to $301 million. As of this writing, Pacific Rim has succeeded in getting its case through the ICSID's jurisdictional hearings, which means that ICSID deemed that the case had enough merit or substance to be heard. Therefore, the case moved into the "merits," or substantive, stage at ICSID, with one insider estimating that the costs alone could exceed $12 million for each side before any actual judgment. And the interest of global mining corporations in the El Dorado claim—and the richness of that vein—was reinforced by events of late 2012 and 2013. In late 2012, when Pacific Rim seemed to be running out of money for its ICSID suit, Australian/Canadian firm OceanaGold infused a significant amount of capital into Pacific Rim. In November 2013, shareholders approved an outright purchase of the company. As a result, Pacific Rim became a wholly owned subsidiary of OceanaGold, with enhanced financial ability to pursue the case at the tribunal.[37]

For this chapter, we attempted to interview Pacific Rim board chair Catherine McLeod-Seltzer in 2011, but her office steered us to the CEO of Pacific Rim's US subsidiary, Thomas Shrake. In a tersely worded e-mail, Shrake "respectfully denied" our request. In 2012, we also attempted to interview PacRim's Salvadoran-based vice president, Ericka Colindres, who represented PacRim at the consultations, but she never responded.

However, McLeod-Seltzer and Shrake did publicize their arguments in other forums. Part of their defense involved maligning the groups and individuals on the ground in El Salvador and the activists' supporters across the globe. As McLeod-Seltzer explained on a Canadian radio documentary, "These people purport to be environmentalists, they're not. They're anti-development. They are not pro-environment, if they were, they would support this mine."[38] Pacific Rim also claimed that the reason it was not granted an exploitation license was, in Shrake's words, "all about corruption" of Salvadoran officials who were, Pacific Rim explained, holding up the licensing since PacRim had not bribed anyone.[39]

Dozens of human rights, environmental, and fair trade groups across North America—from US–El Salvador Sister Cities and the Committee in Solidarity with the People of El Salvador to Oxfam, the Council of Canadians, Public Citizen, MiningWatch Canada, and the Institute for Policy Studies (IPS)–responded to such allegations with outrage and rallied to the defense of both La Mesa and the Salvadoran government. A loose coalition of international allies from across the globe came together to pressure both Pacific Rim to withdraw the case and the ICSID tribunal to rule in favor of El Salvador.

But many believe that even if Pacific Rim/OceanaGold withdraws its case or loses in this tribunal, the very existence of investor-state clauses in trade agreements is an affront to democracy.[40] "For democracy to prevail," Sarah Anderson of IPS told us, "citizens' movements and their allies in governments must work hard to eliminate these clauses from all trade and investment agreements." Beyond supporting El Salvador at ICSID and eliminating such existing investor-rights clauses, there is also a need to make sure that new agreements are not signed that expand such investor rights and further limit the ability of countries around the world to do what El Salvador is trying to do: prioritize its responsibilities to its citizens and future generations.

### A People's Economy

Back in Santa Marta in the province of Cabañas, citizen groups are building sustainable farming as an alternative economic base to mining. Their goal is a "solidarity economy," or, as Vidalina Morales termed it, a "people's economy." Explained Vidalina, her voice deep with passion, "We reject the image of us just as anti-mining. We are for water and a positive future. We want alternatives to feed us, to clothe us."[41]

Elvis Nataren, a philosophy student, led us to the riverbank and pointed to communal land where organic farms will be built. Three towering greenhouses already contained plump hydroponic tomatoes, green peppers, and other vegetables. Together these should make Santa Marta self-sufficient in corn, beans, and vegetables. As Elvis explained, food sovereignty was even more urgent in the wake of CAFTA's passage, given the cheap foreign produce that began to flood the Salvadoran market.

In the nearby province of Chalatenango, in the town of Buena Vista, we met community leaders in an open-air multipurpose building surrounded by dresses made entirely of corn stalks that had been worn in the local corn festival contest. The leaders talked about the alternative local economy they were creating. A key first step was to build a community water system. Then they established a cooperative organization enabling the community to process sugarcane, manage beehives for honey, and oversee a fish hatchery. The women of the town organized themselves to plant corn and beans collectively, and they also produce shampoo, soap, and alternative medicines ("so that we don't have to run to a pharmacy"), while running a small massage business. Their farming is organic.

Elvis, Vidalina, Miguel, Francisco, and others we met in Cabañas and Chalatenango were well aware that as they nurture farmlands and the river vital to this alternative future, their success also depends on struggles and debates in San Salvador and Washington.

## Conclusion

What happens in this small country the size of Massachusetts has importance far beyond its borders. If El Salvador is able to continue to prevent new metallic mining, that precedent opens possibilities for other countries. But if it is forced to allow mining, if PacRim prevails in ICSID, if the Commerce Group is able to cut and run, then this will give greater power to the mining firms.

In addition, in business, academic, and activist circles, there is a larger debate about whether large-scale mining can be done ethically, socially, and environmentally responsibly or whether large-scale mining should, at least in some cases, be banned. Our conclusion from El Salvador is that the country is so environmentally fragile and already so environmentally challenged that the only environmentally and socially responsible policy option is to make the whole country a no-go zone for metals mining. Mining should be banned there.

So too does the evidence from El Salvador suggest that an effective "anti-" campaign must simultaneously be "pro-" something—in this case, pro-water, pro-food sovereignty, and, overall, as people in El Salvador term it, pro-life. In other words, in the struggle to ban gold mining in El Salvador, a politics of resistance has joined a politics of alternatives. This also appears to be the case with other KIIG struggles.[42]

In material terms, this Salvadoran KIIG case study is not really about gold; it is about water (and, for that matter, soil, land, rights, and so on), just as the other KIIG fights are not really about fossil fuels, but about climate destabilization, toxic accumulation, healthy ecosystems, and water.

Meanwhile, the streams and springs of San Sebastian continue to change color from cranberry red to neon orange and yellow depending on the season. The overstressed Rio Lempa watershed continues to be near, if not beyond, its ecological limits. And the violence introduced to Cabañas continues. On June 30, 2012, David Urias, the son of an anti-mining activist, was brutally murdered—his brain crushed by a rock—as he walked alone on a dirt road, three years to the day from when Marcelo Rivera's tortured body was found.

Yet determined social movements and a responsive government have shown the world that it is possible to choose water and life over dangerous mining and destroyed ecosystems. "Keep gold in the ground" joins "keep oil in the soil" as an example that a longer-term vision of a better future is not only possible, it can prevail.

## Acknowledgments

This chapter would not have been possible without the willingness of the many people in El Salvador who kindly allowed us to interview them. In addition, we extend thanks to those who read through all or parts of drafts of this chapter, including Sarah Anderson, Robert Goodland, Jack Manno, Pamela Martin, Jen Moore, Jan Morrill, Marcos Orellano, Thomas Princen, Manuel Perez Rocha, Christina Starr, and Geoff Thale, as well as other participants in an International Studies Association workshop held in April 2013. Research assistance in the United States was provided by Pauline Abetti and Rachel Nadelman. We were accompanied in much of our Salvadoran research by Jan Morrill, who served as everything from local expert, to driver extraordinaire, to translator. Special thanks to this book's editors, Thomas Princen, Pam Martin, and Jack Manno, who not only envisioned this book project but also brought chapter authors and others together at venues such as the International Studies Association to share our research and insights. Robin Broad thanks the School of International Service at American University for funding provided to support this work. We also thank the Wallace Global Fund and editors at *The Nation* for encouraging and supporting our initial writing on this topic.

This chapter expands and updates an article based on our 2011 field research in El Salvador that was originally published as "Like Water for Gold in El Salvador," *The Nation*, July 11, 2011.

# Notes

1. See http://www.itmtrading.com/historical_gold_prices.

2. Note that this refers to a comprehensive, national-level policy. A few other countries have more limited moratoriums or bans. Most notable of these seems to be Costa Rica, which, since November 2010, has had a legislative ban (not just an executive branch moratorium) on a specific kind of mining: open-pit mining and the use of cyanide. In 2013, the ban was upheld by the Costa Rican Supreme Court, which rejected an appeal by Canadian gold mining company Infinito Gold to open its Crucitas open-pit mining site in Costa Rica. The ban continues to hold, despite Infinito Gold's then proceeding to sue the country at the International Centre for Settlement of Investment Disputes. See "Costa Rica Lawmakers Vote to Ban Open-Pit Mining," Reuters, November 10, 2010, http://af.reuters .com/article/metalsNews/idAFN0912629920101110; L. Arias, "Infinito Gold to Move Forward with Billion-Dollar Lawsuit against Costa Rica," *Tico Times,* June 19, 2013, http://www.ticotimes.net/2013/06/20/infinito-gold-to-move-forward -with-billion-dollar-lawsuit-against-costa-rica; and "Calgary-based Infinito Gold Fails to Strong-Arm Costa Rican Judiciary as Final Appeal Rejected by Supreme Court," *MiningWatch Canada,* July 4, 2013, http://www.miningwatch.ca/news/ calgary-based-infinito-gold-fails-strong-arm-costa-rican-judiciary-final-appeal -rejected.

Some other countries have implemented "no-go zones" or specific areas where mining is not allowed. Examples include Argentina, which in 2010 banned mining in glaciers and periglacier areas; Panama, which in 2012 banned mining on indigenous lands; Guyana, which as of 2012 "indefinitely banned all diamond and gold mining in rivers because of pollution, loose regulations and other concerns"; and Zimbabwe, which in 2013 outlawed all alluvial mining near waterways. See Dorothy Kosich, "Mining Companies Must Comply with Glacier Law-Argentine Supreme Court," *Mineweb,* July 4, 2012, http://www.mineweb.com/mineweb/ content/en/mineweb-political-economy?oid=154474&sn=Detail; "New Panama Law Bans Mining on Native Land," March 27, 2012, http://phys.org/news/2012 -03-panama-law-native.html; "Guyana Bans Gold, Diamond Mining in Rivers," *Jamaica-Gleaner,* July 8, 2012, http://jamaica-gleaner.com/gleaner/20120708/ business/business85.html; and Municipal Reporter, "Alluvial Mining Banned," *Herald,* June 11, 2013, http://www.herald.co.zw/alluvial-mining-banned.

In addition, there are examples of temporary mining moratoriums. A notable example is the Philippines. In July 2012, Philippine President Aquino issued an executive order (no. 79) that prohibits acceptance of new mining applications and the approval of new mining permits but does not stop ongoing mining. This temporary moratorium was put in place so that the Philippine Congress could deliberate a new law to replace the Mining Act of 1995. As is typically the case in such temporary moratoriums, the Philippine government's main focus appears to be on negotiating better deals in terms of royalties, taxes, and other payments from mining companies. Among the best sources on this is Kaka Bag-ao, "Extracting Good Policy from Bad Legislation? A Review of Executive Order 79, Series of 2012," legal notes, Akbayan Party, July 13, 2012. See also ABS-CBNnews.com on this including: ABS-CBNnews.com, "Q&A on Executive Order 79," July 9, 2012, and Cathy Garcia, "Mining EO Out; Moratorium on New Mining Deals Stays," ABS-CBNnews.com, July 9, 2012, http://www.abs-cbnnews.com/business/07/09/12/mining-eo-out-moratorium-stays. For updates on Philippine mining deliberations, see the website of the Alyansa Tigil Mina in the Philippines: http://www.alyansatigilmina.net.

For more on no-go zones, see Robert Goodland, "Responsible Mining: The Key to Profitable Resource Development," *Sustainability* 4, no. 9 (2012): 2099–2126, http://www.mdpi.com/2071-1050/4/9/2099. For more on how to assess these various mining reforms, see Robin Broad, "Responsible Mining: Moving from a Buzzword to Real Responsibility," *The Extractive Industries and Society Journal*, March 2014, doi: 10.1016/j.exis.2014.01.001.

3. Unless otherwise noted, the information and quotations are from our interviews in El Salvador.

4. For the 6.2 million figure, see World Bank, "Country at a Glance—El Salvador" (Washington, DC: World Bank, 2011), http://www.worldbank.org/en/country/elsalvador.

5. We conducted field research in El Salvador in April and May 2011, July and August 2012, May 2013, and July 2014. Research ranged from more formal interviews (especially the case with interviews conducted with government officials in various ministries in San Salvador) to informal, multiple-day participant observation outside of San Salvador, especially in the province of Cabañas.

6. The details on Marcelo's death were repeated to us numerous times. For a written source, see, among others, Richard Steiner, "El Salvador: Gold, Guns, and Choice," International Union for the Conservation of Nature (IUCN) Commission on Environmental, Economic, and Social Policy, February 2010), 13, 40–44, http://www.walkingwithelsalvador.org/Steiner%20Salvador%20Mining%20Report.pdf. Marcelo was abducted in mid-June 2009, and his body was found on June 30. For the June 30 date, see Committee in Solidarity with the People of El Salvador, "Body of Missing Activist Found with Signs of Torture; Social Organizations Demand Justice," July 17, 2009, http://wetlands-preserve.org/phpUpload/uploads/CISPES%20on%20Mining.pdf.

7. More specifically, in terms of the #2 ranking: "El Salvador is the second most deforested country in Latin America after Haiti. Almost 85 percent of its forested cover has disappeared since the 1960s" (143); and "El Salvador has the second highest economic risk exposure to two or more hazards, according to the Natural Disaster Hotspot study by the World Bank. The same study also ranks El Salvador second among countries with the highest percentage of total population considered at a 'Relatively High Mortality Risk from Multiple Hazards'"(141). Global Facility for Disaster Reduction and Recovery, *Disaster Risk Management in Latin America and the Caribbean Region: GFDRR Country Notes—El Salvador* (Washington, DC: World Bank, 2011), 141, 143.

El Salvador has also been ranked number one among countries across the globe in terms of environmental vulnerability: "According to the Global Facility for Disaster Reduction and Recovery (GFDRR) and the United Nations Disaster

Assessment and Coordination (UNDAC), El Salvador is recognized as the most vulnerable country in the world. The 2010 UNDAC report, "Assessment of the Capacity for Emergency Responses 2010," states that almost 90 percent of the territory in El Salvador is located in an area of high risk. These areas are home to more than 95 percent of the country's population and are linked with approximately 96 percent of the country's gross domestic product (GDP). According to studies from the Economic Commission for Latin America and the Caribbean (ECLAC), natural disasters have caused 6,500 deaths since 1972, with an economic cost of greater than 16 billion dollars. Of these impacts, more than 62 percent of the deaths and between 87 to 95 percent of the economic losses were related to climatic events." United Nations Development Programme, *Paving the Way for Climate-Resilient Infrastructure: Guidance for Practitioners and Planners* (New York: UN Development Programme, 2011), vi.

For more on El Salvador's environmental degradation, see United Nations Disaster Assessment and Coordination, *Evalucaión de la Capacidad Nacional para la Respuesta a Emergencias* (San Salvador, El Salvador: United Nations, 2010); United Nations Development Programme, *El Salvador, Informe anual PNUD El Salvador 2010* (San Salvador, El Salvador: Programa de las Naciones Unidas para el Desarrollo (PNUD), 2010); United Nations, Economic Commission for Latin America and the Caribbean, *The Economics of Climate Change in Central America: Summary* (New York: United Nations, 2010); and Republic of El Salvador, Ministry of Environment and Natural Resources, El Salvador, *Resultados del Informe de la Calidad de Agua en los ríos de El Salvador* (San Salvador, El Salvador, 2010).

See also the work of Angel Ibarra, including Florian Erzinger, Luis Gonzales, and Angel M. Ibarra, *El Lado Obscuro del oro: Impactos de la mineria metalica en El Salvador* (San Salvador: Caritas de El Salvador and Unidad Ecologica Salvadorena, 2008).

8. Specifics on Pacific Rim plans in El Salvador were found on the website of Pacific Rim: http://www.pacrim-mining.com/s/Home.asp. As of late 2013 and the purchase of Pacific Rim by OceanaGold (discussed later in this chapter), information on Pacific Rim appears on the OceanaGold website: http://www.oceanagold.com.

9. See the website of Goldman Foundation Prize: www.goldmanprize.org.

10. On the six years, see Robert Moran, "Technical Review of the El Dorado Mine Project Environmental Impact Assessment, El Salvador" (October 2005), 2 available at http://www.votb.org/elsalvador/Reports/Technical_Review_El_Dorado_EIA.pdf.

11. Steiner, "El Salvador: Gold, Guns, and Choice," 15.

12. This was from the Pacific Rim website: http://www.pacrim-mining.com/s/Home.asp. Note that, as of late 2013 and the purchase of Pacific Rim by OceanaGold (discussed later in this chapter), information on Pacific Rim appears on the OceanaGold website: http://www.oceanagold.com.

13. For the poll, which was carried out by the University of Central America, see http://www.uca.edu.sv/publica/iudop/Web/2008/finalmineria040208.pdf. The

relevant question is number 43 on p. 54. It asks: "Do you think El Salvador is an appropriate country for metallic mining?" The answer of 62.4 percent is no.

14. This information comes from our interviews, but it is also in Steiner, "El Salvador: Gold, Guns, and Choice," 13.

15. Steiner, "El Salvador: Gold, Guns, and Choice," 13–14.

16. The text message had the name Oscar with a small "o."

17. CISPES, "Not Another Mine, Not Another Death! Release from the National Anti-Mining Group in El Salvador," June 16, 2011, http://www.cispes.org/programs/anti-mining-and-cafta/not-another-mine-not-another-death-release-from-the-national-anti-mining-group-in-el-salvador.

18. Republic of El Salvador, Ministry of Environment and Natural Resources, "MARN confirma presencia de cianuro y hierro en Rio San Sebastian, La Union," July 15, 2012, http://www.marn.gob.sv/index.php?option=com_content&view=article&id=1462%marn-confirma-presencia-de-cianuro-y-hierro-en-rio-san-sebastian-la-union&catid=1%noticias-ciudadano&Itemid=77.

19. For more on Commerce Group, see http://www.commercegroupcorp.com/mines.html, http://www.stopesmining.org/j25/index.php/commerce-groups, and http://www.minec.gob.sv/index.php?option=com_phocadownload&view=section&id=16:commerce-group-vrs-repblica-de-el-salvador&Itemid=63.

20. Goodland, "Responsible Mining."

21. For more on the technical and financial challenges of acid mine drainage (AMD) remediation, see the websites of MiningWatch Canada (http://www.miningwatch.ca) and Earthworks (http://www.earthworksaction.org). English-language sources that directly address the AMD cleanup situation in Central America are more difficult to find than are Spanish-language sources. Sources (available via the MiningWatch Canada site) that address the question of whether AMD cleanup is possible permanently or temporarily include Environmental Mining Council of British Colombia, "Acid Mine Drainage: Mining and Water Pollution Issues in British Colombia," 2006, and Environmental Mining Council of British Colombia, "EMCBC Mining and the Environment Primer: Acid Mine Drainage" (March 31, 2006). The first report asserts that what makes AMD such a threat is the fact that modern technology cannot completely clean up this kind of pollution; thus, "while Acid Mine Drainage is not the only threat to waterways from mining, it is the biggest threat" (8). The second is a brief article that states that AMD is the most significant environmental problem facing the mining industry and the areas in which mining takes place because of technical difficulties and expense: "[AMD] is virtually impossible to reverse with existing technology, and once started, costs millions of dollars annually to treat and can continue for centuries." See also Ata Akcil and Soner Koldas, "Acid Mine Drainage (AMD): Causes, Treatment and Case Studies," *Journal of Cleaner Production* 14, issues 12–13, 2006, 1139–1145.

Useful Spanish-language sources include:

• Comisión para la gestión integral del agua en Bolivia (CGIAB), "Agua, mineria y comunidades locales" (La Paz, Bolivia: Seminario Nacional, November 21,

2008), which includes work presented at a 2008 seminar, Water, Mining and Local Communities, focusing primarily on Chilean and Bolivian case studies—for example, "The characteristics of the [mining] contamination make it very difficult to reuse the [contaminated] water, and the possibility of returning the water to its prior ambient state is zero. Calculations have been done on the possibilities for cleaning or decontaminating a portion of the water contaminated by mining. Particularly difficult is acid drainage, which is not the direct product of the transformation of rock into mineral, but rather collateral contamination that has not been given sufficient attention" (8; translation by Rachel Nadelman).

• Iberoamericana Organización Red for Medio Ambiente Subterráneo Y Sostenibilidad (MASYS), 4° Jornada Iberoamericana De Medio Ambiente Subterráneo Y Sostenibilidad Medio Ambiente Subterráneo: Contaminación De Aguas Subterráneas" [Organization of the Iberoamerian Network for the Sustainability of the Subterraneous Environment, *Proceedings of the Fourth Iberoamerican Conference for Environmental Sustainability and Subterraneous Environment: Groundwater Contamination*] (Oruro, Bolivia, November 2011). This document contains the proceedings from a three-day workshop, with attendees from Argentina, Bolivia, Brazil, Chile, Cuba, Ecuador, Mexico, Venezuela, Spain, and Portugal (translation by Rachel Nadelman).

• Verónica Odriozola, "No todo lo que es oro brilla. Resumen de impactos ambientales de la minería de oro," Campaña de Tóxicos de Greenpeace, Argentina. [Not all that is gold, shines: Summary of environmental impacts of gold mining, Greenpeace's Campaign on Toxics], January 2003. This report on AMD and other environmental impacts of gold mining includes examples from across the world and also goes into the question of which entities end up being responsible for cleanup costs: "Who pays the costs of remediation? One of the biggest problems is who will pay the costs of remediation (when it is possible to remediate) once the problems occur and the mine is no longer active or the company has left" (translation by Rachel Nadelman).

22. Rob Nixon, *Slow Violence and the Environmentalism of the Poor* (Cambridge, MA: Harvard University Press, 2011).

23. For more on the adverse environmental impact of gold mining, including the release of arsenic, see the work of Ohio University geologist Dina L. Lopez (http://www.ohio.edu/geology/lopez/). On arsenic and gold mining in particular, see Jochen Bundschuh et al., "One Century of Arsenic Exposure in Latin America: A Review of History and Occurrence from Fourteen Countries," *Science of the Total Environment* 429 (2012): 2–35; and William Holden and R. Daniel Jacobson, *Mining and Natural Hazard Vulnerability in the Philippines: Digging to Development or Digging to Disaster* (New York: Anthem Press, 2012).

24. Our sources on Butters include Thomas Rickard, *Interviews with Mining Engineers* (San Francisco: Mining and Scientific Press, 1922); "News Releases," Gold Reign Resources September 2009, http://www.goldenreignresources.com/s/NewsReleases.asp?ReportID=365934&_Title=San-Albino-Murra-Gold-Property-Nicaragua-Update; "Proceedings of a Workshop on Development of Mineral, Energy and Water Resources and Mitigation of Geologic Hazards in Central

America" (US Geological Survey, 1987), 14; "Central America in 1913," *Engineering and Mining Journal* 97, n. 2 (New York: Hill Publishing, 1914), 140; and Percy Falke Martin, *Salvador of the Twentieth Century* (New York: Longmans, Green, 1911).

25. CEICOM reports are available at http://www.ceicom.org.sv/index.php/en.

26. For more on the role of the US government in the Salvadoran civil war, see Cynthia Arnson, *Crossroads: Congress, the President, and Central America, 1876–1993* (University Park, PA: Pennsylvania State University Press, 1993, 2nd edition). See also Mark Danner, *The Massacre at El Mozote: A Parable of the Cold War* (New York: Vintage, 1994); James Chace, *Endless War: How We Got Ourselves Involved in Central America and What Can Be Done* (New York: Vintage Books, 1984); and Raymond Bonner, *Weakness and Deceit: U.S. Policy and El Salvador* (New York: Times Books, 1984).

27. See Steiner, "El Salvador: Gold, Guns, and Choice," 6; Economic Intelligence Unit, "Country Report— El Salvador," March 2011, 4.

28. The next legislative assembly election is in 2015.

29. Steiner, "El Salvador: Gold, Guns, and Choice," 16.

30. The $1,500 is from *Washington Post*, April 21, 2011, A12. On September 13, 2012, the price of gold was $1,769 per ounce. See http://www.itmtrading.com/historical_gold_prices/.

31. The quotation and Hind's titles are from Oxfam America, "Metals Mining and Sustainable Development in Central America" (2009), 15, http://www.oxfamamerica.org/publications/metals-mining-and-sustainable-development-in-central-america.

32. Steiner, "El Salvador: Gold, Guns, and Choice," 7–8.

33. Note that the "jurisdictional ruling" allowed the case to proceed based on El Salvador's investment law at the time. This was the version of the investment law that had been revised in 1999, in connection with World Bank structural adjustment lending, to allow foreign corporations to file such suits in international tribunals rather than go through domestic courts. In 2013, it was subsequently revised to put domestic courts back in the loop. Two studies are particularly useful in linking that 1999 domestic law with the World Bank's structural adjustment requirements: Maria Eugenia Ochoa, Oscar Dada Hutt, and Mario Montecinos, "El Impacto De Los Programas De Ajuste Estructural Y Estabilizacion Economica En El Salvador" [The impact of structural adjustment programs and economic stabilization in El Salvador], Structural Adjustment Participatory Review International Network (SAPRIN), December 2000 (see especially chap. 1, pp. 12–14), http://www.saprin.org/elsalvador/research/els_cover_index.html; and Francis Montserrat Sanchez Garcia, Nancy Reyes Yolanda Nunez, and Mabel Denisse Velásquez Leiva, "Evaluación De Políticas De Inserción Laboral Y Su Impacto En Los Jóvenes" [Evaluating labor market integration policy and its impact on youth"), José Simeón Cañas Universidad Centroamericana, UCA, Graduation Work Prepared for the Faculty of Economics and Social Sciences, September 2010 (see especially pp. 5–8), http://www.uca.edu.sv/deptos/econ

omia/media/archivo/b02a67_evaluaciondepoliticasdeinsercionlaboralysuimpac-toenlosjovenes.pdf.

34. See Pamela L. Martin, "What You Don't Know Can Hurt You: Investor-State Disputes and the Protection of the Environment in Developing Countries," *Global Environmental Politics* 6 (2006): 73–100; Sarah Anderson and Manuel Perez-Ro-cha, *Mining for Profits in International Tribunals: Lessons for the Trans-Pacific Partnership* (Washington, DC: Institute for Policy Studies, April 2013); Michael Mortimer and Leonardo Stanley, "Justice Denied: Dispute Settlement in Latin America's Trade and Investment Agreements" (Working Group on Development and Environment in the Americas, Discussion Paper 27, October 2009); and José Reyes, "El Salvador," in *Latin American Investment Protections: Comparative Perspectives on Laws, Treaties, and Disputes for Investors, States and Counsel*, ed. Jonathon C. Hamilton, Omar E. Garcia Bolivar, and Hernando Otero (Leiden: Martinus Nijhoff, 2012), 293–316.

35. Center for International Environmental Law, "Pac Rim Cayman LLC v. Republic of El Salvador, ICSID Case No. ARB/09/12 Submission of AMICI CURIAE Brief," May 20, 2011; Center for International Environmental Law, "Pac Rim Cayman LLC v. Republic of El Salvador, ICSID Case No. ARB/09/12 Application for Permission to Proceed as AMICI CURIAE," March 2, 2011; and International Centre for Settlement of Investment Disputes (ICSID), "ICSID Case No. ARB/09/12 Between Pac Rim Cayman LLC v. The Republic of El Salvador: Decision on the Respondent's Jurisdictional Objections," June 1, 2012. For subsequent details and submissions for this case, see the ICSID website: https://icsid.worldbank .org/ICSID/FrontServlet. Many documents also are posted on the Salvadoran Ministry of Economy's site: http://www.minec.gob.sv/index.php?option=com_ph ocadownload&view=category&id=26:otros-documentos&Itemid=63.

On the issue of reincorporating a subsidiary in the United States, see Public Citizen, "CAFTA Investor Rights Undermining Democracy and the Environment: Pacific Rim Mining Case" (May 2010), 3, http://www.citizen.org/documents/ Pacific_Rim_Backgrounder1.pdf.

36. Sarah Anderson, Institute for Policy Studies, 2011 interview. See also "Commerce Group CAFTA Ruling Highlights Threat of Foreign Investor Rules Also Included in Korea FTA," Public Citizen press release, March 15, 2011, http:// citizen.typepad.com/eyesontrade/2011/03/commerce-group-cafta-ruling-highlights -threat-of-foreign-investor-rules-also-included-in-korea-fta.html.

37. Note that the "jurisdictional ruling" allowed the case to proceed based on El Salvador's pre-2013 investment law. The $77 million is from an e-mail from Meg Kinnear, head of ICSID, to Robin Broad, June 6, 2011; other sources cite a higher figure. The purchase and updated figures are from a press release of Pacific Rim on April 1, 2013: "Statement of Claim Filed in Arbitration Case against El Salvador; PacRim Seeks Damages of US $315 Million," http://www.pacrim-mining .com/s/News.asp?ReportID=578430 and subsequent OceanaGold and Pacific Rim Mining, "OceanaGold Agrees to Acquire Pacific Rim Mining," press release, October 8, 2013. The buyout seems reflective of the relationship between "junior" exploratory ventures and more "senior" mining companies. On this global

trend, see Michael Dougherty, "The Global Gold Mining Industry: Materiality, Rent-Seeking, Junior Firms and Canadian Corporate Citizenship," *Competition and Change* 17 (2013): 339–54.

This information on OceanaGold is based on our research and interviews not only in El Salvador but also in the Philippines in July and August 2013. This included fieldwork at OceanaGold's Didipio gold and copper mine in the northern Philippines. See John Cavanagh and Robin Broad, "The Real Cost of Gold in the Philippines," *YES! Magazine*, September 13, 2013, http://www.yesmagazine .org/blogs/john-cavanagh-and-robin-broad. On the Philippines, see also William Holden and R. Daniel Jacobson, *Mining and Natural Hazard Vulnerability in the Philippines: Digging to Development or Digging to Disaster* (New York: Anthem Press, 2012).

38. Catherine McLeod-Seltzer, chairman of the Pacific Rim board of directors, interviewed in Karin Wells, "High Stakes Poker," *CBC Sunday Edition*, January 11, 2013.

39. From interviews in Wells, "High Stakes Poker."

40. See Robin Broad and John Cavanagh, "A Strategic Fight against Corporate Rule," *The Nation*, February 3, 2014.

41. See Santiago Humerto Ruiz Grandadino, Estudio Comparativo del Impacto Economico de la Explotacion Minero Metalica vrs el Impacto Economico de la Reactivacion Agropecuaria en la Zona Norte de El Salvador, ADES, Cabañas, El Salvador, June 2012.

42. The El Salvador case is not unique, as the case studies in this book reveal. However, it seems that the mainstream media invariably play up the "anti-" side rather than the positive alternatives being pushed.

# 8

## Slowing Uranium in Australia: Lessons for Urgent Transition beyond Coal, Gas, and Oil

James Goodman and Stuart Rosewarne

If keeping fossil fuels in the ground requires a perceptual shift from viewing them as highly valued, net beneficial resources to seeing them as costly, planetary threats, then Australia's relationship with uranium mining illustrates both the possibility and the difficulty—perceptual and, especially, political—of making that shift. In this chapter, James Goodman and Stuart Rosewarne describe the cultural pendulum of legitimization, delegitimization, and, more recently, the possible relegitimization of uranium. They trace the history of ideas and actions of anti-uranium peace activists, labor unions, farmers, and Indigenous peoples who succeeded in convincing the Australian people and their government to impose a moratorium on uranium mining in the mid-1970s, a ban that lasted for almost thirty years. Australia's action demonstrates that a nation can choose to reject at least some of the enticing promises of easy wealth that can be had by taking stuff out of the ground, and do so based largely on nonmonetary ethical, spiritual, security, health, and human rights grounds.

This story adds a complication to the keep-it-in-the-ground (KIIG) argument: when it is implemented, the subterranean riches are still there, essentially free for the taking; a decision to ban is thus inherently unstable. Continuing to say no to extraction thus requires more than ethical righteousness; it requires a politics of co-creation alongside the resistance, building societal relations where people thrive on much lower levels of energy and material consumption, what Australian activists have only begun to engage. Lacking such a transformation, demand for fuel and power and the temptations of cheap extraction and costless externalization (what Australia knows all too well with its other mining operations, including coal) can easily overwhelm a decision to keep them in the ground. So here, although the pendulum swung toward uranium extraction in the early 2000s, it could well swing back to KIIG as the nuclear power industry and citizens, the ones who ultimately pay true costs, face the twenty-first century reality of never-ending long-term waste storage problems, aging plants, and nuclear disasters such as Fukushima (see chapter 9 for Germany's reaction to Fukushima).

In these swings, there is a lesson for fossil fuels and extraction in general: irreversible processes that visit slow violence on both marginalized peoples and the planet as a whole are illegitimate; the time to start stopping is now.

Leaving fossil fuels in the ground currently means locking up what are defined as high-value assets. The International Energy Agency (IEA) states that two-thirds of the world's fossil fuel reserves must be permanently sequestered if climate stability is to be maintained.[1] To achieve a shift of this sort on a global scale means that fossil fuels must cease to be defined as resource assets and instead become recognized as liabilities that do untold damage when they are extracted and burned. Rather than be defined as a fuel, coal, oil, and gas must be recognized as health and climate hazards, a fundamental threat to human life and the environment. There are signs that fossil fuel industries, especially coal-fired power stations, are increasingly recognized as "stranded assets" and that this produces market volatility and investment uncertainty in the sector. The uncertain future for fossil fuels is discouraging risk-averse investment, yet change remains at the margins and piecemeal. The wholesale closure of fossil fuel industries, on the scale required by the IEA, remains elusive.

Uranium offers an example of an energy commodity that has undergone a similar transformation from fuel to poison. The transformation was achieved in large part by the anti-uranium movement, which reached its peak in the 1970s and 1980s. With the first experimental nuclear reactor in 1951, uranium was transformed into a valuable commodity, declared the fuel of the future. Countries with uranium deposits were suddenly enriched as suppliers to the new nuclear industry. Challenging the definition of uranium as a fuel, the anti-nuclear movement sought to recast it as a danger to humanity, a mineral to be left safely in the ground. The military uses of plutonium, the by-product of nuclear power, had been demonstrated to devastating effect at Nagasaki and Hiroshima and were subsequently played out in the Cold War arms race. From its inception, uranium as a commodity was stained by its association with weapons of mass destruction and with the nightmare scenario of nuclear holocaust. Even when distanced from its military uses, the nuclear industry could never shake off concerns about public safety, of nuclear reactors and nuclear waste, and mining for uranium oxide, so-called yellowcake.

In this chapter, we argue that there are many lessons to be learned from the anti-uranium movement for developing strategies to leave fossil fuels in the ground. There are strong parallels between the political struggle in Australia to leave uranium in the ground and current efforts worldwide to phase out fossil fuel extraction. In both cases, the stakes are high. The apocalyptic scenario of runaway climate change, of making the planet uninhabitable, conveys much of the terror felt from the prospect of mutually assured destruction and the ensuing fallout and nuclear winter. Both climate catastrophe and nuclear holocaust are unimaginable and generate similarly visceral responses, including despairing for our planetary future. Also, in both cases the risk of global disaster is created by the consumption of mineral commodities for energy. Political contention centers on arguments against minerals extraction, whether uranium or fossil fuels, and instead for energy conservation and or renewable energy.

The parallels end when we contrast the direct logic of human agency embedded in the prospect of nuclear weapons crisis—the finger on the button—and the more mediated logic of climate crisis. The threats they pose—planetary irradiation or climate catastrophe—are quite distinct and have their own separate logic. Reflecting this, nuclear power and fossil fuels are often presented as alternatives: expanded nuclear generation in India and China is driven in part by the desire to reduce reliance on coal-fired power and minimize greenhouse gas emissions; in contrast, in the aftermath of Japan's Fukushima incident in 2011 expanded use of fossil fuels has allowed a planned phase-out of nuclear power in Japan and in Germany. The dystopias offered by uranium and fossil fuels are set against each other as alternatives, as the only options on the table. More positive-sum possibilities offered for society by renewables, or by steady state and de-growth models, are ignored or foregone or postponed.

For this book's purpose of seeking a fossil fuel phase-out, there are strong similarities between the two issues and how they may be politically addressed. The merits in removing fossil fuels from the commodity chain and keeping them in the ground is the premise of our argument in this chapter, where we explore the Australian experience of leaving uranium in the ground. Australia's partial success was achieved principally through a government moratorium on expanded uranium mining, enforced nationally from 1983 to 1996 and at the State level until 2007. The process that led to the moratorium, its limits, and how it subsequently unraveled, holds important lessons for instituting a similar ban on new mining and drilling for fossil fuels at the national level.

At the outset, a note of caution is required: sequestering fossil fuels, giving them up for safekeeping, presents much more of a challenge than sequestering uranium. Even in the scale of operations, we are not comparing like-with-like. The nuclear industry is a minor player in global energy when compared with fossil fuels. Nuclear energy accounts for almost 6 percent of global energy production; fossil fuels account for at least 80 percent.[2] World electricity production from uranium stabilized at about 2,700 terra watts in 2004 and remained at this level in 2012. Nuclear power accounts for about 12 percent of world electricity output, but this is concentrated in just six countries that together account for three-quarters of output: the United States, France, Japan, Russia, Korea, and Germany.[3] In contrast, reliance on fossil fuels for electricity production is the global norm. The burning of fossil fuels accounted for three-quarters of global electricity supply in 1971, falling to two-thirds in 2010 (the fall being mainly due to the emergence of nuclear power). The three-fold growth in global electricity output from 1971 to 2009 translated into a tripling of coal-fired power output.[4] Ideologically there are also important differences. Coal, gas, and oil are historically embedded as industrial fuels, positioned as normal (chapter 3); nuclear is presented as a successor fuel, but is also linked to immediate safety concerns and to nuclear weaponry and proliferation, and hence is relatively easy to delegitimize.

With these important caveats in mind, this chapter charts the origins of the national ban in the mid-1970s to its demise with the fall of the Federal Labor government in 1996. It traces the legacy of the ban at State level and in social movement mobilizations through to 2007 when the Labor Party (the ALP) returned to national power and officially repudiated the policy and instead relegitimized uranium as an export fuel. The discussion contrasts with the experience after 2007, that is, after the ALP ban was lifted, with the moratorium period from the 1970s. The aim is to highlight the effectiveness of the moratorium in keeping uranium in the ground. During the period of the moratorium, several studies sought to demonstrate its negative impacts in terms of lost export earnings.[5] In 2007, commenting on the lifting of the moratorium, the director of the Australian Uranium Association stated, "It is perhaps difficult for those not from Australia to appreciate the significance of the change this represents."[6] The conclusion discusses this kind of government action as a strategy for halting the extraction of fossil fuels.

## The National Moratorium

Australia's anti-uranium movement developed into a robust political force over the course of the 1970s and 1980s to have a decisive impact on government policy formulation. The movement secured broad popular support across the nation, and its momentum was maintained because of its focus on the community health effects of uranium mining and radioactive waste, and the role of uranium in the global arms race and in creating the prospect of nuclear war. The movement was also significant because it brought indigenous issues into the political frame. Indeed, the idea that Australia should leave its uranium in the ground became a national political issue in the early 1970s only with the discovery of large uranium deposits in relatively remote regions inhabited by indigenous peoples, in the Northern Territory (at Nabarlek and Ranger in 1969) and South Australia (at Roxby Downs, with copper and gold, in 1975). Deposits at these three sites accounted for up to 40 percent of worldwide economically viable uranium deposits at the time and were of added significance as the other large deposits were either already devoted to military uses, such as in the United States and Russia, or were located in unstable countries, such as South Africa.[7] Australia offered a large-scale reliable supply of uranium for the world nuclear industry, said to be on the cusp of a boom in the wake of the 1973 oil crisis. This positioning makes it all the more remarkable that the anti-uranium movement was able to gain such political traction.

A small number of small-scale uranium mines had been operating in Australia from the 1950s, supplying uranium for the British and US weapons programs.[8] Between 1955 and 1963, the British government operated nuclear test facilities in Australia, with twelve major explosions of atomic bombs and a further estimated 700 experiments with radioactive material.[9] By the 1960s, the military market for Australian uranium had dried up. The Mary Kathleen mine in Queensland, for instance, opened in 1958 but was closed by 1963. Nonetheless, in 1972, in the closing days of the conservative Liberal-National Coalition government, new contracts for the export of uranium were approved. In 1973, the newly elected Federal Labor government, the first in office since 1949, imposed a ban on new exports of uranium primarily to gain leverage for an Australia-based uranium enrichment industry.[10] As part of this strategy, in late 1974 the federal Labor government assumed a 50 percent stake in the Ranger project (through the Australian Atomic Energy Commission) and in 1975 announced a public inquiry. This

Ranger Environmental Inquiry was to be the first inquiry under the 1974 Environmental Protection Act and would be chaired by a senior judge, Justice Russell Fox (hence, dubbed the Fox Inquiry).

The announcement of the Fox Inquiry and the consequent prospect of large-scale uranium mining in Australia sparked extensive political mobilization. French nuclear weapons testing in the South Pacific in 1972 and 1973 had already begun to position Australian uranium as an environmental hazard, with unions imposing work bans on uranium exports to France.[11] Occupations against nuclear power stations in Germany, India's use of Canadian uranium in its first atomic bomb test, and in March 1975 a nuclear accident at Brown's Ferry in the United States set the international context. The Fox Inquiry itself was to proceed for more than a year, and its public hearings, where more than 300 people gave evidence, provided an important focus.[12] By 1976 a strong and broadly based movement had emerged, initiated by environmental organizations (notably Friends of the Earth and the Australian Conservation Foundation), encompassing churches, State branches of the Labor Party, some blue-collar trade unions and regional labor councils, and a range of professional bodies, including teachers, doctors, and scientists.[13]

The Australian Council of Trade Unions (ACTU) and the national conference of the ALP became central sites for advancing the campaign to stop uranium mining. As Cupper and Hearn outline, opposition to uranium mining cascaded through the labor movement from 1975, with an appeal from the environmental movement that the 1973 ban be maintained, and later from community-based coalitions such as the Movement against Uranium Mining and from indigenous groups affected by the proposed new mines.[14] The uranium debate assumed central significance for the Labor government: some at the time likened the issue to that of the Vietnam War.[15] In 1977, the ACTU stated that "no issue since conscription during the First World War—with the possible exception of Vietnam—has so split society."[16] The movement quickly injected environmental concerns into a national minerals policy agenda. At the time, Australia was undergoing a minerals boom, and there was an intense political debate about the public benefits of expanded mining, especially in terms of revenues.[17] The Labor government commissioned a report into mining finance, the Fitzgerald Report, that confirmed the drain on the economy and, acting on this, sought to buy back mining resources and support minerals processing in Australia. The government recognized the economic costs of dependence on extractive industries and that uranium mining was a politically contentious issue that, because of the popular opposition to mining and the broader questions of human health and safety, could dominate the policy debate.

Under pressure from several unions, the ACTU became concerned about the health effects of mining and transporting uranium and the impacts of the nuclear cycle more generally. In September 1975 the ACTU resolved to proscribe trade unionists from working in the industry, pending the outcome of the Fox public inquiry. In May 1976, the issue was forced further up the political agenda when a Queensland railway supervisor was fired for refusing to permit the transport of material for the Mary Kathleen mine.[18] Unions responded by launching a one-day national railway strike, and the industrial action could well have precipitated a national general strike had not the ACTU appealed to unions to postpone further action in line with its previous undertaking to await the assessment of the Fox Inquiry.

However, the labor movement's growing determination to block any further expansion of uranium mining was overshadowed by a more immediate pressing political challenge. In November 1975, the conservative Liberal-National Coalition used its control of the upper house to block the federal government's finance bill. With finance blocked, the business of government policy was completely frustrated, and this sparked a constitutional crisis. The governor-general dismissed the Federal Labor government from office and appointed an interim caretaker government led by the Coalition. With public sentiment running against Labor, a subsequent election resulted in the return of the Coalition to government. This proved to be highly significant because the Federal Labor Party, out of government, became more receptive to arguments against uranium.

The first report of the Fox Inquiry, published in October 1976, began by refusing to treat uranium as any other mineral, stating in the first sentence of its Preface, "Uranium is a very special metal: it contains fissile atoms."[19] It recognized the risks associated with mining and exporting uranium, including in terms of the arms race, and urged that any decisions on approving mining be approached with caution and be subject to further public debate. Interestingly for today's debates about fossil fuels, it was the democratic argument that brought the inquiry closest to recommending a complete ban:

One of the arguments which has been used against any mining development is that, once it is started, no government will have the strength to resist pressures for its continuance and even its expansion. We believe this is a serious consideration. If the argument is a sound one, the proper course is to recommend against commencement.[20]

The Coalition government claimed there had already been sufficient debate and immediately renewed the existing export contracts. The Fox Inquiry's second report, of May 1977, focused on the prospect for uranium mining at Ranger and reiterated its earlier concern on the need for caution in considering project approval. While it highlighted the risks of uranium mining, it also recommended mechanisms to minimize safety risks, and in the process, it effectively legitimized new mines.[21]

In response to the second Fox Report, the Coalition announced conditional support for expanded uranium exports and the opening of several new mines subject to the consent of local indigenous landholders. The Labor Party opposition opposed new mines while stating, "Existing contracts for uranium mining should be honored." Given that the coalition government had stated it would approve export contracts for the new mines, the Labor opposition acknowledged it would be required to honor them while committing to not approving any new operations.[22]

Meanwhile, the anti-uranium movement had gathered momentum. By 1977, the link between environmental issues and peace concerns against uranium mining had become well established. Publications by the Australian branch of Friends of the Earth, such as *Ground for Concern,* but more especially the much-reprinted collection *Redlight for Yellowcake,* charted these linkages and articulated much of the urgency.[23] Although linked, the anti-uranium movement quickly subsumed and superseded the antiwar movement. By 1977, State-based groups had formed a national Uranium Moratorium movement behind a petition with a single demand for a five-year delay on uranium mining to enable a national debate. The petition was explicitly based on the Fox Report and foregrounded its security concerns about the risk of nuclear war; about theft, sabotage, or blackmail using radioactive material; and about the threat of nuclear waste. Concerns about broader environmental impacts and implications for indigenous peoples were cited but were not central.

The petition gathered 250,000 signatures, and in August 1977 more than 80,000 people demonstrated against the Coalition's decision to proceed with exports.[24] From June 1977, the movement sought to directly halt the export of uranium yellowcake that had been newly authorized by the federal government. Police violence at the dockside in Sydney and Melbourne led to work bans by waterside workers and an increased profile for the issue.[25] The anti-nuclear movement engaged local communities throughout much of Australia, and many local government shires and councils were enjoined to declare their local areas nuclear-free zones (and the Local Government Association was to maintain its opposition to uranium mining into the new millennium). However, while such declarations remained an enduring symbol of the struggle, anti-uranium activism was not sustained. Dozens of local and workplace anti-uranium groups emerged through 1977, with most winding up by 1978. The vast majority had passed into history by 1980.[26] Notwithstanding the ephemeral nature of groups opposed to uranium mining, the anti-uranium movement was later rechanneled from the 1980s into the peace movement, centering on the intensifying arms race and nuclear weapons proliferation.[27]

While short-lived, the anti-uranium upsurge had a direct and lasting impact on ALP policy. In 1977, reflecting the growing movement, the ALP position hardened into outright opposition to uranium exports. In July 1977 the ALP national conference declared a "moratorium on uranium mining and treatment in Australia" and repudiated all existing uranium contracts.[28] The stronger conference position reflected public concerns about nuclear proliferation and also about issues of mine safety, the safety of nuclear plants, and the issue of nuclear waste. This set of concerns had been raised at the Fox Inquiry and remained unresolved. In addition, there were concerns about the political impacts of the industry in terms of concentrated economic power and national security regulations that defined uranium mines as defense projects, where anyone threatening to boycott or advocating obstruction could be imprisoned for twelve months.[29] Despite these restrictions, at a special conference in 1978 the ACTU banned union labor from new uranium mines until "adequate safeguards" were in place for workers and for local indigenous people affected by the proposed mines (although it resolved not to impose work bans on existing contracts). In September 1979, following the March 1979 Three Mile Island nuclear meltdown in the United States, the position was reaffirmed with an ACTU Congress resolution against all uranium mining and exports (although not for enforceable work bans, enabling existing mines to operate and export). Given the political context, the resolution was then translated into the Labor Party's platform for government.

### The Persistent Moratorium

In 1977 Mary Elliott and Friends of the Earth stated presciently that the ALP conference of July 1977 was a "decision of principle" that would "ensure the uranium debate goes on."[30] In many respects, the debate it initiated continued for thirty years, until the ALP fully reversed the party policy in 2007. Yet even by the late 1970s, the ALP and ACTU positions had reached their high-water mark. With Nabarlek opening in 1979 and Ranger in 1980, the ACTU formally abandoned work bans from 1981, and the ALP's conference in 1982 resolved to "phase out Australia's involvement in the uranium industry" rather than repudiate it, preventing any need to legislate for closure. The ALP now was "not [to] allow any new uranium mines" (beyond Narbalek and Ranger) but would consider approval where uranium was "mined incidentally to the mining of other minerals" (thus enabling approval of Roxby Downs). Mindful of its experience in government over 1972 to 1975, when big business constantly challenged the direction of Labor's management of the economy and the threat of capital flight engendered considerable economic uncertainty, the Labor Party sought to position itself as a responsible economic manager in the lead-up to the 1983 federal election. Caution framed the policy debate within the party conference, and the debate on uranium centered on the consequences of repudiating uranium contracts, said to be worth over $400 million and whether investors should be compensated.[31]

While the new ALP policy sought to embrace a sense of caution as the premise of responsible management, the Coalition government had no such pretensions and was determined to preempt any moves to block the expansion of uranium mining by granting mining licenses and approving export contracts before the 1983 election. Here local indigenous land councils played an important initial role in delaying mining approvals. Indigenous consent to uranium mining quickly became a bargaining chip in the land rights debate. The Northern Territory Land Rights Act of 1977 (which, in 2013, still afforded the strongest rights for customary indigenous owners of any land rights legislation in Australia) gave veto power for local Indigenous people over mining development but then removed it for "national interest" projects, such as uranium mines. Reflecting this, the Act explicitly exempted the Ranger uranium mine from any local veto power, forcing the local land council to negotiate a deal.[32]

In 1984, with Labor in government, the ALP Congress moved to endorse a "three mines policy," explicitly protecting Narbalek, Ranger, and Roxby. The aspiration to phase out uranium mining was deleted, but new mines were prohibited, thus establishing a moratorium on the number of mine sites.[33] The policy was then entrenched as federal policy for the successive Hawke- and Keating-led Labor governments from 1984 to 1996. The national three mines policy also played out after a Coalition government was elected in 1996 because the federal

Coalition government could not override State rights, and, with Labor in office in key States, and these governments refusing to permit uranium exploration, the moratorium remained effectively in place until 2007. The Northern Territory was the exception because, unlike the State jurisdictions where governments possessed constitutional responsibility with respect to the issue of exploration and mining licenses, the granting of mining licenses in the Territory was the constitutional prerogative of the federal government.

The longevity of the moratorium is in itself remarkable, especially given the shift of the ALP in government to a more pro-market policy framework through the 1980s, and given the uranium industry's ongoing campaign for expanded mining. One explanation for this lies with the new, more dangerous nuclear standoff following the end of détente in the late 1970s. With the Soviet invasion of Afghanistan in 1979 and the rise to power of the Republican US presidency led by Ronald Reagan in January 1981, the world embarked on a series of new hostilities, dubbed the Second Cold War.[34] With Reagan in office until 1989, the 1980s were marked by rapid growth in the nuclear disarmament movement, not least in Australia.

During this period, the environmental concerns of the anti-uranium movement became more embedded in debates about the nuclear stand-off.[35] In 1981 People for Nuclear Disarmament was founded in Melbourne, and yearly Palm Sunday peace rallies began in 1982, attracting close to 400,000 people by the mid-1980s. As political scientist Camilleri reported in 1986, by the mid-1980s Australia had caught "the same nuclear allergy that had struck much of the Western world," with upward of two-thirds of the population stating that the use of nuclear weaponry could never be justified.[36] Public concerns centered on nuclear weaponry and proliferation, defining a special responsibility for Australia as a uranium supplier. The nexus was expressed in the 1984 Independent Committee of Inquiry into Nuclear Weapons and Other Consequences of Australian Uranium Mining, which sought to update the Fox Inquiry process in the context of the new federal Labor government.[37] The 1986 Chernobyl accident confirmed the already firm public opposition to Australian participation in the nuclear cycle.

## Nuclear Boosterism

As long as the Cold War proceeded and concerns about nuclear safety prevailed, it was difficult for uranium to be treated as just any other mineral commodity. With all its human and environmental risks as a radioactive mineral with vast destructive capacity, potentially casting a pall over millennia to come, uranium could not easily be treated as a mineral on a par with, for instance, iron or gold. The struggle for the industry, and for the government, was to shed uranium of its risky associations. One way to achieve this was to establish a distinction between peaceful and military uses of uranium and impose requirements that Australian exports of uranium only be used for peaceful ends. This approach was pursued with an approvals and monitoring framework established in the 1980s. As early as 1984, the Hawke Labor government was arguing that the export of uranium from Australia contributed to world peace by empowering Australia to argue for disarmament. Speaking in Moscow in 1984, the Australian foreign minister stated:

The simplest way to ensure that no Australian uranium is ever used in a nuclear weapon—so the argument goes—is to keep it in the ground. But ... cutting off the supply of uranium will not have any effect in reducing the number of nuclear weapons in the world. It will seriously damage arms control and disarmament and it could deal a serious blow to the single most effective arms control and disarmament measure in effect at the moment—the Nuclear Non-Proliferation Treaty.[38]

Another way of normalizing uranium was to diminish the salience of military uses and delegitimate other sources of energy. The end of the Cold War reduced the intensity of the military threat, although the issue of nuclear proliferation remained. The intensification of concern about global warming as caused by the burning of fossil fuels repositioned the nuclear industry, once again, as the fuel of the future. Nuclear energy could now be presented as clean energy, entailing lower risks in comparison with fossil fuels. This was a political gift for nuclear advocates.

Despite the end of the Cold War, the ALP maintained its moratorium at the national level. Pro-uranium advocates sought to have the moratorium lifted in 1994 and 1998, but they failed due to concerns about maintaining ALP political unity. Public concerns remained in place and could resurface unexpectedly, for instance, in 1995 when France resumed its nuclear testing in the Pacific. Here the Labor government was forced to respond to public disquiet but did so claiming that Australian uranium sold to France was used only for peaceful ends.[39] As noted, when Labor lost the federal government in 1996, State Labor governments continued to maintain a ban on new uranium mines. The State bans frustrated the development of new mines.[40] More important was the expansion of uranium mining at the approved mines, especially at Roxby, later renamed Olympic Dam. With known deposits of 1.5 million tons, Olympic Dam is the world's largest uranium deposit and has expanded production.[41] Indeed, it was the expansion of output at Olympic Dam that was largely responsible for the doubling of uranium exports between 1996 and 2007, to about 10,000 tons per year.[42]

Meanwhile, the federal Coalition government focused its efforts on promoting the expansion of uranium mining in the Northern Territory. With the constitutional authority to make determinations with respect to mining in the Territory, the federal government was able to override local-level political priorities, although in practice, the incumbent conservative Country Liberal Party had been promoting uranium for many years. However, the Coalition government's moves served to revitalize the anti-uranium movement. With new purpose, anti-uranium campaigning focused on reengaging popular opposition and on leveraging with local indigenous communities and through international institutions. The key point of contention was the proposed Jabiluka mine, to be located near the Ranger uranium mine in what had become the Kakadu National Park (partly as a result of the Fox Report). With local indigenous people able to make claims through land rights legislation and with Kakadu listed as a World Heritage site under UNESCO, plans to mine uranium were forced into abeyance by Indigenous and environmental campaigners, driven by a mixture of dogged determination and strategic maneuvering. The Jabiluka campaign began in 1996 and ended only in 2002 when mining giant Rio Tinto acquired the site and resolved that the cost of overriding Indigenous opposition was too high in terms of reputational capital.[43]

### The New Millennium Uranium Putsch: An Energy Supply Nation

From the mid-2000s, a succession of governments of both political persuasions embraced the economic promise of uranium mining. They joined the chorus of industry in seeking to rehabilitate uranium mining, arguing the case for nuclear power as a source of clean energy as well as contributing to medical advances. With Australia as the world's preeminent exporter of coal, rivaling Saudi Arabia in terms of the value of exports of fossil fuels,[44] successive governments contended that adding uranium to the export mix would be critical to positioning Australia as the foremost source of the world's energy resources.

Toward this end, support for a range of proposed new mines and the establishment of a waste dump that would receive imported uranium waste prefigured a uranium renaissance in Australia in the 2000s. From 2005 the conservative Coalition government signaled that it would relax restrictions on the exploration and mining of uranium. In February 2006, the Coalition hosted the G20 meeting and nominated energy security as one of the key issues to be addressed. Chairing the event, the Coalition Treasurer declared Australia's pivotal role in "an energy and minerals freeway linking suppliers and consumers across the globe."[45] The government was clearly seeking to position Australia as an energy superpower, as potentially the world's largest uranium exporter as well as the largest coal exporter. World leadership in the export of gas was also part of this agenda. The government's clean energy alternative to the Kyoto Protocol, the 2006 Asia-Pacific Partnership on Clean Development, neatly expressed this geo-energy ambition. Alongside the establishment of a Low Emissions Technology Development Fund that would underwrite research and investment in clean coal technology, the growth of nuclear power—and the expanded mining of Australian uranium—was claimed by the Coalition government, and also supported by the successor Labor government's minister for energy and resources, to be a critical aspect of the "clean energy" mix.[46]

In the process of promoting nuclear power, the Coalition government commissioned a number of federal public inquiries, all headed by nuclear power advocates: a joint industry-government committee, the Uranium Industry Framework, in August 2005, chaired by John White who had a direct interest in companies that invested in uranium enrichment and had lobbied the government to support a nuclear enrichment plan at BHP-Billiton's Olympic Dam uranium mine; the Uranium Mining, Processing and Nuclear Energy Review in May 2006 chaired by Ziggy Switkowski, who himself was chair of the Australian Nuclear, Science and Technology Organisation; and the House of Representatives Standing Committee on Industry and Resources Inquiry into Australia's Uranium, chaired by well-known pro-uranium politician Geoff Prosser. One obvious consequence was a rush of expressions of investment interest from energy companies and state-owned enterprises from Canada, China, France, India, and Japan to explore and mine uranium in Australia.

The Coalition government also sought to secure a place on the "energy and minerals freeway" by signing onto the Global Nuclear Energy Partnership. The Bush administration had proposed a partnership of suppliers and users with a view to developing a "worldwide consensus on enabling expanded use of economical, carbon-free nuclear energy," in effect bypassing Nuclear Non-Proliferation Treaty (NPT) obligations and the International Atomic Energy Agency (the Global Nuclear Energy Partnership was subsequently renamed the International Framework for Nuclear Energy Group).[47] The Coalition had already acceded to the terms of this alternative institutional framework, outside the NPT, by endorsing the sale of uranium to Taiwan. This was in line with and followed the US administration's decision to supply nuclear technology to Taiwan, despite Taiwan's not being covered by the NPT (as a nonmember of the United Nations, Taiwan cannot be an NPT signatory, although it is subject to an International Atomic Energy Agency [IAEA] safeguard agreement). When Australia joined the Global Nuclear Energy Partnership in 2007, it was also signing onto the Partnership's commitment to creating a closed fuel cycle, where suppliers would not only provide new fuel but would also agree to contaminated waste being returned to the uranium-source country.

Surprisingly, the Taiwan uranium export deal attracted little opposition within Australia. This was perhaps because the contract between BHP Billiton and Taipower was organized as an indirect sale arrangement: uranium would first be shipped to the United States for enrichment before being transhipped to Taiwan, and the arrangement was sanctioned through an Australian-US bilateral agreement.[48] The lack of debate may also be explained by the prior resolution of an earlier debate about the sale of uranium to China.[49] The prospect of Australian uranium being sold to China had aroused considerable concern even though China is a signatory to the NPT (whereas Taiwan is not). There were lengthy deliberations in the parliament, and some of the concerns were dispelled with the Coalition determining that a firm condition of the export contract would be that uranium could be used only for peaceful purposes and this condition would be monitored by the IAEA.[50] Officials in the Australian Safeguards and Non-Proliferation Office called into question this monitoring arrangement, as the IAEA does not have the authority to track the movement of uranium, although the parliamentary committee that endorsed the arrangement to export uranium to China recommended that Australia fund the IAEA to police the contractual obligations.[51]

A more politically contentious situation emerged following the Bush administration's decision to provide India with nuclear technology for civilian purposes. The refusal of India to accede to the NPT represented a real challenge to successive Australian governments' commitment to the terms of the Treaty and whether it should consider taking advantage of the US rapprochement with India and approve the export of uranium to

India. Initially the Coalition declared that it would stand by its commitment to the NPT and not entertain the possibility of supplying uranium to India. But this position was soon abandoned on the grounds that the controls proposed in the US-India bilateral agreement provided the necessary safeguards and that planned inspection arrangements would draw India under an NPT-like umbrella.

The election of the Rudd-led Labor Party government in November 2007 heralded a potential retreat from the Coalition's rehabilitation of uranium mining. Rudd was committed to the NPT and suspended the agreement to export uranium to India. However, at its April 2007 National Conference, the Party abandoned its three mines policy, the culmination of an unfolding retreat from the Party's previous commitment to the moratorium on any expansion of uranium mining. In 2011, and in line with this policy retreat, the ALP set aside its opposition to the sale of uranium to India following the Nuclear Energy Suppliers Group resolution to exempt India from a ban on the supply of uranium. Rudd's successor as prime minister, Julia Gillard, argued that there was little logic in prohibiting exports to nonsignatories to the NPT when other uranium producers could supply India, and the Labor Party's 2011 national conference fell into line, abolishing the policy proscribing the supply of uranium to nonsignatories to the NPT.[52] A not-altogether-different set of circumstances shadowed deliberations over whether to ratify an earlier Coalition agreement to permit uranium exports to Russia: in the face of criticisms that Russia was being politically singled out, the Labor government had approved the agreement in 2010.[53]

While diplomatic and strategic concerns were critical in this export drive, economic considerations had also been important. The commodity boom had captured the Labor Party, and the Rudd Labor government's minister for energy and resources, in particular, was an enthusiastic advocate for adding uranium to the resource export mix. Minister Martin Ferguson commissioned the Bureau of Resource and Energy Economics to review the potential of the global uranium market in anticipation of a nuclear power boom in the Asia-Pacific region.[54] The potential markets included nonsignatories to the NPT, and from 2011, uranium supply deals were being explored wherever there was an opportunity to outcompete rival exporters. In 2012, for instance, the Labor government agreed to export uranium to the United Arab Emirates (UAE)—ahead of other potential suppliers such as Kazakhstan and Canada—and the government anticipated that this would become a springboard for negotiating other uranium supply agreements in the Middle East (the UAE had concluded agreements with Korea to construct two nuclear power reactors with plans for two more).[55]

A second critical element in this campaign for a nuclear energy future was the Labor government's decision to affirm the Coalition plan to establish a radioactive waste dump in Australia. This, it was argued, would integrate the import of nuclear waste as part of the uranium trade cycle and would greatly strengthen the attraction of Australia as a uranium supplier. Just as the commitment to promoting uranium mining and exports was argued to reflect Australia's comparative advantage in uranium, so the argument that nuclear waste could be deposited in a geologically stable and remote location was represented as another dimension of the nation's comparative advantage. Imports of nuclear waste were to complement Australia's international trade closing the circle of the uranium commodity chain.

## Post-2007: Local Mobilization Disrupting the Uranium Rush?

The mounting federal enthusiasm for uranium mining and export forced the environment movement to rethink its strategic focus. With almost no traction in the federal political sphere, the environment movement turned its energies to testing the scope for stopping mining development through the environment and conservation legislative framework. Resources were directed to evaluating planned projects to expose the local environmental and social impacts of new mines and to scrutinizing and questioning the merits of the environmental impact assessments. Governments were also pressured to extend the frames of reference of impact assessments to include cultural concerns, particularly as these relate to Indigenous communities. Since 2012, impact assessments have been required to provide greater reflection on cultural issues, and this has been especially important given that uranium deposits are generally located in remote areas on traditional Aboriginal lands.

Anti-uranium movements have also turned their attention to lobbying State governments. With the exception of the campaign to block the uranium waste facility being established at Muckaty in the Northern Territory, a federal government responsibility, campaigns against uranium mining have concentrated on lobbying State governments when State governments are assessing applications for exploration and mining approvals. This, however, has not been particularly successful, especially with the election of pro-uranium Coalition governments in place of Labor governments across the country. In 2008, for instance, the incoming Liberal government in Western Australia lifted a previous ban on uranium exploration and three new uranium mines—Comeco's (formerly BHP Billiton's) Yeeliree, Lake Maitland, and Lake Way—were scheduled to start production by 2014. The Australian Conservation Foundation (ACF) held out the hope that it might stop an incoming government in Queensland from following the lead of the Western Australian government, securing assurances from the two competing parties that they would not lift the ban on mining, but the trouncing of the Labor Party at the election handed the conservative Campbell Newman government such a large majority in 2012 that it immediately announced it would not honor its written undertaking to the ACF. The Coalition government in New South Wales, which had held the line against uranium mining, quickly followed suit and lifted restrictions on exploration.

Still, the lack of traction in lobbying efforts of State governments has not been confined to instances where the Coalition party governs. This was demonstrated in the struggle against the approval of BHP Billiton's plans to develop the Roxby Downs mine to establish what would be the world's largest open-cut uranium mine, the Olympic Dam project. The environment movement was pitted against a State Labor government in 2008 when BHP Billiton announced that it wanted to invest $30 billion to expand the Roxby Downs underground mine into a massive undertaking, expected to extract and export up to 19,000 tonnes of uranium as well as recover copper, gold, and silver.[56] Both the federal and Labor-led South Australian governments indicated their support for the proposal, and in 2009 BHP Billiton submitted a draft environmental impact statement (EIS) for planned substantial expansion. There were 4,000 public submissions in response to the EIS, and the great majority of these identified innumerable negative environmental impacts. The new mine would require an additional 200 million liters of water per day to be drawn from the Great Artesian Basin, which itself feeds Lake Eyre, and flow rates are declining. A proposed desalination plant would release saline brine into the Upper Spence Gulf, potentially damaging the marine ecosystem and threatening unique breeding grounds. Energy consumption would draw on 20 percent of South Australia's electricity capacity and result in a dramatic increase in greenhouse gas emissions.[57] The environmental concerns were well documented; even the company acknowledging the leakage of radioactive waste into the underlying rock and the aquifer (BHP Billiton 2011). Although the 2009 EIS understated the company's ambitions, the company issued a revised version, and both the federal and South Australian Labor governments approved the project on 10 October 2011.[58]

The reorientation in the political focus of the anti-uranium movement is not new. It builds on longstanding practices of mobilizing locally in the face of governments' pro-uranium stance. Indeed, when the pro-uranium Coalition government was elected to office in 1996 and approved applications for new mines and the expansion of existing mines, anti-uranium campaigns concentrated on canvassing support within local communities to lobby government. The campaigns were more often than not initiated by State or Territory-based organizations, principally conservation councils.

A key feature of these campaigns was their attention to environmental and community safety concerns and the failure of mines to abide by set environmental guidelines. Campaigns were focused on exposing the repeated incidences of leaks, spills, and accidents, emphasizing the fact that uranium could not be produced safely. In the process, activists examined the records of some of the uranium mining companies' overseas operations to highlight systemic shortcomings in their safety records.[59]

Coalition building was also a crucial feature of these endeavors to reignite the anti-uranium movement, as were decisions to inject a more confrontational dimension into campaigns through direct action and acts of nonviolent civil disobedience. As well, engaging Indigenous communities figured prominently in these endeavors. In 1998, for instance, a coalition of forces, initially coordinated by the Mirrar people, the customary Aboriginal landowners, and the Australian Conservation Foundation, launched national protests to block approval for the proposed Jabiluka mine in the Northern Territory. Notwithstanding Coalition support for the project, the sustained campaign forced the indefinite postponement of the project, although the collapse in commodity prices was a contributing factor because lower prices made the project less viable.[60] In fact, this success set the pattern of the movement against uranium mining for the next decade. Steeled by the success at blocking Jabiluka, another direct action campaign was launched against the proposed Beverley uranium mine in northeast South Australia.

Direct action protests have been designed to rouse support among the general population. The Sleepy Lizard Revenge march, from July 14 to July 20, to convene at the Olympic Dam site, drew on traditional Indigenous iconography to signal the common purpose of local Indigenous communities and the activists drawn from urban centers.[61] The potential for legal action had been explored, with traditional elder Uncle Kevin Buzzacott pursuing action to halt the expansion in the federal court, on the grounds that the federal minister for the environment had failed to consider crucial environmental factors.[62] That action proved unsuccessful, although, as outlined below, subsequent uncertainty about the future of the nuclear industry later prompted the company to shelve its expansion plans.

Indigenous communities and activists played a key role in these campaigns, which in the main centered on halting the mining, the occupation, and contamination of traditional lands in remote Australia. Indigenous groups began to assume a more prominent, if not leading, role in defining campaigns against uranium mining. A key concern has been the potential impact on traditional lands, on Indigenous country. For instance, following government approval for the Canadian company, Uranium One, to develop the Honeymoon mine in South Australia, the Adnyamathanha traditional owners appealed for greater transparency in the approval and oversight process of what was happening on Yarta land. At the Olympic Dam uranium mine, elders from the Arabunna community were resolute in their campaign to protect their country from the expansion of mining. The Arabunna nation, for instance, petitioned the government to have Lake Eyre declared a World Heritage site as a strategy to block further development of the Roxby Downs mine. Indeed, Indigenous leadership and involvement in the movement against uranium mining became an essential feature of the contemporary political focus for movements against uranium mining. As a result, Indigenous cultural concerns were woven into the anti-uranium mining narrative, alongside ecological and safety considerations.

The assertion of Indigenous rights has also been at the forefront of efforts to block the federal Labor government plans to establish a radioactive material waste repository in the Northern Territory. Local Indigenous landowners at the Muckaty cattle station, where the proposed dump is to be located, launched a national campaign in 2009 against the proposal. With support from various environmental activists and trade unions, the Muckaty people toured major urban centers canvassing support to stop the proposal. They questioned the legitimacy of the negotiations and the authority and right of those with whom the government had been negotiating to speak on behalf of all of the Muckaty traditional owners. Following a petition by five groups of traditional owners, the government proposal became the subject of federal court action, and the matter was heard in 2013. In the meantime, popular opposition to the

Muckaty waste repository proposal continued to build, steeled by the successful campaign by the Kupi Piti Kungka Tjuta to check the idea of a nuclear dump being established in South Australia, and by strengthening union support, with leading Northern Territory unions, notably the Maritime Union of Australia, declaring their intention to stop the project from proceeding. Confronted by this concerted campaign and facing possible defeat in the court, the government announced on June 19, 2014 that it would not proceed with the proposal.

While environmental movement campaigns against uranium mining have been buoyed by engagement with traditional custodians of the land, differences of opinion within Indigenous communities on the economic benefits, or otherwise, of mining or waste dumps could deliver to these communities have highlighted the divisive impact of uranium mining. Several national-level Indigenous councils, frustrated with the impoverishment of Aboriginal communities in remote Australia, regard the development of uranium mining projects or the establishment of waste sites as bringing considerable economic advantage to these communities. The Central Lands Council, an Indigenous representative body at the regional level, for instance, supported the Angela Palmer uranium mine near Alice Springs in central Australia, setting aside the opposition of local traditional landholders to the project. Likewise, the Western Desert Land Aboriginal Corporation has supported Cameco's Kintyre Rocks project in Western Australia in the face of the reticence of the local traditional custodians. The Northern Land Council has endorsed to the Muckaty waste site proposal in opposition to the local people.

Governments and mining companies have been quite effective in selling the message that Indigenous endorsement of mining will provide remote Aboriginal communities with a sustained income flow and a potential source of employment. Where they have not been successful in persuading Indigenous landholders, they have had few qualms in riding roughshod over local opposition. One example was the West Australian government's refusal to meet Indigenous elders' demands for a more thorough environmental and cultural impact assessment of the Yeelirrie uranium mine project. Notwithstanding these defeats, there has been one successful campaign to stop any development of a uranium deposit at the Koongarra deposit, with the site being incorporated into the Kakadu National Park in 2011 in line with the wishes of the traditional land custodians.[63]

Such successes are limited and localized, but are remarkable given that the full force of the state and the corporate sector has been mobilized in support of uranium. As such, the successes are highly symbolic and have revitalized opposition to uranium mining. The Muckaty campaign in particular has proved a catalyst in bringing fresh impetus to the movement. By linking the dangers of embracing the uranium cycle with the assault on Indigenous rights, the campaign has steeled a new determination to campaign against the thrust of the Labor government's policies. One aspect of this has been a strengthening of coalition building across different social movements and faith groups in order to bolster local opposition to the waste dump. A new grouping, the Choose Nuclear Free project, for instance, was formed in 2011, bringing twenty-seven nongovernment organizations together, including leading environmental and public health and State conservation councils. Member groups have declared their determination to maintain the fight against uranium mining and, working with the Muckaty campaign, block the establishment of radioactive waste facilities.[64] In the process, concerns with securing and maintaining the cultural integrity of Indigenous communities, of defending connection with Indigenous lands, have become as important as the defense of ecological integrity and environmental and safety issues in challenging the notion that uranium is a commodity like any other.

This campaign to block the establishment of a nuclear waste storage facility has proved important in other respects. It has exposed the dangers of uranium mining and of the long-term consequences of the government's support for mining and its decision to accept the import of nuclear waste as part of the ambition of selling Australia's comparative advantage in the international trade in uranium. This commitment to the storage and sequestration of nuclear waste, as another stage in the uranium commodity chain, refocused the attention of the environmental movement, trade unions, and civil society organizations on ending Australia's involvement in this international trade. The many groups that have endorsed the New South Wales Uranium Free Charter, for instance, are set on working toward stopping uranium mining, to keeping uranium in the ground, and, in effect, to abolishing the commodification of uranium.[65] Government efforts to bolster uranium sales by welcoming waste to Australian shores had clearly started to backfire.

## The Movement against Uranium Mining and the Force of the Market

The ALP's 2007 decision to renege on its 1977 moratorium on new uranium mines, now backed by Labor parties in all State jurisdictions and the Northern Territory, has led to a significant expansion in uranium mining in Australia. The Australian government has begun to openly compete with other uranium-exporting countries by abandoning the NPT and offering the prospect of establishing a new import trade for nuclear waste, destined for Indigenous communities in central Australia.

Paradoxically, as the movement against uranium mining struggles to influence government policy, the uncertainties in the global economy could prove a more powerful obstacle to the industry's future expansion. In August 2013, BHP Billiton announced the postponement of the Olympic Dam project. BHP Billiton also disposed of its Yeeliree mine, which it sold to the Canadian company Comeco. The company blamed escalating costs in getting projects off the ground and softening uranium demand and prices. It has also become apparent that the company was overly optimistic in assuming that the leaching technology that was being tested, as a substitute for the more expensive smelting process, could be scaled up to the required production level. There was no guarantee that the technology would be available in the immediate future, let alone what the cost would be.[66] Nor is the postponement of the Olympic Dam project a unique event. There are, in fact, a number of projects that have been put on hold, including Comeco's Yeeliree mine.

However, while growing uncertainty in the global uranium market may prove the undoing of the Olympic Dam expansion project, it would be wrong to simply attribute the decision not to proceed to erroneous accounting forecasts. Indeed, the softening uranium demand and prices reflect mounting global concerns about the risks associated with nuclear power. The Fukushima meltdown has been a critical factor in this. As well, reports of leaks and mishaps at other nuclear installations in Europe and the United States have underwritten concerns about the risks of nuclear power and strengthened the political voice of the antinuclear movement. Global campaigns to halt the engagement with nuclear energy have in fact resonated on international commodity markets.

Despite these international developments, the Australian government has remained resolutely committed to the expansion of uranium mining. The Labor government's resources minister continued to champion uranium as part of the resources export mix and even as a source of energy within Australia.[67] When BHP Billiton extended its Olympic Dam indenture agreement through to October 2016, its chief executive officer, Marius Kloppers, was reported as saying that the Federal and South Australian governments had been "fabulous" in backing the project, presumably in the hope that global demand and market prices for uranium will rebound.[68]

Notwithstanding government support, uranium and the nuclear industry remain high risk. In 2007, Michael Angwin, the director of the Australian Uranium Association, delivered a triumphalist speech to the World Uranium Association declaring an upcoming nuclear renaissance in Australia with the lifting of the ALP moratorium, claiming public support for uranium in every State and Territory. BHP Billiton's decision to postpone development of uranium mining at Olympic Dam because of the collapse in commodity prices suggests that the promise of the nuclear renaissance might well be wishful thinking. The Fukushima meltdown and the concerns in Japan that another nuclear plant is sitting on a fault line and will probably have to be shut down, along with the regularity of accidents reported at nuclear power stations throughout the world, have raised serious questions. The viability of nuclear power as an alternative to fossil fuels is now put seriously in question, and confidence in a nuclear future is severely shaken. Even if the federal government is able to proceed with the Muckaty waste disposal site, which now looks increasingly unlikely, Muckaty would be far from an adequate solution to the problem of nuclear waste. The withdrawal of federal funding to develop waste storage facilities in the United States indicates that the problem of safe and effective waste management is virtually, if not actually, insoluble.

## Conclusion

Campaigns against uranium mining in Australia successfully revalorized uranium, defining it as a hazard rather than as an asset. The result was that for several decades, some of the world's most strategic and accessible deposits of uranium were not mined. The lessons for campaigns to leave fossil fuel in the ground are manifold. First, it is clear that a national-level legislated moratorium can be highly effective even if it allows the expansion of existing mines. Better still, and certainly necessary, would be a legislated phase-out of fossil fuel industries. Second, it is worth emphasizing that the uranium moratorium in Australia arose from a nationally organized anti-uranium campaign that was strongly linked with Indigenous peoples, environmentalists, the labor movement, and key elements in the Labor Party. Third, what linked these groups was a shared rejection of mining and exporting uranium principally due to its cultural, environmental, health, and safety impacts, impacts that were directly felt by local Indigenous peoples and other communities and by workers in the industry. These segments of the population have to varying degrees provided the foundation for a continuing campaign against uranium in Australia. Fourth, beyond those immediately affected by the industry, the campaigns were able to draw in constituencies concerned about the broader implications of nuclear power, nuclear weaponry, and nuclear waste.

We can find similar themes in the contemporary climate change movement. The movement requires a strategic vision that can engage legislative power to produce structural transformation. Lacking the capacity to start stopping across all fossil fuels and in all contexts, rather than to stop here only to expand elsewhere, is critical. Alliance building in a campaign to halt mining and extracting, rather than simply to mitigate impacts, is also critical: as with uranium, the first problem is how to halt extraction, and this has to be the condition for any alliance building. Indeed, it is clear that case-by-case contestation of exploration and mining, with action enacted on the ground to draw together people most directly affected by the fossil fuel sector, has to be a high priority. Indeed, in many ways, the emergent climate justice movement has shifted away from abstract debates about climate policy to focus more on these material contexts in which fossil fuels are mined and burned.[69] The corollary of this shift to the material, though, is the requirement to embed such local struggles in the broader questions of climate stability and endless expansion, as it is only in the context of this broader frame that the local issues gain salience and traction. Without the wider context, the movement may become sidetracked into a series of fragmented efforts at self-protection, not-in-my-backyard efforts that miss the wood for the trees.

Finally, it is worth reflecting on the importance of addressing climate change across all energy sources. Nuclear energy has undergone something of a global renaissance as a clean source of energy. Climate change becomes an opportunity for the nuclear industry as coal, oil, and gas-fired electricity are devalorized. The obvious options, renewable energy and reduced use and greater efficiency, are left unexplored. The vested interests that drive the nuclear sector reconfigure our energy future in their interests. Yet the inherent risks associated with uranium continue to haunt the sector. Despite claims to the contrary from industry and government, what is remarkable in recent years in Australia is the story of strengthening public opposition to uranium exports, rising from a low of about 25 percent in 1982 to about 39 percent by 2007.[70] More recently the so-called nuclear renaissance has been dramatically truncated. Here the combination of public outrage and government-led phase-outs (e.g., in Japan and Germany) have dramatically undermined investor certainty in the sector, forcing a retreat. Again, there are parallels with the fossil fuel sector, where the rush to gas as a relatively low-emissions fuel has dramatically unraveled in the face of large-scale environmental impacts. While the Fukushima incident reverberated globally, dealing a blow to the nuclear renaissance, abrupt weather events associated with climate change are having a similar impact, allowing fossil fuels to be framed as inherently dirty and dangerous. The revalorization process combines the direct experience of degradation with a broader crisis of confidence in the commodity. The challenge for keeping them in the ground is to secure this revalorization of fossil fuels as a global hazard. The experience of leaving uranium in the ground demonstrates that the end game for coal/oil and gas is feasible as well as necessary. It would be a legacy of the planet's energy and carbon cycle that these substances are left undisturbed.

**Notes**

1. International Energy Agency, *World Energy Statistics* (Paris: IEA, 2012).

2. Ibid.

3. Ibid.

4. OECD, *Factbook 2011: Environmental and Social Statistics* (Paris: OECD, 2011).

5. Carlos Sorentino, "Uranium Mining Policy in Australia: One Step Forward and Two Steps Backwards," *Resources Policy* 16, no. 1 (1990): 3–21; Andrew Ferguson and Peter Lam, "U308: Assessing the Wealth Effects of the Three Mines Policy" (Presented at the 32nd Annual Conference of the Accounting and Finance Association of Australia and New Zealand, 2007), http://www.afaanz.org.

6. Michael Angwin, "The Policy Challenges for Australia's Uranium Industry" (London: World Nuclear Association, 32nd Annual Symposium, 2007): 30, http://www.world-nuclear.org.

7. Brian Martin, "The Australian Anti-Uranium Movement," *Alternatives: Perspectives on Society and the Environment* 10, no. 4 (1982): 26–35.

8. Gavin M. Mudd and Mark Diesendorf, "Sustainability of Uranium Mining and Milling: Towards Quantifying Resources and Eco-Efficiency," *Environmental Science and Technology* 42 (2008): 2624–30.

9. Martin, "The Australian Anti-Uranium Movement."

10. R. A. Panter, "Uranium Policies of the ALP, 1950–1990" (Department of the Parliamentary Library, research service background paper, 1991).

11. Sigrid McCausland, "Leave It in the Ground: The Anti-Uranium Movement in Australia 1975–82" (PhD diss., University of Technology Sydney, 1999).

12. Les Dalton, "The Fox Inquiry: Public Policy Making in Open Forum," *Labour History* 90 (2006): 137–54.

13. James Camilleri, "Nuclear Disarmament, an Emerging Issue in Australian Politics" (Australian Studies Centre, University of London, Working Paper 9, 1986), 40.

14. Les Cupper and June Hearn, "Unions and the Environment: Recent Experience," *Industrial Relations* 20 (1981): 221–31.

15. See Don Chipp, former coalition government minister, cited in Ashley Lavelle, "'Conflicts of Loyalty': The Australian Labor Party and Uranium Policy 1976–82," *Labour History* 102 (2012): 177–96.

16. McCausland, "Leave It in the Ground," 402.

17. Sorentino, "Uranium Mining Policy in Australia."

18. Camilleri, "Nuclear Disarmament."

19. Russell Fox, G. Kelleher, and C. Kerr, *Ranger Environmental Inquiry First Report* (Canberra: Commonwealth of Australia, 1976).

20. Ibid., 183.

21. Dalton, "The Fox Inquiry."

22. Lavelle, "Conflicts of Loyalty."

23. Mary Elliott, *Ground for Concern: Australia's Uranium and Human Survival* (London: Penguin and Friends of the Earth Australia, 1977); Denis Hayes, Jim Falk, and Neil Barrett, *Redlight for Yellowcake: The Case against Uranium Mining* (Melbourne: Friends of the Earth Australia, 1977).

24. James Camilleri, "Nuclear Controversy in Australia: The Uranium Campaign," *Bulletin of Atomic Scientists* 35, no. 4 (1979): 40–44.

25. McCausland, "Leave It in the Ground."

26. Ibid.

27. Verity Burgmann, *Power and Protest: Movements for Change in Australian Society* (St. Leonards, NSW: Allen and Unwin, 1993).

28. Panter, "Uranium Policies of the ALP."

29. Martin, "The Australian Anti-Uranium Movement"; Camilleri "Nuclear Controversy in Australia."

30. Elliott, *Ground for Concern,* 4.

31. Lavelle, "Conflicts of Loyalty."

32. Jim Falk, *Global Fission: The Battle over Nuclear Power* (Melbourne: Oxford University Press, 1982); Richie Howitt and John Douglas, *Aborigines and Mining Companies in Northern Australia* (Sydney: Hale and Iremonger, 1983).

33. Panter, "Uranium Policies of the ALP."

34. Fred Halliday, *The Making of the Second Cold War* (London: Verso, 1986).

35. McCausland, "Leave It in the Ground."

36. Camilleri, "Nuclear Disarmament," 3.

37. Independent Committee, *Australia and the Nuclear Choice: Report of the Independent Committee of Inquiry into Nuclear Weapons and other Consequences of Australian Uranium Mining* (Sydney: Total Environment Centre, 1984).

38. Bill Hayden, "Uranium, the Joint Facilities, Disarmament and Peace" (Australian Government, Canberra: Publishing Service, 1984), 3.

39. Damian Grenfell, "Environmentalism, State Power and National Interest," in *Protest and Globalisation: Prospects for Transnational Solidarity*, ed. James Goodman (Vancouver: Pluto Press and Fernwood Publishing, 2002).

40. Ferguson and Lam, "U308."

41. Jim Falk, Jim Green, and Gavin Mudd, "Australia, Uranium and Nuclear Power," *International Journal of Environmental Studies* 63 (2006): 845–57.

42. Greg Baker, "Australia's Uranium" (Department of Parliamentary Services, Research Paper 6, 2009).

43. James Goodman, "Leave It in the Ground! Ecosocial Alliances for Sustainability," in *Nature's Revenge: Reclaiming Sustainability in an Age of Corporate Globalism*, ed. Josée Johnston, Mike Gismondi, and James Goodman (Peterborough, ON: Broadview Press, 2006), 155.

44. Guy Pearse, "Land of the Long Black Cloud," *The Monthly* (September 2010), 20-25.

45. Peter Costello, "Resources are the issue," *Australian Financial Review,* November 16, 2006, 62.

46. Commonwealth of Australia, *Australia's Uranium—Greenhouse Friendly Fuel for an Energy Hungry World* (Canberra: House of Representatives Standing Committee on Industry and Resources, 2006), http://www.aph.gov.au/house/committee/isr/uranium/report/fullreport.pdf Mark Diesendorf, *Greenhouse Solutions with Sustainable Energy* (Kensington, NSW: University of New South Wales Press, 2007); Ian Lowe, *Reaction Time: Climate Change and the Nuclear Option* (Melbourne: Black Ink, 2007); Leslie Kemeny, "The Right Road to Clean Energy," *Australian Financial Review*, June 27, 2012, 55.

47. Jim Falk and Domenica Settle, "Australia: Approaching an Energy Crossroads," *Energy Policy* 39 (2011): 6804–13.

48. Katherine Murphy and Misha Schubert, "Government rejects nuclear power," *The Age*, 5 April, 2006: 12. . The machinery to bypass obligations under the Nuclear Non-Proliferation Treaty to support an arrangement to export uranium to Taiwan had been negotiated in 2001, according to an exchange on National Interest Notes between the United States and Australian governments, http://www.austlii.edu.au/au/other/dfat/nia/2001/28.html.

49. The government approved BHP Billiton's 2007 moves to negotiate the export of uranium to China, despite concerns expressed by the Australian Conservation Foundation that China was not meeting its Nuclear Non-Proliferation Treaty obligations. Australian Conservation Foundation, "BHP to Ship Uranium as China Celebrates One Party State" (October 1, 2009).

50. Richard Leaver, "The Economic Potential of Uranium Mining for Australia," in *Australia's Uranium Trade: The Domestic and Foreign Policy Challenges of a Contentious Export*, ed. Michael Clarke, Stephan Frühling, and Andrew O'Neil (Farnham, Surrey: Ashgate, 2011), 87.

51. Matthew Franklin, "Uranium to China within Months," *The Australian*, January 6–7, 2007.

52. Michael Clarke, "The Third Wave of the Uranium Export Debate: Towards the Fracturing of Australia's 'Grand Bargain,'" in *Australia's Uranium Trade: The Domestic and Foreign Policy Challenges of a Contentious Export*, ed. Michael Clarke (Surrey: Ashgate, 2011), 109; Rory Medcalf, "Powering Major Powers: Understanding Australian Uranium Export Decisions on China, Russia and India," in Clarke et al., *Australia's Uranium Trade*, 67; Ben Doherty, "Gillard Ready to Discuss Indian Nuclear Exports Despite Fears over Safety," *Sydney Morning Herald*, October 15, 2012, 6.

53. Clarke, "The Third Wave."

54. Diesendorf, *Greenhouse Solutions*.

55. Greg Sheridan, "Mid-East Uranium Sales on Cards," *The Australian*, August 2, 2012, 8.

56. The expansion plans would make Olympic Dam the world's largest uranium mine, as well as increase copper output almost fourfold to 750,000 tonnes per year and boost gold production eightfold and uranium production almost fivefold.

57. Friends of the Earth Australia, "Submission to Joint Standing Committee on Treaties: Inquiry into Nuclear Non-proliferation and Disarmament" (Melbourne: Friends of the Earth, 2009), http://www.foe.org.au/anti-nuclear/issues/oz/u/isl.

58. The Roxby Dam operates under the Roxby Downs Indenture Act (1982), which provides for wide-ranging exemptions from several South Australian laws, including the Aboriginal Heritage Act, the Freedom of Information Act, and the Natural Resources Management Act. The Indenture Act was amended to 2011 to enable the expansion of the mine. Australian Conservation Foundation, News and Media: "Olympic Dam Economics: Do the Benefits Outweigh the Costs?" August 6, 2009.

59. For instance, the safety record of the operators of the Honeymoon mine in South Australia, Heathcote Resources, a subsidiary of General Atomics, has been highlighted with reports of a number spills at the General Atomics plant in Oklahoma; Jim Green, "General Atomics, Heathgate and the Beverley Uranium Mine," *Punch*, August 2, 2012, http://www.foe.org.au/chain-reaction/edition/116/heathgate.

60. Burgmann, *Power, Profit and Protest*.

61. The Sleepy Lizard Revenge campaign takes its lead from the Arabunna dreamtime story of Kalta, a sleeping lizard that "has in its belly yellow poison … [and] should never have been woken," http://www.helencaldicott.com/2012/07/the-lizards-revenge.

62. WGAR News, "Court Throws Out Olympic Dam Challenge," April 20, 2012, http://indymedia.org.au/2012/04/22/wgar-news-court-throws-out-olympic-dam -challenge-abc-pm.

63. Commonwealth of Australia, Completion of Kakadu National Park (Koongarra Project Area Repeal) Act 2013—C2013A00031 (Canberra: 2012), http:// www.comlaw.gov.au/Details/C2013A00031.

64. See Beyond Nuclear Initiative, http://beyondnuclearinitiative.com/tag/ muckaty.

65. See Uranium Free NSW, http://www.uraniumfreensw.org.au.

66. Sarah Martin and Michael Owen, "Olympic Dam May Never Proceed," *Weekend Australian*, August 25–26, 2012, 4. However, BHP Billiton has since announced that it intends to experiment with a new processing technique—heap leaching—using chemicals to separate minerals from waste ores. Peter Ker, "BHP in fresh bid to unlock Olympic Dam," *The Sydney Morning Herald* July 28, 2014, 1.

67. Leonore Taylor, "Study Rates Nuclear a Cheap Source of Energy," *Sydney Morning Herald*, August 1, 2012, 5.

68. Mathew Dunckley and Ayesha de Kretser, "BHP Recommits to Olympic Project," *Australian Financial Review*, November 14, 2012, 7; Peter Ker, "BHP Gets Four More Years to Expand Olympic Dam," *Sydney Morning Herald Business Day*, November 14, 2012, 5; Sarah Martin, "Olympic Threat as BHP Puts Brakes On," *Weekend Australian*, July 28–29, 2012, 1, 6.

69. Stuart Rosewarne, James Goodman, and Rebecca Pearse, *Climate Action Upsurge: The Ethnography of the Climate Movement* (London: Routledge, 2013).

70. Clarke, "The Third Wave," 141.

# 9

## The Future Would Have to Give Way to the Past: Germany and the Coal Dilemma

Tom Morton

Germany is in the forefront of Europe's ecological modernization move, the technological transformation of industrial society toward cleaner and greener energy and a sustainable economy. Yet even as it does, plans are underway in the heart of former Communist East Germany, to expand the mining of brown coal, the most polluting and inefficient as well as the cheapest and most available form of coal. Here Tom Morton explores Germany's coal dilemma through the words and struggles of farmers and villagers whose lands and homes are threatened once again by a coal juggernaut, a juggernaut many Germans thought would have been long ago abandoned, the dirty legacy of the Communist era with millions of tons of brown coal safely left in the ground.

The dilemma derives in part from Germany's decision to abandon nuclear power in the wake of the Fukushima nuclear disaster in Japan, itself a step toward leaving uranium in the ground. But that policy, along with a commitment to renewable energy, has put even more pressure on Germany's energy supplies to maintain its status as a leading industrial country. Fearing energy shortages with serious economic consequences, powerful interests are now promoting brown coal as a transition fuel; it is, after all, domestically abundant, cheap, and available, they say. What they do not say, though, is that entire villages will have to give way, along with people's livelihoods. Nearly powerless in the larger scheme of national and international energy politics, these villagers are speaking out, calling not just for compensation but claiming the entire project of burning more coal is not legitimate.

Morton captures in his chapter title these peoples' struggle and, for that matter, Germany's dilemma as it tries to transition out of both nuclear power and fossil fuels. To use coal, the future would have to give way to the past, a notion that suggests that Germany's much-heralded future of a society powered by alternative energy would be giving way to a checkered past of dirty coal. What's more, communities of the former East Germany would once again be sacrificed against their will for a powerful system's vision of power from coal—in this case, ironically, to exit fossil fuels. What's more, it is a vision that insists on energy source replacement, unable to entertain the possibility of reducing overall energy use, which in this book we have argued will occur one way or another anyway. As with the other cases in part 2, the outcome is contested as this book is published.

It's actually an environmental crime. When I say an environmental crime, I mean the 120,000 hectares of land in Lusatia, which are covered in signs saying, "Trespassers prosecuted," because the land belongs to the coalmines. You can't go on the land and you can't use it for anything. It's a catastrophe, and so is everything that goes with it. Brown coal is uncontrollable, and what it has done to this region is immense.
—Günter Jurischka, small-business owner and activist, Proschim

On the very piece of ground where they want to dig the new pit (Welzow Part 2) there are solar panels installed right now. They would have to give way to the mine. So you see, the people in Proschim are very open to renewable energy. But if the new mine goes ahead, renewable energy would have to give way to coal. The future would have to give way to the past. We'd be making the energy transition go backwards.
—Petra Rösch, local magistrate, farmer, managing director of Proschim Agricultural Enterprises .[1]

In May 2013, I visited the village of Proschim in Lusatia, a region of eastern Germany not far from the Polish border. Proschim appears to be a small, sleepy rural village, surrounded by farmland and forest; the nearby countryside was a patchwork of brilliant yellow fields of rapeseed, lush green pasture, the white blossom of plum and cherry trees, and woodlands bearing the leaves of a late spring.

Proschim and its inhabitants are facing a precarious future. One of the largest energy companies in Europe, the state-owned Swedish energy concern Vattenfall, wants to expand its coal-mining operations in Lusatia. If Vattenfall's plans go ahead, Proschim will be wiped from the face of the earth. Its population, and those of the neighboring villages of Welzow and Lindenfeld, around 800 people in total, will be uprooted to make way for an open-cut brown coal mine.

The proposed mine is an extension of an existing open-cut pit to the north and east of Proschim, known as Welzow-Süd. Günter Jurischka, who lives in Proschim, drove me to a vantage point close to the mine entrance, at its lower southwest corner. The mine is already massive; it covers approximately 11,000 hectares and produces 20 million tons of brown coal every year.[2] If approved, the extension, Welzow Part 2 (Welzow Teilfeld II), will allow the company to expand the mine by a further 1,900 hectares and mine an additional 200 million tons of brown coal over the next twenty-five years.

Jurischka and I looked out over a uniform gray-brown landscape of craters and ridges stretching north all the way to the horizon and east to the coal-fired power plant Schwarze Pumpe ("Black Pump"), which the mine supplies. I had seen a similar landscape more than twenty years before. In March 1990, just a few months after the fall of the Berlin Wall, I toured the coal-mining areas of Saxony, close to the East German city of Leipzig, as a radio reporter for the Australian Broadcasting Corporation. At the time, I was shocked by the sheer scale of the environmental devastation I saw: huge swathes of the countryside had been gouged and torn by many years of open-cut coal mining.

Since that first encounter, I have visited the former East Germany a number of times as a journalist, documentary producer, and now as an academic researcher who still practices as a journalist. Over two decades, I have followed how the people of the former German Democratic Republic (GDR) have adapted to the challenges of reunification. I knew that many of the coal mines I had seen around Leipzig in 1990 had been closed and an extensive program of rehabilitation carried out, funded by the German federal government. I was all the more shocked, then, to discover that new open-cut brown coal mines were being proposed in Lusatia, and I became interested in the struggle of local people to stop the mines from going ahead.

A loose alliance of individual activists and local organizations has been fighting the expansion of Welzow-Süd for six years—among them Günter Jurischka and Petra Rösch, the two individuals quoted at the start of this chapter. Welzow-Süd, however, is not the only proposed new mine in the region. Vattenfall is seeking to open five new open-cut brown coal mines in Lusatia. If all five mines go ahead, six villages—Proschim, Lindenfeld, Welzow, Kerkwitz, Grabko, and Atterwasch—would be demolished or rendered uninhabitable and around 1,500 residents relocated. According to Matthias Berndt, a Protestant pastor in the village of Atterwasch, uncertainty about the future of their homes, their farms, their villages, and their familiar surroundings has led to a "great depression," particularly among the older residents of the affected villages: "People are threatened with being cut out by the roots. The uncertainty is making them sick. ... They feel that the value of their lives is being called into question by the government and the company. They've worked their whole lives long for their farms and their families, and all of a sudden all that could be swept away."

The people of Proschim, Lindenfeld, Welzow, Kerkwitz, Grabko, and Atterwasch are fighting first and foremost to save their homes and villages from destruction. However, environmental groups in the region, such as the Green League Network (Grüne Liga Netzwerk ökölogischer Bewegungen), also point to climate change, air quality, and impacts on local water resources as important concerns for those working in campaigns to keep brown coal in the ground. These groups stress their historical roots in environmental activist groups in the GDR and argue that they have consistently opposed coal mining and coal-fired power stations since reunification.[3] These local environmental groups are now gaining support at a national level from organizations such as Greenpeace and Bund (Friends of the Earth Germany). The German Climate Alliance has designated Lusatia as a focus of campaigning against brown coal.[4]

The local struggle in Lusatia over coal and the future of coal mining reflects a much larger struggle over the future of energy policy and climate action in Germany. The German government has taken bold policy action through a range of laws known as the German Energiewende (energy transformation) to shut down its nuclear power stations and rapidly expand the use of renewable energy. At the same time, however, Germany faces an extremely challenging policy dilemma in carrying through this energy transformation without becoming more, rather than less, dependent on coal-fired power in the short term.

Germany is already the world's largest producer of soft brown coal, accounting for 17.2 percent of total global production. Brown coal already accounts for around a quarter of electricity generation, and its share has actually increased since the beginning of the energy transformation.[5] It is no exaggeration to say that what happens in Lusatia—whether the new coal mines go ahead or are stopped by their opponents—will have an important bearing on the future of coal mining in Germany, and campaigns within Germany itself and Europe more broadly to leave coal in the ground. Thus, Lusatia is a particularly rich and complex local context within which to analyze the micropolitics of the coal dilemma and the motivations and commitments of individuals and social actors who are fighting to leave coal in the ground.

Lusatia is the second-largest brown coal mining region in Germany. According to company figures released in March 2013, in 2012 Vattenfall produced 62.6 million tons of brown coal from its open-cut pits in Lusatia—3 million tons more than in 2011. Three coal-fired power stations in Lusatia operated by Vattenfall, together with a plant half-owned by Vattenfall in Saxony, and its gas turbine power plants, produced 58 billion kilowatt-hours of electricity—enough to provide power to around 16 million households. Overall, Vattenfall's coal-fired power stations produced almost one-tenth of every kilowatt-hour of electricity used in Germany.[6]

There is also a long history of forced relocations in Lusatia. Since the late 1960s, thirteen villages have been partially or totally demolished and around 3,500 people relocated to make way for the expansion of the Welzow-Süd mine alone.[7] Many of these relocations occurred under the former Communist regime in the GDR, better known as East Germany, a historical legacy that continues to play an important part in shaping the political consciousness and local identity of people in Lusatia.

## Analyzing the Coal Dilemma: A Transnational and Transdisciplinary Approach

This chapter is based on interviews gathered in Lusatia in March and May 2013. Gülseren Ölcum, a German freelance journalist, and I interviewed a number of local activists and residents of the affected villages, as well as the head of the works council (Betriebsrat) at the Jänschwalde mine, and a representative of Vattenfall. These interviews represent the initial phase of a longer-term research project, in the course of which we will follow developments in Lusatia over the next three years and document and analyze their impact on individuals and communities in the region.

Our project starts from two basic premises. The first is that ending reliance on coal is not simply a technological question; it is also a sociopolitical question. As Ortwin Renn has noted, "A better understanding of the human drivers for initiating, promoting, or hindering political change [in the arena of climate action] is as crucial to effective decision-making as are the findings of the natural and climate sciences."[8] It is crucial to our project to investigate how minds are changed: how the contestation of coal arises at a local level, how material and social conditions inform opposition, and how local conflicts over coal are reflected in national (and international) public debates and their impact on decision makers.

Our second premise is that individual countries, their coal and energy industries, and the localities where coal is mined and consumed are deeply enmeshed in a global energy economy. In order to understand the interaction of the local, national, and transnational dimensions of the current "coal rush," our project will compare the contestation of coal in three locations—Lusatia in Germany, the Hunter Valley in Australia, and Andhra Pradesh in India—drawing on methodologies and perspectives from political economy, environmental sociology, documentary production, and ethnographic practice.[9] (These comparisons will be explored in forthcoming publications, but are not addressed here.)

Before I explore the material from the initial interviews in Lusatia in more detail, a brief biographical and historical sketch will provide some further context.

### Coal in the GDR: A Brief (Geo-)Political Biography

The impact of coal mining on the people and landscape of the former GDR has made a deep impression on me since my first visit as a radio reporter in 1990. I began that visit by covering the first (and last) free elections in the GDR. At that point, the GDR was still officially a separate country but would not remain so for long. The elections were a landslide for the Alliance for Germany, a coalition dominated by the Christian Democrats (CDU), and paved the way for German reunification later that year. After the elections, I spent three weeks traveling around the GDR, gathering material for a radio documentary about how people in the East were experiencing the dramatic changes in their country. In the course of my travels, I visited Leipzig, a major industrial center in the GDR, and met a young freelance journalist, Peter Scholz, who offered to take me on a tour of the coal mines outside the city.

After approximately a thirty-minute drive outside Leipzig, we entered a landscape of huge craters and slag heaps, over which diggers crawled like massive predatory insects. Especially striking was the absolute destruction of nature: I saw only a trace of plants, trees, or any other living thing—apart from the miners themselves. I was shocked by the sheer scale and extent of the mining. Scholz spoke to me passionately about the dust storms, which often swept through the region, and the high levels of respiratory disease in the population, which he attributed partly to coal dust and partly to the smoke and smog from the coal-fired power stations surrounding the city. Such claims were, of course, based entirely on anecdotal evidence; the GDR authorities published no official epidemiological data or official statistics on air quality, and until the very last months before the Communist regime collapsed, they refused to acknowledge that pollution, contamination, or other environmental problems existed.

During the 1980s, however, researchers and environmental groups in West Germany (the Federal Republic of Germany) had become well aware of the environmental problems in the GDR. A report published in 1988 described the GDR as "one of the worst air polluters in Europe," pointing specifically to the 4 million to 5 million tons of sulfur dioxide emitted there every year by brown coal–fired power stations.[10] Evidence gathered before and after the collapse of the Communist regime in 1989 shows that sulfur dioxide levels in the air in the GDR were the highest in Europe and that levels of airborne particulate matter (dust and ash) were especially high in the regions around Leipzig, Halle, and Cottbus, the center of the brown coal mining district in Lusatia.[11]

Coal mining had been an important economic activity in Saxony since the middle of the nineteenth century. Its importance as an energy source in the GDR declined briefly in the 1960s, when the GDR began to import massive quantities of cheap oil from the Soviet Union. In the 1970s, however, the international oil crisis and high world oil prices led the Soviet Union to increase exports to Western Europe and increase prices for Soviet bloc countries such as the GDR. The SED regime (the SED was the Sozialistische Einheitspartei Deutschlands, the East German Communist Party) recognized that it would need to become less reliant on Soviet oil, and expanded brown coal mining. In 1981, the West German newspaper *Die Zeit* reported that the GDR was the world's largest producer of brown coal, with an annual yield of 256 million tons, or 28 percent of total world production.[12] In 1988, when production reached its peak, 310 million tons of brown coal were mined in the GDR. Hermann Wittig describes the consequences of this heavy reliance on brown coal as "an almost unimaginable destruction of the countryside and burdening of the environment; inefficiency in the generation of energy; wastefulness in the use of energy, encouraged by misguided subsidies; and a lack of measures to compensate [dislocated] populations and rehabilitate the landscape."[13]

Wastefulness in the use of energy was certainly clearly apparent when I visited the GDR in March 1990: although the weather was relatively mild, every factory, government building, and office I visited was overheated to such an extent that, in some at least, the temperatures were almost tropical.

In the years following reunification, many of the open-cut brown coal mines I saw on my tour around Leipzig were closed. According to Wittig, local populations were no longer willing to accept the social and environmental impacts of mining and burning brown coal. Brown coal production in the former East fell by 33.8 percent between 1989 and 1994 as a result of this shift in social attitudes and the arrival of the market economy.[14] A far-reaching program of ecological rehabilitation was undertaken, financed by the Federal government: abandoned coal pits were flooded, and a new, artificial countryside of man-made lakes, fields, and green belts replaced a landscape ravaged by coal mining.[15] I saw one example of this transformation in 2010 when I visited the eastern German city of Bitterfeld. Bitterfeld was once an important center for the GDR's chemical industry, and by the late 1980s it had become notorious as one of the most heavily polluted cities in the whole of Eastern Europe. Twenty years later, Bitterfeld was home to the world's largest manufacturer of solar cells. The open-cut brown coal mine adjacent to the city, known locally as *die Goitzsche*, filled with water from the River Mulde during the floods of 2002 and now has an artificial beach, a marina filled with sailing boats, and a modest ecotourism industry.

In light of these developments, it seems all the more ironic that a substantial expansion of brown coal mining is again being proposed in Lusatia. This irony is all the stronger when seen in the specific sociopolitical context of the former GDR. For ordinary citizens of the GDR and for the underground environmental groups that sprang up in the 1980s, the exploitation of the natural world and the exploitation of human beings in the "workers' and peasants' paradise" went hand in hand. The environmental devastation, which had been allowed to continue unchecked in their country over decades, was both a symptom and a symbol of the regime's complete lack of regard for the well-being of its citizens.

On my first visit to eastern Germany in March 1990, I interviewed Monika Maron, who had been a reporter for the East German newspaper *Wochenpost* and had visited Bitterfeld in the 1970s. In 1981, her first novel, *Flugasche* (Flying Ash), was published in West Germany. It was set in a city identified only as "B" and described in graphic detail the pervasive contamination of the environment by the chemical industry and the appalling conditions in which people lived and worked.[16] In my interview, I asked Maron to reflect on the circumstances that had led her to write the novel. She told me that her visit to Bitterfeld had been a turning point in her attitude to the East German socialist state; it had convinced her that the GDR was nothing other than a state of "modern slavery":[17]

When I wrote the book, I was thinking less about environmental issues and more about the impossibility of democracy in the GDR. I wrote a novel about a journalist who believes she has to do something, who gets involved in a situation, and who fails. I didn't formulate this in my head as an ecological problem. Rather, I thought to myself, there has to be something wrong when the process of production destroys human beings and destroys the basic conditions for a human existence.

Maron stresses that for her, the ecological devastation of the GDR was primarily a problem of democracy. As we shall see, more than twenty years later, this formulation has a particular resonance for the experiences of those opposing coal mining in the eastern German region of Lusatia. We shall now consider their struggle and its motivations in greater depth.

### Brown Coal Mining, Climate Change, and the Energy Transformation

Most of the interviewees do not argue directly for an immediate and total stop to coal mining in Lusatia or in Germany as a whole. The views of Petra Rösch are broadly representative. Rösch is fifty-eight years old and has lived in the village of Proschim, one of those slated for demolition, all her life. She describes herself as having very deep roots in the place. She is a local magistrate and general manager of Proschim Agricultural Enterprises, the parent company for a number of small businesses, including the local butcher. She is married, with two adult sons, one of whom works installing solar panels. Rösch sees the recent history of brown coal mining in their region as a history of broken promises:

In the old East Germany, brown coal was the only energy source, and the open-cut mine, Welzow South, swallowed up a whole lot of villages and forest and arable land. In 1991, after the fall of the Berlin Wall and reunification, we were told it wouldn't go any further. There would be an extension to Welzow South, Welzow South Part 1 (Welzow Süd Teilfeld 1), but after that no more new open-cut mining, and no more destruction of villages.

The existing Welzow South mine (including the extension that Rösch refers to) still contains approximately 360 million tons of coal. Rösch and most other interviewees argue that this will be enough to supply the local power plant *Schwarze Pumpe* (Black Pump) for another thirty years. They are unequivocal in their opposition to the proposed new mines or, indeed, any other expansion of coal mining in Lusatia:

Then all of a sudden, in 2007, they say, we've changed our minds, we want to dig up coal again. And now they want to start Welzow II, and mine until 2042, and destroy our villages and their natural surroundings, and we say, "That simply isn't compatible with the way things have developed here. And we strongly oppose it."

In 2007 Petra Rösch's company, Proschim Agricultural Enterprises, installed a biogas (methane) power plant at a cost of 1.7 million euros. The plant runs on dung and silage and can generate up to 536 kilowatts.[18] Large banks of solar cells occupy a number of surrounding fields. According to Rösch, these renewable energy sources between them can generate 1 megawatt at peak load, more than enough to supply the energy needs of the village. In Rösch's view, Proschim is a living embodiment of the energy transformation put into practice. Yet she does not explicitly describe herself as an environmental activist: "No, I am not an environmental activist. I am a farmer. I have a lot of time for nature and the environment because I grew up with them. For me agriculture and the environment are the same thing."

Rösch's son, Hagen Rösch, operates both the biogas and solar power plants. In a recent press article, he describes the brown coal industry as "like a dinosaur looking for its old habitat."[19] His mother says that her involvement in activism and opposition to new coal mines has been a gradual process and has come at a cost:

People have been very critical of me. You have to realize that as farmers with pasture right next to Welzow South Part 1, we are confronted with open-cut mining every day. We have to deal with it, and with the political situation, and we've been dealing with them ever since reunification. Then I became the local magistrate. I would say that I only started to become active and to speak out publicly quite late, because I'm responsible for nearly ninety people in our firm. I thought that as soon as I started to speak out, they'd make trouble for us, and that's what's happened.

Rösch's views are broadly similar to those of Erhard Lehmann, who worked for twenty-six years as a coal miner in the Welzow South mine. Lehmann was also born in Proschim and grew up in the early years of the GDR. He is a member of the conservative Christian Democrat Party. Now sixty-two, he describes how he saw the villages in the district disappear to make way for coal mining. He believed, however, that it would never happen to his own village:

In the local newspaper they're always describing us as anticoal. But what I want to say is that my colleague Jurischka, and Mrs. Rösch and myself, we're fighting for our homes and our property, and that's supposed to make us anticoal. I'm not anticoal in principle, because I worked in the mines for a long time, but you have to leave the villages and the people where they are, and if you're going to mine coal, then not the way they're doing it here. People say to me, we can't leave Proschim where it is, that would be uneconomic. In other words, don't reduce our profits. ... We don't seem to be able to mine coal sensibly and keep the landscape in order here in Welzow.

Kathi Gerster, official spokesperson for Vattenfall, believes that opponents of the mine extension are overstating the resistance to relocation among local people:

What is relocation? If we get approval for all the proposed new mines we will be relocating a total of 3,000 people. We will be doing so with contracts that will limit the social impact. If the 8,000 people who are directly employed here by Vattenfall, and the further 8,000 who are indirectly dependent on Vattenfall all lose their jobs, how much of a social impact will that have? ... There are some local residents who can't imagine living anywhere else, and there are some who come to us as soon as there's a whiff of relocation and ask if they can sign a contract straight away.

Many of the local residents, however, are concerned not only with their homes and villages but also with the natural world. Although Lehmann states that he is not anticoal in principle, he and his colleague Günter Jurischka, a former small business owner and fellow activist, are surprisingly blunt in their descriptions of what brown coal has done to the countryside:

That's actually an environmental crime. When I say an environmental crime, I mean the 120,000 hectares of land in Lusatia, which are covered in signs saying "Trespassers prosecuted," because the land belongs to the coal mines. You can't go on the land and you can't use it for anything. It's a catastrophe, and so is everything that goes with it. Brown coal is uncontrollable, and what it has done to this region is immense.

Lehmann also recognizes the contribution coal makes to climate change and is skeptical about the future of coal:

I see people on the television who are already suffering, who are already affected by coal and climate change. They haven't lived their lives the way we have, and they won't have to live like us, and they're content. Then they say that we're arrogant because we need brown coal. We don't need brown coal any more, at least not for electricity.

Like Lehmann, Petra Rösch stresses the wider impact of coal mining and its contribution to climate change: "I have to say that I think brown coal has a drastic effect on the environment. And it really doesn't matter where that is. This impact on the environment isn't making climate [change] any better."

A younger interviewee, Julia Albinos, is opposed to coal mining in general "because people elsewhere are affected as well, and Germany is a very densely populated country, and other people could be forced to move. When you see what it's like to be affected by coal mining, you don't want them to keep on doing it."

Albinos, twenty-three years old, is a former Miss Brandenburg who works in the payroll department at eBay in Berlin. She is from Atterwasch, another village threatened by demolition. Albinos is also more forthright than the older interviewees in stating that climate change is an important factor in her motivations for opposing coal mining: "Of course [I believe in] … the devastation it causes. The landscapes look like lunar landscapes. The water supply is shrinking. The summers are getting hotter all the time here. The ground is burning up."

Interestingly enough, both Lehmann and Rösch, and a number of the other interviewees, see the push to expand brown coal mining as out of step with the ecological modernization that has already taken place in the immediate vicinity of Proschim and in Lusatia more broadly. Rösch says:

The energy transformation has to be supported by deeds, and not just words, and we've done a lot in this place already. … You can see for yourself, the energy transformation is going ahead. Brandenburg could live from renewable energy. We're exporting energy.

We've installed a biogas plant here where our business is. We're an agricultural producer, and we've built up a good production chain. All the dung and animal waste, which comes out of our farms, is converted to fertilizer and methane and the biogas plant converts the methane into energy. That is a virtuous circle. We get good fertilizer and the smell becomes energy! We've installed solar panels on the roofs of all our buildings, which will generate up to 1 megawatt at peak. On the very piece of ground where they want to dig the new pit [Welzow Part 2], there are solar panels installed right now. They would have to give way to the mine.

So you see, the people in Proschim are very open to renewable energy. But if the new mine goes ahead, renewable energy would have to give way to coal. The future would have to give way to the past. We'd be making the energy transformation go backward.

Lehmann, the former coal miner, expresses similar views: "That's the crazy thing: we're supposed to give up clean energy for dirty energy. It's a disgrace. I just don't understand it."

Another interviewee who strongly supports the energy transformation and its aims is Rainhard Jung. Like Petra Rösch, Jung is a local farmer and secretary of the Alliance for Home and Future in Brandenburg.

He describes how the Alliance for Home and Future grew out of the Farmers' Federation in Brandenburg:

We started looking at the issues of CCS [carbon capture and storage] technology and new open-cut brown coal mines quite early on. We came to the conclusion that as farmers, we are fundamentally in favor of the energy transformation, the energy transformation that has been declared by our chancellor, Angela Merkel. In order for this energy transformation to succeed, the opposing forces have to be driven back. That's the decisive issue. What that means in concrete terms in Brandenburg is stopping any new brown coal mines. In order to achieve this, we founded the Alliance for Home and Future with many politicians who are sympathetic to us, including politicians from the Christian Democrats, the Greens, and the Left Party. The Landowners Association and a number of other socially oriented organizations are also on board. We are a broad spectrum of groups fighting to save the regions of Brandenburg, which are threatened by coal mining.

Although most of the interviewees hesitate to describe themselves as environmentalists or environmental activists (or, as Petra Rösch does, actively reject this description), both Günter Jurischka and Erhard Lehmann are involved with the environmental organization BUND (Friends of the Earth Germany). Jurischka and Lehmann say they are "strongly involved in working together with Greenpeace. Their 750,000 members worldwide mean that we have some financial clout behind us because Greenpeace is saying quite unambiguously, it can't go on like this. We're grateful for their advocacy on our behalf."

All of our interviewees identify two main obstacles to their campaign to stop the new coal mines in Lusatia. The first of these is the social and political legacy of the East German past. Rösch believes that Vattenfall's plans would have been nipped in the bud quickly if the proposed new coal mines were in Bavaria or Baden-Württemberg:

What's happening here couldn't happen there [in Bavaria or Baden-Württemberg]. We have a different history here. In the old East Germany, there was no "yes" or "no." It was a dictatorship. There was the Stasi, and if you stuck your head out, you were dealt with. People in this region, whether it's in Saxony or Brandenburg, they are still really shaped by that, especially the older people.

Rösch believes the citizens of the West German states Bavaria and Baden-Württemberg would be more likely to resist mining in their communities because their entire lives have been lived in a liberal democracy. Her fellow citizens in Brandenburg, by contrast, and especially the older generation, are still shaped by their experience of the authoritarian socialist state:

We've had a different system for more than twenty years, but still they [Vattenfall and the state government] know that if you tell people often enough, "There's no other way, this is how it will be, we're going to get our way," then people who've grown up in the old system say, "Oh well, someone else has always made decisions for us; I suppose they must know what's right."

In other words, Rösch echoes Monika Maron's view that the environmental devastation of both the natural and human worlds in the old East Germany are primarily a problem of democracy—one that continues into the present. However, she and her fellow activists identify a different kind of democratic deficit in the German federal system. This has its roots in what they perceive as the very close political relationship between Vattenfall and the ruling Social Democratic party in the state government of Brandenburg. Rainhard Jung of the Alliance for Home and Future describes this relationship as follows: "We have a state government which is a loyal vassal of the large Swedish concern Vattenfall, and which in the final analysis will do anything that Vattenfall asks of it."

According to Rösch, the influence of what she describes as "the brown coal lobby" is so strong that no politician in office can resist it, regardless of which party he or she belongs to:

I am dumbfounded and dismayed at what politicians say before an election and what they do afterward. Before the elections, a lot of promises were made, especially in the state of Brandenburg. The then premier, Manfred Stolpe (SPD), gave his word back in the 1990s that Horno was the last village that would be demolished to make way for brown coal. He was here in Proschim, and he said that Mr. Platzeck, who is our state premier now in Brandenburg, was a founding member of the Green League. Today he's a member of the SPD, and now he's totally in the pocket of the brown coal lobby. I have really been shaken by the way decisions have been taken over the heads of people in this region, but those decisions haven't been finalized yet, and we say, "Not with us; we won't let that happen here."[20]

This close relationship between Vattenfall and the political establishment is openly acknowledged by Kathi Gerster, press spokesperson for the company:

We have good contacts with politicians, especially in the state government of Brandenburg. The state premier, Platzeck, is a strong supporter of the coal industry, and he's said so again and again in his public appearances. Of course, there are some politicians from the region, maybe some of them are even Christian Democrats, if they're from the villages that are threatened, they might have a different opinion from the party leadership. But fundamentally, all the ruling parties have expressed their support for the coal industry and for a secure source of electricity generation, and they don't see what else could fill the gap.

At a local level, all the interviewees believe that Vattenfall has been very successful in dividing the local community through financial patronage and by co-opting or infiltrating local organizations, as Rösch describes: "They're playing off villages and village councils against each other. They've driven a wedge through the community, through the Association for Brown Coal in Lausitz and the Association for the Future of Lausitz." The latter two organizations are community organizations that support the proposed expansion of mining, and that Rösch believes are supported directly or indirectly by Vattenfall.

Matthias Berndt also believes that proposed new coal mines are creating deep divisions in communities in Lusatia. Berndt is the Protestant pastor in the village of Atterwasch, to the northeast of Proschim, very close to the Polish border. Atterwasch is one of three villages in the vicinity whose populations are facing relocation to make way for another open-cut brown coal mine, Jänschwalde North, which would supply coal to the Jänschwalde power plant. Both the new mine and the power plant are operated by Vattenfall. If Jänschwalde North goes ahead, Atterwasch and its church, which was consecrated in 1220 AD, will be demolished. Berndt is responsible for the pastoral care of congregations throughout the region who are affected by brown coal. He believes that uncertainty about the future is having a corrosive effect on the mental health of individuals and whole communities:

On the one hand, people are suffering from insomnia and depression, and on the other, they're angry that this is being done to them. This anger and aggression often gets directed against other people in the community, even though it should really be directed at the government and the coal industry.

Former coal miner Erhard Lehmann argues that many people in the community are reluctant to voice their opposition to the new mines because members of their own families work in the industry:

Some people [who would otherwise be protesting] keep quiet because their son is working for Vattenfall. Or the son is hoping to get an apprenticeship, and for that, the parents have to be prepared to keep their mouths shut and give up their land if the mine comes, or sell it. Lots of people say to us, "It's good that you're fighting, but we have to keep out of it, because it could affect our kids' prospects of getting a job."

Generational differences contribute to a lack of community cohesion in another way. Many young people have moved away from Lusatia because of a lack of job opportunities, a structural economic problem that affects large parts of the former East Germany. Julia Albinos is twenty-three years old, and originally from Atterwasch but is now working in Berlin. She says some members of her generation don't feel such a strong sense of belonging to their home villages as their parents: "I think there are young people who are saying, 'If my village disappears, well, that's not so bad, I want to leave anyway.' It's not as bad for them as it is for their parents who are still living there. ... The villages are really suffering; people are fighting more and more."

Albinos herself says she would feel a strong sense of personal loss if Atterwasch were to be demolished:

My home would be gone. The place where I grew up, where lots of things happened in my life, and where I'm always happy to go back to. It would make me very unhappy. I remember watching, more or less, when the people in Horno were relocated. I thought it was terrible, how the people from the surrounding villages would go and take the tiles from the roofs of the houses that were left. It was like they were just taking stuff from a junkyard. People had built something up over years, and all of a sudden it was gone. It was awful.

These initial interviews suggest a number of provisory conclusions and avenues for further research and analysis. Perhaps the most important of these is that although the interviewees do not necessarily describe themselves as anticoal in principle—in other words, they do not believe that coal mining should cease immediately and all coal should be left in the ground—the logic of their arguments appears to lead inexorably in this direction. In other words, their concrete experiences of struggle against coal mining in their own communities and localities, what might be described in German as their concrete subjectivity, leads them to embrace the energy transformation and to distance themselves from the idea that brown coal can be a bridging technology in this transition. Their experience with the transition to renewable energy at a local level leads them to the view that to return to a reliance on brown coal would mean, in Petra Rösch's words, that "the future would have to give way to the past." It is worth noting here that this lived experience of the energy transformation in Germany places Rösch and others like her in Lusatia in a very different context from the Appalachian groups that oppose mountaintop removal (chapter 6, this volume). As Bozzi writes, "Gaining public support to keep coal in the ground in central Appalachia would likely require compensating that same region with benefits (e.g., funds to kick-start renewable energy production)." This has already happened in Germany, through the incentives given to solar and wind energy and the strong strategic support from the German Federal government for the policies of ecological modernization described later in this chapter.

Put simply, the people of Lusatia already have direct experience of what the foundations of a just transition might look like.

## Contesting Coal: The Policy Context

The local struggle over coal and its future in Lusatia reflects a much wider debate in Germany over the future of energy policy and climate action.

In 2011, Germany embarked on what Ottmar Edenhofer, co-chair of working group III of the Intergovernmental Panel on Climate Change, described in the German press as "one of the greatest social experiments there has ever been in Germany, comparable with the process of reunification."[21] This social experiment is the German energy transformation (*Energiewende*), a package of laws passed by the German parliament on June 30, 2011, in the aftermath of the Fukushima nuclear accident.[22] The laws provide for a phase-out of all nuclear power plants currently operating in Germany by 2022 and a major expansion of renewable energy. They set an ambitious target of 35 percent of total energy use to be provided by renewables by 2020 (compared with 20 percent at present). By 2050, according to the targets set in the energy transformation legislation, all but a fraction of Germany's energy needs should be provided by renewables.

The German energy transformation sets out a policy framework and a schedule for transition to a post-emissions economy that is one of the most ambitious in the industrialized world. Germany already has a highly developed renewables sector, with solar (photovoltaic) becoming an increasingly competitive energy source for electricity generation.[23] The energy transformation has the support of both major political parties in Germany and was greeted with general approval by environmental organizations such as Greenpeace and the German Climate Alliance (Klima Allianz Deutschland). According to Mark Lewis, an analyst at Deutsche Bank quoted in the *Economist*, Germany is alone in the industrialized world in possessing "the means and will to achieve a staggering transformation of the energy infrastructure."[24]

The energy transformation continues an approach to environmental and energy policy in Germany that a number of academic commentators have described as ecological modernization. According to Rainer Hillebrand, ecological modernization "emphasizes the ʽwin-winʼ opportunities of technological progress in industrialized countries … in contrast to environmental approaches which stress the negative ecological impact of economic activities and the physical boundaries of economic growth."[25]

As Hillebrand shows, a commitment to ecological modernization has underpinned German climate protection policy since the 1990s, and this approach has been widely perceived as a success.[26] At the same time, however, ecological modernization faces "severe limitations": industries such as aluminum, steel, and coal-based electricity generation, which stand to lose from the process of modernization, have been able to water down climate policies, and play "an important (veto) role in German politics."[27]

Hillebrand's analysis helps to illuminate the challenging policy dilemma that Germany confronts in bringing about the energy transformation. Coal-fired power plants currently generate 42 percent of total electricity used in Germany, and the country would need to burn an extra 3 million to 4 million tons of coal a year to meet the shortfall from the nuclear phase-out if current levels of energy use are to be maintained.[28] In the short term, this would boost total current emissions by around 10 percent.[29] The exit from nuclear energy has given proponents of coal extraction and coal-fired power generation a powerful pretext to frame coal, and brown coal in particular, as a transitional energy source or bridging technology, and a vital component of the transformation of German energy infrastructure. Typical for this position is the view expressed in a media interview by Claudia Kemfert, an energy expert with the German Institute for Economic Research, that "we cannot get out of nuclear energy and coal-fired power simultaneously."[30]

Thus, on the one hand Germany can rightly be seen as leading the industrialized world in its commitment to action on climate change and nuclear risk and its pursuit of policies to end fossil fuel dependence by midcentury. On the other hand, the short-term impact of the energy transformation, especially with the nuclear exit, may be to increase reliance on coal, and brown coal in particular.

In April 2012, Hannelore Kraft, state premier of North Rhine–Westphalia, officially opened the largest brown coal–fired power station in the world.[31] Operated by the German energy concern RWE, one of the four largest energy companies in Germany and a major player in the European energy market, the new power plant at Grevenbroich is close to a large open-cut brown coal mine also owned by RWE. The new plant cost 2.4 billion euros to construct and will generate 2,200 megawatts at peak load. Describing the opening of the plant as "the right step at the right time" and "a special day for RWE, for the state of North Rhine–Westphalia, and for Germany as a whole," Kraft was at pains to portray coal-fired and renewable energy as partners in the energy transition. She had previously stressed the risk of "deindustrialization" if the energy transition were to proceed too fast and promised that her government would not support a hasty exit from coal.[32]

This is a theme stressed repeatedly by state and federal politicians in Germany from both major political parties. The former German environment minister, peter Altmaier, stated in 2012 that "for many years, we will need conventional fossil energy in addition to renewables. For 2020 the aim is to have 35% renewables, but that means 65% fossil electricity."[33] In speeches, press releases, and official documents, the current German chancellor, Angela Merkel, her ministers, and government agencies have described fossil fuels as an important bridging technology within the framework of the energy transition. Citing the need to maintain the consistency and reliability of the energy supply, the government's own policy documents argue that in the immediate future, "it will be necessary to invest in highly modern coal- and gas-fired power stations which will function as conventional power plants with fossil fuels. This course of action is necessary to secure Germany's position as a strong economy with a large number of jobs."[34]

The new RWE power plant at Grevenbroich is a particularly potent symbol of the contradictions inherent in the German energy transition. Brown coal is the most polluting and inefficient form of thermal coal. Its share of the German energy mix is increasing, while the extraction and burning of black coal (otherwise referred to as metallurgical or coking coal) is decreasing. In 2007 the German federal government announced that Germany would end the mining of black coal by 2018 as part of a European agreement to phase out subsidies to uneconomical coal mines. However, black coal accounts for only a fraction of total coal mining and coal use in Germany. In 2009 German mines produced 14 million tons of black coal compared with around 175 million tons of brown coal. Germany is already the world's largest producer of soft brown coal, accounting for 17.2 percent of total global production. In a postnuclear dash for coal, pressure is increasing for an expansion of brown coal production.[35]

According to recent press articles, brown coal has made a dramatic comeback as an energy source since the passing of the energy transformation laws in 2011. Based on figures published by the Working Group for Energy Accounting (Arbeitsgemeinschaft Energiebilanzen), brown coal now accounts for approximately a quarter of electricity generation in Germany and increased its share of the total energy mix by 3.3 percent in 2011.[36] According to the working group, this trend would have led to an increase of 0.8 percent in greenhouse gas emissions in 2011 had it not been for an exceptionally warm winter the previous year and the dampening of energy demand due to the economic crisis in Europe.

The economic attractiveness of brown coal has been strengthened by the low price of carbon credits within the European Union (EU) emissions trading scheme and by the falling cost of electricity generated from renewables, which are now undercutting gas, making brown coal the only cheaper option.[37]

However, the German federal government has assured the EU that rising dependency on coal-fired power will not affect its target of a 40 percent reduction in greenhouse gases by 2020. New coal-fired stations are to be highly efficient next generation and "CCSready" at the same time, older coal-fired power stations are being decommissioned.[38] The technology of CCS, otherwise known as clean coal technology, holds out the promise of capturing carbon dioxide emissions from coal-fired power plants and storing the captured carbon dioxide in deep geological formations underground.[39] However, serious doubts have been cast on the future of CCS technology in Germany. According to a recent study by the German Institute for Economic Research, numerous CCS pilot projects have been abandoned or postponed. According to the study's author, Christian von Hirschhausen, CCS technology will not play a role in the German energy sector for the next twenty years: the technology has proved "too technologically demanding and too expensive to implement."[40]

At the time of writing, both the German and the European energy markets are extremely volatile. In 2012, the four largest energy companies in Germany, E.ON, RWE, EnBW, and Vattenfall, which account for 89 percent of electricity production, reported heavy losses in their annual financial results. These losses were attributed to falling electricity prices, high gas prices, and the beginning of the nuclear phase-out.[41] In March 2013, RWE reported a 28 percent drop in net profit for 2012. The *Wall Street Journal* ascribed this result to falling electricity demand, lower wholesale prices for electricity, and the growing competitiveness of renewables in the German energy market.[42] At the time of writing, Vattenfall itself has indicated that it believes the prognosis for energy markets across Europe is poor: in a press release issued on July 23, 2013, the company stated that "like other European energy suppliers, Vattenfall is affected by a bleak outlook in the market. Our company assumes that this outlook will not alter in the foreseeable future."[43]

Despite these bleak prospects, German energy companies are continuing to assert publicly that they will invest in brown coal extraction and new brown coal–fired power plants in the future. This trend toward an expansion of coal-fired power is strongly contested by environmental groups and policy analysts in Germany, with some arguing that gas is preferable as a transition fuel.[44] Greenpeace Germany has outlined a strategy and a legal framework for an exit from coal by 2040. Under the Greenpeace plan, the last brown coal–fired power plant would close in 2030 and the last plant burning black coal in 2040.[45]

It falls outside the scope of this chapter to evaluate these proposed alternatives to an expansion of coal. In the final part of this discussion, we return to Lusatia to examine the impact of opposition to new mines and the likely prospects for the future.

### Will the Future Give Way to the Past?

Thus far, it would appear that local opposition groups have been successful in delaying the opening of the proposed mines but not in stopping them altogether. Vattenfall's plans were originally announced in 2007 in a joint press conference with the state government of Brandenburg, which has remained a strong supporter of the mines, and of coal mining and coal-fired power in general.[46] As already noted, the company itself openly acknowledges the close relationship between the company and political parties in Brandenburg.

Vattenfall submitted the initial documentation for government approval in 2008 and undertook a scoping study in 2011. Planning authorities considered the application for a full year before it was referred to the environmental planning authorities. So far, full planning authority has not been granted. In the latest developments, the Brown Coal Commission of the state of Brandenburg (Braunkohlenausschuss) met in late May 2013 in the city of Cottbus to consider Vattenfall's application to extend the Welzow South mine. The meeting was accompanied by vigorous demonstrations from both supporters and opponents of the mine extension. A majority of the commission voted to approve Vattenfall's application. In June 2014, the state parliament of Brandenburg gave formal legal approval for the mine to proceed.

Given the uncertain economic prospects for the energy industry, however, and Vattenfall's own sober assessment of the short- and long-term outlook, speculation about the company's commitment to its operations in Germany, and the possibility of a legal challenge through the German courts, it is extremely hard to predict whether the new mines will indeed go ahead. What does seem clear, however, is that the uncertainty that has hung over the future of communities such as Proschim and Atterwasch for six years has had a corrosive effect on the communities themselves and on the mental health of many individuals in those communities.

Most of the people interviewed have lived in the region for much or all of their lives. All feel a strong sense of connection to their home towns or villages and to the region as a whole. The older interviewees also express a strong sense of history, often relating current developments to the unique history of Lusatia during the Communist period and its aftermath. They do not necessarily see themselves as environmentalists, but all see coal mining in their region as damaging to the local environment, and most relate it to broader themes of climate change and environmental risk, which transcend its impact on their own communities.

They see the pressures for an expansion of coal mining as a reversal of the process of ecological modernization over the past two decades, to which some have made a direct practical and economic commitment themselves. Finally, they express strong skepticism about the capacity of mainstream political parties and parliamentary politics to deliver good policy outcomes for their region and environmental justice for their communities. As this book goes to press, an estimated 7,500 people from twenty countries gathered in Lusatia in August 2014 to protest against any expansion of coal mining. The protesters formed a human chain eight kilometres long from Kerkwitz, one of the threatened villages close to the border with Poland, to Grabice on the other side of the border, also threatened by plans for new brown coal mines in western Poland. In so doing, they aimed to draw attention not only to the transnational nature of the coal industry, but the possibilities for transnational solidarity in contesting coal.

In closing, I offer one final observation, based partly on my conversations with people in Proschim and partly on my reading of Stephen Gardiner's work on the ethical dimensions of climate change.

As I have noted earlier, while many of the activists in Proschim do not describe themselves as anticoal, joining a political struggle against the coal industry has led them to question its legitimacy. While their initial motivation in opposing new coal mines springs from a desire to protect their homes, livelihoods, and community identity, their lived experience of the energy transformation at a local level, and of the clash between future and past, has led some to engage with wider issues of climate change and environmental justice. This process already gives us insight into what Ortwin Renn describes, in a phrase quoted earlier, "the human drivers for initiating, promoting, or hindering political change" in the area of climate action. In other words, it reveals, how individuals may begin to frame their own efforts to leave coal in the ground not merely as a local, primarily pragmatic struggle, but rather as an unfolding ethical commitment strongly oriented toward the future. As Gardiner puts it, "The dominant discourses about the climate threat are scientific and economic. But the deepest challenge is ethical. What matters most is what we do to protect those vulnerable to our actions and unable to hold us accountable, especially the global poor, future generations and nonhuman nature."[47]

## Notes

1. Except where otherwise indicated, all quotations from Günter Jurischka, Petra Rösch, Erhard Lehmann, Reinhard Jung, Matthias Berndt, Julia Albinos, and Kathi Gerster are from personal interviews conducted in Germany by the author and freelance journalist Gülseren Ölcum in April and May 2013.

2. Brown coal, also known as lignite, has a "low energy and high ash content." Australian Bureau of Agricultural and Resource Economics and Sciences, http:// adl.brs.gov.au/data/warehouse/pe_aera_d9aae_002/aeraCh_05.pdf.    Although definitions of brown coal vary from country to country, it generally refers to poorer-quality coal that produces less heat than black coal (thermal or metallurgical coal) and generates more carbon dioxide and other greenhouse gases when burned. The International Energy Agency defines brown coal as "nonagglomerating coal with a gross calorific value less than 5700 kcal/kg (23.9 GJ/t) containing more than 31 percent volatile matter on a dry mineral matter free basis." International Energy Agency, "Coal Information" (2012 edition), http://wds.iea.org/wds/pdf/Documentation%20for%20coal%20information%202012.pdf.

3. See, for example, http://www.lausitzer-braunkohle.de/index.php: "Wer wir sind: In der DDR unterdrückte Umweltgruppen, sogenannte "Umwelt- und Friedenskreise," waren es, die vor und während der Wendezeit 1989/90 als erste die Probleme des Braunkohlenbergbaus öffentlich machten. Aus diesen und weiteren Gruppen gründete sich 1990 der überparteiliche Umweltverband GRÜNE LIGA."

4. See Climate Alliance, http://www.die-klima-allianz.de/braunkohle-in-der-lausitz-2.

5. According to the Working Group for Energy Accounting (Arbeitsgemeinschaft Energiebilanzen), brown coal increased its share of the total energy mix by 3.3 percent in 2011. See http://www.stern.de/wirtschaft/news/folgen-der-energiewende-braunkohle-erlebt-schmutziges-comeback-1796042.html.

6. See   http://corporate.vattenfall.com/about-vattenfall/vattenfall-in-brief,   and
http://www.freiepresse.de/NACHRICHTEN/SACHSEN/Vattenfall-foerdert-63
-Millionen-Tonnen-Lausitzer-Braunkohle-artikel8234521.php.

7. See Braunkohlenplan Tagebau Welzow-Süd, Weiterführung in den räumlichen
Teilabschnitt II und Änderung im räumlichen Teilabschnitt I (brandenburgischer
Teil) Gemeinsame Landesplanungsabteilung Berlin-Brandenburg, Referat GL
6: 15, http://gl.berlin-brandenburg.de/imperia/md/content/bb-gl/braunkohle/bk
_welzow_sued_entwurf_20_07_11.pdf.

8. Ortwin Renn, "The Social Amplification/Attenuation of Risk Framework: Application to Climate Change," *Wiley Interdisciplinary Reviews: Climate Change*
2 (2011): 154–169.

9. There is now a lively and extensive scholarly literature drawing connections
between the ethnographic methodologies of anthropology and journalism (Vesperi 2010; Singer 2009; Hannerz 2004). As Boyer (2010) has noted, "long-form
investigative reporting, for example, seems to share much with the rich tradition
of critical public anthropology." Both, Boyer argues, have a *"translocal* and *epis-
temic* orientation as practices of making and communicating knowledge about
the world across social and spatial distance." Both are narratives discourses, both
are involve a sustained engagement with particular individuals and their communities and life-worlds, and both are engaged in social analysis. Dominic Boyer,
"Divergent Temporalities: On the Division of Labor Between Journalism and Anthropology," *Anthropology News* 51, no. 4 (2010), 6–9. See also See also: Ulf
Hannerz, *Foreign News. Exploring the World of Foreign Correspondents* (Chicago: University of Chicago Press, 2004); Jane B. Singer. "Ethnography" *Journalism and Mass Communication Quarterly* 86, no. 1 (2009): 191–198; Maria D.
Vesperi, "Attend to the Differences First: Conflict and Collaboration in Anthropology and Journalism," *Anthropology News* 51, no. 4, 7–9.

10. See, for example, "Die Energiepolitik der DDR: Mängelverwaltung zwischen
Kernkraft und Braunkohle," Friedrich Ebert Stiftung, Bonn (1988): 33.

11. See, for example, Ökologie und Ökonomie: Umweltpolitik in der DDR. Stiftung Haus der Geschichte der Bundesrepublik Deutschland, http://www.hdg.
de/lemo/html/DasGeteilteDeutschland/NeueHerausforderungen/OekologieUn-
dOekonomie/umweltpolitikInDerDDR.html; See also U. Krämer, R. Dogner,
and, H.-J. Willer, Auswirkungen der Luftverschmutzung auf die Gesundheit von
Schulanfängern—Eine Vergleichende Studie aus Ost- und Westdeutschland, *Gesundheit und Umwelt Medizinische Informatik, Biometrie und Epidemiologie* 75
(1992): 126–30.

12. Joachim Nawrocki, "Wenn Braunkohle zu Eis wird," *Die Zeit* 13., no. 3
(1981), http://www.zeit.de/1981/12/wenn-braunkohle-zu-eis-wird.

13. Hermann Wittig, "Braunkohlen- und Sanierungsplanung im Land Brandenburg," in Wolfram Pflug, ed., *Braunkohlentagebau und Rekultivierung: Landschaftsökologie. Folgenutzung. Naturschutz* (Berlin: Springer, 1998): 475–486,
here 475.

14. Ibid.

15. See, for example, Wolfram Pflug, ed., *Braunkohlentagebau und Rekultivier-ung: Landschaftsökologie. Folgenutzung. Naturschutz* (Berlin: Springer, 1998).

16. I also visited Bitterfeld in 1990. Conditions then were still much as Maron had portrayed them in her novel.

17. Monika Maron, interview with author, March 1990, subsequently broadcast on *Books and Writing*, Radio National, Australian Broadcasting Corporation, May 1990.

18. See "Proschimer Landwirte füttern Biogasanlage," Lausitzer Rundschau, December 13, 2007, http://www.lr-online.de/regionen/weisswasser/Proschimer -Landwirte-fuettern-Biogasanlage;art13826,1871758.

19. See Benjamin von Brackel, "Renaissance der Kohle," *Klimaretter,* July 23, 2013, http://www.klimaretter.info/energie/hintergrund/14143-renaissance-der -kohle.

20. The SPD is the Social Democratic Party. The term *state premier* is a transla-tion of the German Ministerpräsident, the head of a federal state government. The equivalent office in the United States is that of governor.

21. http://www.focus.de/politik/weitere-meldungen/atom-energiewende-als -grosses-experiment-_aid_636032.html.

22. I have chosen to translate the German *Energiewende* as "energy transforma-tion." The German expression contains resonances not immediately apparent in English, alluding as it does to d*ie Wende* (literally "the turn"), the phrase common-ly used in German to describe the political and economic transformation of the former GDR. While *Energiewende* is sometimes translated as "energy transition," we prefer the more forceful "energy transformation," coined by Mark Lewis in the *Economist* article cited in note 24. For the actual laws passed by the German Federal government, see German Bundesregierung, http://www.bundesregierung .de/Content/DE/Artikel/2011/08/2011-08-05-gesetze-energiewende.html.

23. http://www.ftd.de/politik/:braunkohle-energiewende-verhilft-braunkohle-zu -comeback/70003723.html.

24 Energiewende, "Germany's Energy Transformation," *Economist* 28, no. 7 (2012), http://www.economist.com/node/21559667.

25. Rainer Hillebrand, "Climate Protection, Energy Security and Germany's Poli-cy of Ecological Modernization," *Environmental Politics* 22 (2012): 666.

26. Ibid., 668.

27. Ibid., 669.

28. Brigitte Knopf et al., *Scenarios for Phasing Out Nuclear Energy in Germany* (Bonn: Friedrich-Ebert-Stiftung, 2011).

29. See Henning Gloystein and Jackie Cowhig, "German Nuclear U-Turn Means Jump in Emissions," Reuters, April 4, 2011.

30. Ibid.

31. See Daniel Wetzel, Ein Kohle-Koloss soll die Energiewende sichern, Welt-Online, August 8, 2012, http://www.welt.de/wirtschaft/article108636663/Ein -Kohle-Koloss-soll-die-Energiewende-sichern.html.

32. See "The Energy Transition Must Not End in a De Industrialization," accessed October 1, 2013, http://www.braunkohle-forum.de/index.php?article_id=131.

33. Quoted in Michael Birnbaum, "German Energy Balancing Act," *Guardian Weekly*, August 31, 2012.

34. Die Bundesregierung, Energiekonzept: Moderne Kraftwerke schlagen Brücke ins regenerative Zeitalter, March 21, 2013, http://www.bundesregierung.de/ Webs/Breg/DE/Themen/Energiekonzept/Energieversorgung/ModerneKraftwerke/ _node.html.

35. See Michael Pahle, "Germany's Dash for Coal: Exploring Drivers and Factors," *Energy Policy* 38 (2010): 3431–42.

36. http://www.ftd.de/politik/:braunkohle-energiewende-verhilft-braunkohle-zu -comeback/70003723.html.

37. http://www.ftd.de/unternehmen/industrie/:strommarkt-braunkohle-gewinner -der-energiewende/70111891.html.

38. B. Knopf, *Scenarios for Phasing Out Nuclear Energy in Germany* (Bonn: Friedrich Ebert Stiftung, 2011), 5–6.

39. See World Coal Association, "Carbon Capture and Storage Technologies," http://www.worldcoal.org/coal-the-environment/carbon-capture-storage/ccs -technologies.

40. See German Institute for Economic Research Berlin, "CCS Technologie ist für die Energiewende gestorben," February 8, 2012, http://www.diw.de/ de/diw_01.c.392660.de/themen_nachrichten/ccs_technologie_ist_fuer_die _energiewende_gestorben.html.

41. See, "Large Financial Losses Incurred by German Energy Companies, Centre for Eastern Studies," *CE Weekly*, March 21, 2012, http://www.osw.waw.pl/en/ publikacje/ceweekly/2012-03-21/large-financial-losses-incurred-german-energy -companies.

42. Jan Hromadko, "RWE Warns Profits to Fall amid Low European Power Prices," *Wall Street Journal*, March 5, 2013, http://online.wsj.com/article/BT -CO-20130305-700442.html.

43. http://www.mynewsdesk.com/de/vattenfall-gmbh/pressreleases/vattenfall -nimmt-umfangreiche-abschreibungen-vor-und-teilt-unternehmen-auf-889121.

44. Knopf, *Scenarios for Phasing Out Nuclear Energy in Germany*, 5–6.

45. See Kohleausstiegsgesetz. Verteilung der Reststrommengen und Folgenabschätzung für den Kohlekaraftwerkspark. Eine Studie durchgeführt im Auftrag von Greenpeace von Ecofys Germany GmbH, 2012.

46. Groups like Grüne Liga also point to what they claim are very close connections between the Social Democratic Party in Brandenburg, trade unions, and the coal industry.

47. Stephen M. Gardiner, *A Perfect Moral Storm: The Ethical Tragedy of Climate Change* (New York: Oxford University Press, 2011), xii.

# 10

## Heating Up and Cooling Down the Petrostate: The Norwegian Experience

Helge Ryggvik and Berit Kristoffersen

At one time, Norway was the world's fourth largest oil producer. Before that, and before North Sea oil was discovered, Norway was known mostly for its fiords and cross-country skiing, not to mention an acute environmental ethic. If any country could handle an oil boom and avoid the resource curse, it was Norway. But as Helge Ryggvik and Berit Kristoffersen argue in this chapter, even if Norway avoided the classic symptoms of the oil curse, it nonetheless has been deeply affected, first by the sheer wealth, then by the power of its own national oil company. Political realism in Norway includes fossil fuel dominance—economic and political.

While this case focuses on the difficulty of maintaining a moderate pace of extraction, the larger picture is indeed the legitimacy of a fuel or, perhaps better put, one society's relations to that fuel. Now voices are being heard within Norway questioning the net benefit presumption of continued extraction, let alone expansion. They are saying that enough is enough. Fossil fuels, some are beginning to argue, are no longer legitimate nationally or globally as Norway's well-being is tied to that of the rest of the planet. What is more, if boom conditions have been difficult to moderate, then bust will be as well; it is time to start stopping. The politics, we will see, are fraught with difficulties. But if a moderate pace was a bold measure back in the 1970s when environmental problems were highly localized and manageable, then it may be that Norway, along with Ecuador and others here in the globally constrained twenty-first century, is crafting another bold measure: deliberately leaving oil in the ground.

### A Moderate Pace of Extraction

This chapter takes up the challenge of confronting the industrial, end-of-pipe view (chapter 1) of Norway's oil sector. We analyze the political economy of an increasingly, but not absolutely, oil-dependent Norway: the pace at which oil and gas resources have been and should be extracted. In the early 1970s, after the discovery of the Ekofisk field, a self-imposed policy of a moderate rate of extraction was chosen to cushion Norway against the expected shocks of sudden wealth. This was notably before knowledge of the impact of fossil fuels on the atmosphere. As the science became clear and with Norway being internationally recognized as a forerunner on environmental issues, one might have expected that the new science on climate would have provided the political ground for an even slower pace of extraction, that is, deliberately leaving a large part of the resources in the ground for a long time. Instead, the industry and its supporters responded with a set of arguments and policies that managed to increase production at the same time as climate change mitigation became part of Norway's state agenda.

When the Norwegian oil fund in August 2014 reached 900 billion dollars (officially called the Government Pension Fund Global), it was a direct consequence of a breach in this key premise of the original oil policy, that is, that oil would be extracted at a moderate pace. In fact, at this time, Norway was near a world record in its extraction rate in relation to proven reserves.[1]

Norway's oil production peaked in 2000. This coincided with the Norwegian oil industry's push to acquire access farther north in Norway's Arctic territories, as well as closer to the coast, in environmentally vulnerable areas that play key roles in ecosystems and fisheries. Since 2005 the government's strategy has been to refocus interests in the Norwegian High North as the most important domestic and foreign policy issue. With prospects for continued oil and gas extraction there, the government has mobilized international law and scientific knowledge about the Norwegian North to reimagine the region as an energy frontier and geopolitical center, tying into international scenarios of opening up the Arctic for economic activity.[2]

Over four decades, what can be termed the Norwegian oil industrial complex came to dominate both the framing of Norway's energy and economic policies and its policy prescriptions.[3] This does not mean, in our view, that there is something inevitable and deterministic in Norway's increasing dependence on fossil fuels. Norwegian oil history can be characterized as a constant struggle not only between different interest groups but also between different visions of Norway's future.[4] So in spite of the strong momentum in favor of the oil industry, the outcome of these struggles was never predetermined. The movement to moderate the extraction rate employed a number of ideas and strategies to change the direction of Norway's oil-dependent path. From the 1990s on, this movement was to some degree trapped by the consumer-oriented international climate regime and its focus on carbon trading as a means to fulfill Norway's obligations in the Kyoto Protocol.

Since 2001, a major political struggle in Norway has been whether the petroleum resources off the coast of the Lofoten and Vesterålen islands (henceforth Lo-Ve) in northern Norway can be opened for extraction, as premapping through seismic surveys has proved to be promising.[5] This is the area where the northeast Arctic cod, the biggest cod stock in the world, comes down to spawn from the northern Barents Sea and the Atlantic Ocean every winter. The political opposition has kept drillers away to date. Today the potential of challenging the oil industrial complex to reduce investment and the pace of extraction, and therefore to leave oil in the ground, is increasingly reemerging as a vibrant unifying argument by various strands of the Norwegian political landscape. For example, some unions fear the offshore oil industry would displace other onshore economic activities. Economists fear an overheating of the economy, causing Norway to lose its competitive strength in European markets. Reflecting international debates on the carbon bubble, both academics and journalists increasingly ask what this might mean for Norway.[6] And as we discuss toward the end of this chapter, climate concerns and uncertainty for what will happen with an oversized oil industry when the easy oil runs out are now forming new alliances for an exit strategy. In 2013, labor unions, the environmental movement, and the Norwegian Christian church came together to launch a campaign in which the two central themes were creating green jobs and slowing the pace of oil and gas extraction.[7]

## Motivation

As academics who have been working with oil-related issues for many years, we have observed how the Norwegian oil experience has been met with great respect outside Norway, especially in countries where oil is a major issue. We have found, however, that outsiders' knowledge of the Norwegian experience often is limited. In March 2008, the lead author (Ryggvik) was invited to talk about the Norwegian experience at Ecuador's Constitutional Assembly in the small town of Monte Cristi near the Pacific Ocean. There he was asked by one of the Indigenous representatives, "Do you, in Norway, have any area you have decided to leave the oil in the ground for all future?" The question was an eye opener. It motivated him both to write a book on the issue and take part in the struggle against oil drilling in Lo-Ve.

Whether it is the financial press boasting about the oil fund, industrialists in the Global South being impressed by the development of a technologically competent offshore service industry, or environmentalists looking for a responsible oil company to model others on, there always seems to be an element of imagining what you would like to see in outsiders' view of Norwegian oil policy. Our motivation for writing this chapter is to paint a more sobering picture of Norway's experience with oil, seen from a climate perspective. The history of Norway's approach to the world of oil is, in fact, full of contradictions, including struggles about Norway's future and, for that matter, the future of the world.

## History Revisited

To track the first debate about the pace at which oil and gas resources should be extracted, we have to go back to when oil was first found. This discovery in the middle of the North Sea on the Ekofisk field in December 1969 spurred the remaking of Norwegian society. Reading the early documents and public debates forty years later, it is striking how important it was to Norwegian policymakers and the public at large never to lose control over its development and how much Norwegians feared the power of an unrelenting foreign oil industry. Many pivotal decisions were made in these first years.

A parliamentary committee formulated in 1971 a ten-point list designed to ensure that "natural resources in the Norwegian continental shelf are exploited in a way that benefits the whole society."[8] The list was later referred to as the "ten oil commandments." First on the list was a general aim to secure "national governance and control" for all activities in the Norwegian continental shelf. Further down the list one could read that while the establishment of a state oil company and a national oil industry was important, all this should take place with appropriate concerns for "existing industry as well as environmental protection."[9]

The early 1970s was marked by a general radicalization of Norwegian society, clearly expressed in struggles against Norwegian membership in the European Economic Community (now the European Union, EU), where a majority in 1972 voted against membership. When the "yes side" and the "no side" came together after the referendum to formulate an oil policy, national governance and control was central in the rhetoric of all sides. A pivotal document in the following period was a governmental white paper, "The Role of Petroleum Activities in Norwegian Society," presented by the Ministry of Finance in February 1974.[10] Wealth from oil should be used to develop a "qualitatively better society," it read. This referred to a society committed to great equality in living standards; to strengthening local communities (skepticism toward centralization and regionalism has long been a strong current in Norwegian politics) and developing its welfare state further. Such commitments were to take place without "swift and uncontrolled growth in the use of material resources."[11] Here, environmental considerations were even more important than in the 1971 oil policy document. At the same time, and crucially, the decisive factor to achieve all this was control of the pace of development. In fact, the 1974 white paper's first point states, "Wishing for a long-term perspective in the exploitation of resources, and after a comprehensive evaluation of its social aspects, the Government has concluded that Norway should take a moderate pace in the extraction of petroleum resources."[12]

By keeping to a moderate pace, the ministry felt it would be easier to ensure that the oil and gas would be extracted safely for both the workers out on the platforms and the environment. It would also prevent the conversion costs of adaptation to a completely new industry from growing too fast. This gave a long-term emphasis to oil development—an epoch, not an episode. Contrary to general perceptions, the ministry assumed that the great price increase experienced while the white paper was being written (a quadrupling, toward nine dollars a barrel) would diminish over time. The assumption was that a strong increase in prices would lead to energy-saving measures. Nevertheless, the white paper also stated that the oil crisis, then at its height, could lead to external pressures from both the oil companies and other Western countries to expand production on the Norwegian continental shelf. In other words, it was important for Norway to steel itself against these pressures. Despite all these factors, Norwegian policymakers, supported by the public, considered it rational for Norway to aim for a moderate pace. Norway should enter the age of oil with good intentions for a better society and not become too dependent on oil.

After establishing a political consensus around the goal of a moderate pace, the question then was how to define and implement the goal. It was soon decided that the best approach would be to put a cap on production. In parliament, an industry-responsive alliance between the Norwegian Conservative Party and the Labour Party finally accepted a proposal from the Ministry of Finance that 90 million ton oil equivalents (o.e.) per year was the upper limit for production on the Norwegian continental shelf.[13]

## Latitude 62

A group of smaller center and left parties supported by the environmental movement wanted a much slower pace of extraction. The political opposition to uncontrolled growth in the Norwegian petroleum sector was based on two rather different approaches, both typical for the first wave of environmentalism in Norway in the early 1990s. In the first approach, environmentalists tied into debates on the limits to growth a neo-Malthusian understanding that no process, especially material growth, can continue forever without change.[14] They considered economic growth and consumerism unsustainable, especially if developing nations were to reach the same level of development as industrialized countries. Petroleum was one among many scarce resources that made such growth impossible in the long run. In Norway, for example, when one of the "ten commandments" stated that there should be no flaring of gas from oil platforms, it was not based on climate concerns (still not a major issue) but for wasting a limited resource that aroused environmentalists and others. They feared that shortsighted self-interest might lead international oil companies to extract only the resources that gave the greatest immediate profit. The second characteristic of environmentalism at the time was a strong focus on pollution and conservation. Regarding the oil sector, the threat of oil spills and the consequences for the coast, fisheries, and birdlife were the major concerns.

As it turned out, environmentalists did not gain sufficient political support for their idea of an even lower production rate. As a result, they joined with the fishing community to fight for limiting the geographical extent of petroleum activities to the North Sea, succeeding in the first four Norwegian concession rounds to allow no blocks north of the latitude 62. Latitude 62 crosses Norway on the west coast just south of Aalesund, in the southern part of Norway. It was initially the Norwegian Institute of Marine Research that had suggested this geographical limit. It meant that approximately two-thirds of the Norwegian coast would be closed for drilling activities.

In the North Sea, petroleum activities were concentrated midway between Norway and Britain. Thanks to geology and geography, both countries had, in effect, a buffer zone in case of major oil spills. The

Ekofisk field was approximately 300 kilometers from shore. The Stat-fjord field was 170 kilometers west of the Sognefjord. When 20,000 barrels spilled during a blowout from the Ekofisk Bravo platform in March 1977, wind and heavy sea dissolved the oil before it reached the coast.[15] Farther north, geological areas of interest to the oil industry were often located closer to the coast and important fishing grounds and, in the case of the area around Lo-Ve, in the midst of the spawning grounds for the northeast Atlantic cod.

When the government in a 1976 white paper first discussed opening areas north of latitude 62, safety issues were crucial.[16] The paper stated that there was 1 blowout per 500 wells drilled globally. With over 200 wells drilled offshore in Norway at the time, the probability of a blowout might have been considered to be very great indeed. When the white paper concluded that the likelihood of a blowout on the Norwegian shelf was minor even though the natural environment was among the harshest in the world, it was based on the assumption that the petro-leum activities were performed more safely in Norway than in other oil-producing regions.[17] At this time in history, there was certainly no evidence that supported this claim. The same international companies on whose activities the international statistics were based performed nearly all operations. It was not before well into the 1980s that Norwegian regulations moved ahead of those of other countries. Importantly, as is generally the case in Norway, the public's trust in state institutions and national performance was strong.

When the Ekofisk field had a blowout in 1977, one of the immediate responses was to postpone the opening of areas for drilling farther north. In 1980, the dangers of operating offshore installations in the harsh environment of the north were demonstrated again when the Alexander Kielland platform capsized, killing 123 oil workers. However, the long-term effect of the Ekofisk blowout was somewhat different from what environmentalists had hoped for. The incident proved that blowouts could happen and that oil spill contingency plans were more or less useless if weather conditions were tough. But after the first major winds a few weeks after the accident, it was difficult to find even a trace of the oil. In the public mind, though, it was as if the worst thing had happened and yet there were no long-term measurable consequences.[18] Everything would of course have been different if a blowout had happened closer to the shore, like the *Exxon Valdez* accident in Alaska eleven years later. When the gov-ernment opened the north for exploration drilling during the early 1980s, the term *acceptable risk* was used to justify the activities.[19]

## From Cap on Production to a Limit on Investments

Into the 1980s, the politically self-imposed constraints of a moderate pace of extraction had to confront an oil industry with an increasingly stronger inner dynamic. From the late 1980s, Norwegian oil companies became dominant as operators of petroleum fields. It started with Statoil taking over responsibility for running the large Statfjord field from the American company Mobil. Soon two other Norwegian oil companies, Norsk Hydro and Saga, gained responsibilities to run oil fields. Norwegian companies became the largest employers by far of offshore workers. The tendency was the same in the related supply and service industry. The Gullfaks A platform that started production in 1987 had the strongest local content ever, with 80 percent of all work hours during the construction phase performed in Norway.[20] This "Norwegianization" was seen as a positive development, supported and promoted especially by labor unions. Going further, Norwegian firms were considered to be more amenable to strict social, safety, and environmental regulations. However, with its "Norwegianness," it became easier for the industry to use periods of unemployment to increase activities, including exploration and production (figure 10.1).

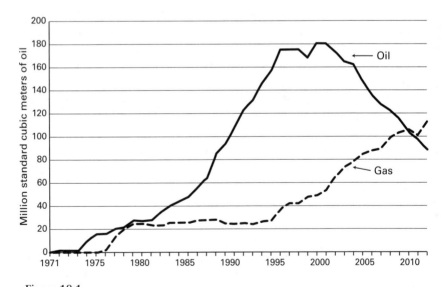

**Figure 10.1**

Norwegian historical production of oil and gas, 1971–2012 (*Source:* Norwegian Petroleum Directorate)

In the first half of the 1980s, the annual rate of extraction was still far below the 90 million target. This was not because the rate of investment had been held back by regulatory measures. Indeed, the challenges and explosion of activities had been so great that the whole sector could be described as a vast technological laboratory. Rather, it took time to get oil onstream. At the same time, representatives from the oil industry feared they would approach that target when the large Statfjord, Gullfaks, and Oseberg fields were ready for production later in the decade. In 1983, the government-commissioned Tempo Committee presented a report that proposed abandoning calculations based on what the committee described as "magical figures," and instead aiming for an objective of a steady level of investments, not of extraction itself. The argument was as follows: if you install new technologies in a field, it is foolish not to increase production by using those technologies.[21] Investments continued to increase powerfully until about 1985 when they reached around 25 billion kroner (about 4 billion USD), and remained stable until 1988. Based on the proposals from the Tempo Committee, the parliament finally agreed in January 1988 that total investments should be limited precisely to 25 billion kroner.[22] At this point, production was nearing the earlier upper limit of 90 million standard cubic meters ($Sm^3$) o.e.[23] The new regulation was to be administered on a first-come, first-served basis. The Ministry of Oil and Energy was to withhold permission to start major construction projects if necessary.

For anyone familiar with the Norwegian oil industry in the 1980s, it was evident that 25 billion kroner in annual investments would lead to a very high level of activity. The long list of major fields that were being developed, with giant installations like Statfjord B, three enormous Gullfaks platforms, and much more, was a concrete expression of such expansions. But in spring 1988, soon after the upper limit on investment was agreed on, the global housing bubble, supported by cheap oil, burst. Falling prices in real estate and the housing market were followed by a financial crisis, and the state had to take over the largest banks. In response, parliament agreed to a series of extension measures, which opened the way for an intense growth in oil investments. The large offshore projects represented both employment and income in the future. In this extraordinary economic situation, all constraints on the pace of development of the industry were put aside. In 1993, investment had shot up to 57 billion kroner annually.[24]

The shift in orientation reflected the growing influence of a worldwide trend of the dominance of market ideology and neoliberal economic reasoning. In 1993, unconstrained growth in oil investment was justified in this way in a white paper by the Ministry of Petroleum and Energy on conditions within the oil industry: "Activity levels in the petroleum industry are to a considerable extent dependent on conditions we cannot control."[25] The starting point for Norwegian oil policy had been a strong desire to secure national governance and control of the industry. Here, however, the Ministry of Petroleum and Energy effectively claimed that this was impossible. In 2013, the overall investment in the Norwegian oil sector climbed above 200 billion kroner annually when new drilling activities are included. Calculated in 1988 prices, this would be around 115 billion kroner in 2013, a fourfold increase. In sum, from 1988, when the 90 million cap or ceiling was broken, until 2000, production increased nearly threefold, reaching around 260 million $Sm^3$ o.e.[26]

### Norway's Climate Dilemma

A timely question is what happened with the 1970s movement that mobilized against drilling in the north and for a slow pace of extraction. The massive increase in Norwegian oil investments and production coincided more or less exactly with the growing awareness of the climate consequences of carbon-based emissions. Why didn't that lead to intensified opposition toward the expansion in the oil sector? It was the UN-appointed Brundtland Commission's report that, more than anything else, popularized climate change and put it on the public agenda.[27] When the report was published in 1987, the commission's leader, Gro Harlem Brundtland, had started her second term as prime minister in Norway. At the same time, there was often talk about Brundtland as the "world environmental minister." Certainly the Brundtland Commission's report influenced the convention written during the UN Conference on Environment and Development in Rio de Janeiro in 1992. The Rio Convention was followed by the Kyoto Protocol in 1997. Hence, one might think that Brundtland in her important political position in Norway would try to restrict the petroleum sector's further growth and put the aim of a moderate pace of extraction back on the agenda.

In the beginning it could seem as if the call from the Brundtland Commission was about to change Norwegian politics. As early as spring 1989, the Norwegian parliament voted on ambitious goals to reduce emissions in Norway. Another important initiative was the introduction of a carbon dioxide emission tax in 1991, directed against offshore installations, which represented a third of Norwegian carbon dioxide emissions. This was in accordance with the aim of the coming convention, which entailed all countries that ratified it to stabilize emissions of climate gases at a level corresponding to that of 1990 in the year 2000. Norway's political parties were undergoing a general greening of their politics, at least in their use of language and symbolism. This second wave of environmentalism in Norway during the late 1980s was key to incorporating environmental issues into formal state activities in Norway. This second wave of environmentalism, however, also contributed to a substantial weakening of grassroots-level action on green issues.[28]

This weakening happened while production of oil and gas reached new heights every year. From the time of the Brundtland Commission report until the year 2000, production more than tripled. In the ten years following the Rio meeting, it doubled. Fortunately for the Norwegian oil industry, the regime that was created for reduction of carbon emissions was based on the principle that countries had to count the emissions where carbon was *consumed*, not where the resources were extracted. This meant that Norway had to count only the emissions from the platforms where the oil was extracted, not from all the oil that poured from Norway into the global market. Since the emissions from the burning of one barrel of oil could be as many as forty times greater than emissions from the extraction phase, this had enormous consequence for how Norway performed in international statistics. And because Norway at the time was blessed with a lot of hydropower, the emission per capita seemed low. Even when Norwegians imported consumer products like cars from Europe and electronics and other products from Asia, partly financed by revenues from oil, the production of these did not increase the Norwegian carbon footprint, at least not as measured by the international climate regime. The same was the case with an increasing number of oil installations, as most of the steel structures from the late 1990s were constructed in Asian shipyards.

The introduction of a carbon dioxide emission tax from 1991 made an impact on oil company practices. Because many oil companies had secured an exemption from the general goal of reducing gas flaring early in the 1970s and then flaring had a considerable cost (0.47 kroner per $Sm^3$ gas in 2010), firms had strong economic incentives to reduce them. A much-talked-about example was the Sleipner gas field where Statoil managed to send carbon dioxide down and store it in an aquifer 800 meters below the seabed. For a time, this carbon capture-and-storage technology created some optimism in the industry, especially because carbon dioxide pumped back into the ground could also be used to increase production. This of course could be seen only as a success with a narrow focus on local emission related to the quotas in the Kyoto regime, not with a focus on Norway's total contribution to global emissions (as discussed in chapter 1, this is yet another effect of end-of-pipe management, focusing on emissions from consumption, not on extraction from production).

But even with this very advantageous system for accounting for carbon dioxide emissions seen from Norway's official standpoint, the Norwegian civil servants who took part in the first international quota negotiations knew that Norway's increased oil production might be a problem. Thus, Norway, in accordance with the country emission targets set in Kyoto in 1997, was actually allowed to raise emissions by 1 percent when compared to the 1990 emissions of carbon dioxide (which the treaty had as its reference point). However, because of the quite literally free flow of investment and production in the petroleum sector, Norway never came near the target of stabilizing emissions at the 1990 level. Flaring during production was only one source of carbon dioxide emissions from an oil platform. The many new installations in the ocean needed power. In practice, establishing local gas-fired power stations solved this. As a consequence, total Norwegian emissions of carbon dioxide rose from around 35 million tonnes in 1990 to over 40 million tonnes in year 2000. Importantly, the oil industry more than swallowed up reductions made in other sectors. Between 1990 and 2008, emissions from the Norwegian oil industry rose from around 8 million tonnes of carbon dioxide equivalent to over 14 million tonnes.[29]

This reflects only emissions according to the international community's consumption-based regime. If the extraction of oil and gas had to meet the same moral criteria used for other kinds of production with unwanted consequences (e.g., narcotics), Norway would be considered one of the largest per capita polluters in the world.

The many internal contradictions in Norway's approach to petroleum policies and environmental politics are reflected in the career of Gro Harlem Brundtland herself. Brundtland rose as a star in the Norwegian Labour Party when she, as minister of environment in 1976, opposed drilling north of the 62nd latitude. Her main objection was poor contingency plans. This gave her the moral high ground when a blowout occurred a year later. Then, in 1980, as prime minister, she was responsible for opening up drilling in northern latitudes (Barents Sea). As prime minister from the late 1980s to the mid-1990s, she supported the carbon dioxide emission tax.[30] Having been one of the first nations to enforce strict environmental regulations domestically, Norway was now recognized as an international leader on the issue. At the same time, however, Brundtland was responsible for the greatest leaps forward in petroleum-related investments and production.

## A Movement Trapped in Quota Discussions?

As in other industrial countries, many social movements in Norway experienced a decline during the 1980s. However, in the late 1980s, the awareness of environmental issues actually grew there. In a second wave of environmentalism, a new generation of environmental activists was radicalized because of the new knowledge about the effect of fossil fuels on the biosphere. This generation of activists certainly put pressure on politicians regarding the formulation of targets for lowering carbon dioxide emissions in the following years. At the same time, however, the all-embracing, actively inclusive state during the Brundtland years was tying the interests of environmental nongovernmental organizations closer to the state's, resulting by the early 2000s in an environmental movement lacking in numbers, radicalism, activism, and autonomy according to political scientists Christian Hunold and John Dryzek.[31]

Since it soon became clear that the production of oil and gas accounted for most of the real increases in Norwegian greenhouse gas emissions (i.e., total emissions from oil extracted in Norway), the petroleum industry naturally got a lot of attention. However, during the 1990s, there were no strong political incentives to control the industry's increasing investment level or the rate at which petroleum was taken out of the ground. Many environmentalists focused on technoscientific solutions and how to reduce emissions in accordance with the Kyoto regime. Norway's official approach moved from "national action" to "thinking globally" in an attempt at becoming "carbon neutral," which allowed for high oil and gas production levels by paying for cuts in other countries through carbon trading.[32] Some environmentalists accepted carbon trading as a means of fulfilling this "neutrality" objective. Besides a carbon dixoide tax, environmentalists demanded that the offshore industry use hydropower electricity in cables from ashore instead of gas-powered aggregates on platforms as a technological fix.

The main political battle between the environmentalist and the oil industrial complex throughout the 1990s and into the early 2000s was whether Norway should develop gas-powered plants for electricity exports. The core argument against such investments was that this would increase Norwegian emissions even further. The counterargument from the industry was that it did not matter for the environment whether the emissions came from gas that was exported in pipelines and then made into electricity in other European countries or, as the industry wanted, made into electricity in Norway and then exported in cables. Environmentalists pointed out that this was opportunistic reasoning when it came from an industry that at the time did not accept the climate dangers of emissions from its main product. The real issue has always been total emissions into a global commons. But the controversy illustrated how the environmental movement, from the departure point of a rational green state, had become a defender of a quota system based on consumption, not production. Had there been a stronger focus on upstream activities (chapter 1), a wider critique of the underlying makeup of Norway's increasingly oil-dependent economy and its national responsibilities globally might have been more likely.[33] Instead, the fact that Norway increased its production manifold didn't become a key issue within the environmental movement, nor the relationship between energy production and consumption.

### Increase Production, Become Less Dependent

When representatives of the official "Oil Norway" travel abroad to promote Norway's oil experience, they rarely mention the extent to which the country broke with all attempts to restrict the pace of extraction, let alone the fact that Norway was, and still is, nowhere close to meeting its emission targets set in Kyoto.[34] Instead, what they highlight is the oil fund. Seen from an environmental point of view, the fund's size is a consequence of two patterns: increased rates of extraction and total (yet uncounted) emissions. As figure 10.2 makes clear, it is worth noting that it was not before the late 1990s that the fund started to expand.

However, both government and industry rationalized their breaches of the former political goals to the environmentally concerned public. The rationale for the oil fund itself was often presented as an instrument to achieve the original goals, a moderate pace of extraction and investment. How could this be? As described in chapter 3, those wanting to boost fossil fuel extraction usually find a way, rhetorical and political, to gain access and expand production.

Representatives of the state and the industry argued that Norwegian production of oil and gas prevented oil dependence and was "clean and green" compared to other countries' petroleum production. Their reasoning started with the economic argument that extracting petroleum

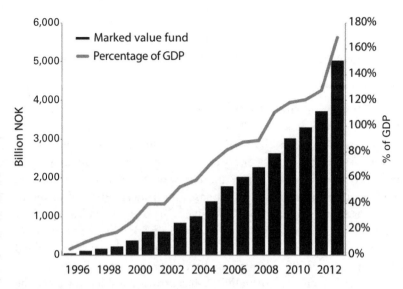

**Figure 10.2**

Size of the oil fund. (*Source:* Statistics Norway, Central Bank of Norway)

resources as fast as possible was the best way to avoid petroleum dependence. The logic was as follows. Placing the oil revenue in portfolio investments and spread in markets all over the world would be less risky than having to rely on the price of only one commodity, oil. At the same time, by keeping petroleum revenues outside the ordinary economy with strict administration of the fund, Norway could avoid the so-called Dutch disease: by limiting spending in the national budget from the fund to a maximum of 4 percent of its size per year, the country should avoid inflation and other negative results of an overheating of the economy. As such, income from the petroleum sector was to be kept separate from spending. The concerns were strictly economic, not social or environmental.

The success (again, strictly economic) of leaving oil and gas in the ground for production, on the one hand, or extract as much as possible, on the other, would be due to future prices. Betting that prices would go up therefore had an element of speculation. As it turned out, Norway effectively lost the bet by concentrating so much of its oil production in the 1990s and early 2000s, a period when oil prices were relatively low. Between 2003 and 2010, the price of oil rose far more than share prices or securities, precisely where, following economists' advice, the petroleum fund's administration put its oil revenues. Politically, however, the major problem with the rationale behind the as-fast-as-possible strategy and the oil fund is that Norway did in fact become an oil-dependent country.

Norwegian oil production peaked in 2000 with 181.2 million $Sm^3$.[35] In 2013 production was less than half of that (84.9 million $Sm^3$). The production decline was partly compensated by an increase in the production of less valuable gas. Still, overall petroleum production has been falling, from a peak of 263.4 million $Sm^3$ o.e. in 2004, down to 213 million $Sm^3$ o.e. in 2013. Given this decline, the massive increase in investment in the same period can be seen as a contradiction. One reason was the rising oil price; another was the fact that production came from large fields in their end phase and from smaller and medium-sized fields that are more labor intensive. This effect is typical for most countries where oil production has reached its peak (chapter 2). So while production has been falling, oil-related employment has nearly doubled in Norway since 2000. This means that the break-even point—the oil price required to ensure profitability—has grown considerably since the turn of the century. Once again, economically and politically, this leaves Norway vulnerable to falling oil prices, even temporarily. Norway has more and more oil workers, oil engineers, and service suppliers dependent on less and less oil.

At the same time, in the years after the 2008 financial crisis, there have been strong signs that typical Dutch disease effects are in fact becoming evident in Norway. Between 2003 and 2012, Norwegian wages were on average 38 percent higher than its Nordic neighbors and 64 percent higher than the Eurozone average.[36] The combination of increased costs and more intense international competition has had a crowding-out effect on investments in other Norwegian industries, including the renewable energy sector.

Given Norway's environmentalist image, there was a lot of talk in the country about increased investment in sustainable energy sources like windmills and wave power, as Norway has completely used its hydropower potential (producing almost all of Norway's electricity).[37] With its geographic location, Norway has had comparative advantages in wind and wave power, and yet renewable energy development from these sources has not yet taken off. The government, led by social democrat Jens Stoltenberg (2005–2013), gave itself the term "red-green." It was a coalition between the Labor Party, the Center Party and the Socialist Party. Trying to live up to that image, the government increased allocations for basic research on climate change and green energy significantly, but very few renewable energy projects have materialized. At the same time, several universities have increased education on environment, climate, and clean energy. Nevertheless, unlike Norway's neighbors such as Denmark, Sweden, and also Germany, there has been little new development of sustainable energy sources.[38] In 2012, Sweden produced more than five times as much wind energy as Norway, and Denmark produced nearly six times as much. The main underlying reason, we conclude, is the very profitable oil industry that crowds out other industrial activities and investments, as seen in the boosterist economy Princen describes in chapter 3.

## The "Clean" Oil and Gas Argument

When the Norwegian minister of petroleum and energy in 2010 argued, "It is not right that we who have the cleanest production of oil and gas should reduce our production," he repeated a line of argument that representatives of the oil industry had used for many years.[39]

In the years after 2000, carbon dioxide emissions from the Norwegian offshore production phase were around 10 kilograms per produced barrel of oil. The lowest measured level was 8 kilograms in 2009. In the same statistics, presented by the Norwegian oil industry, emissions from the Middle East were between 12 and 13 kilograms per produced barrel of oil.[40] In other parts of the world, it was even higher.

However, the "clean production" argument is problematic in many ways. When emissions from Norway were rather low from the actual production phase, it was not only because of the carbon dioxide tax. As noted, the international system of accounting for emissions gave Norway an advantage. Because a lot of the production on the Norwegian shelf is close to the British border, only a small part of the emissions from transportation is on Norway's emission account. The Statfjord oil, for example, is pumped to large tankers in the field. Most of the emissions from those tankers when they sail to the markets are accounted for in the countries the tankers have to pass. At the same time, although the oil market in periods has been unique in the sense that some large producers were able to influence prices by increasing production, it seldom or never has been an automatic adjustment. The most likely effect on reduced production in Norway would mostly be a slight price increase, though sometimes a substantial one. The effect on consumption on such a price hike does not have to be large to have a positive climate effect if one takes into account that the difference between Norway and the Middle East was marginal compared to the 400 kilograms of carbon dioxide per barrel for consumption of both Norwegian and Middle Eastern oil.

The opportunistic nature of the "clean oil" argument becomes even more evident if one looks closer at the numbers from the Middle East. Saudi Arabia, the only country in the Middle East that has had the capacity to play the role of swing producer (and thus change world prices), has produced "cleaner" oil than Norway has.[41] And in 2010 the clean oil argument collapsed when even the aggregate numbers from the Middle East were better than Norway's, with 6.9 kilograms carbon dioxide per barrel, compared to Norway's 8.7 kilograms per barrel.[42]

The same kind of argument was used regarding Norway's increasing production of gas. Production of gas was often presented by governmental and industry representatives as if it was more or less clean. "Clean" Norwegian gas was supposed to replace coal as the major fuel in European power plants. However, given that even gas releases large emissions it cannot be part of a final solution. The massive increase in Norway's export of gas from the mid-1990s has never been coupled to any scheme in Europe for a full replacement of fossil fuels.

The "cleaner" oil argument has perhaps most clearly been used in an opportunistic way when it comes to production from tar sands in Canada. The Norwegian state-controlled oil company Statoil (the state holds two-thirds of the shares) used revenues from oil production in Norway to invest heavily in Alberta. It was argued that Statoil, because of its historical "clean" record and because it comes from an environmentally oriented country, would be better than other international companies. In the case of tar sand, emissions from the production phase are so great that it makes a real difference when compared with other sources of fossil fuels: 70 to 100 kilograms carbon dioxide per barrel. Here, however, Statoil, along with other oil companies operating in Canada, points to the fact that emissions from production of oil from tar sand are "only" 25 percent or less than the emissions when the same resources are finally burned. Emissions from one barrel of oil from tar sand exported to and burned in the United States will be the same 400 kilograms as for ordinary oil. Importantly, one can note how Statoil, like others in the industry, twist arguments depending on what it likes to achieve. In Norway, when the issue is a massive increase in gas production, the argument is that it "reduces" emission abroad. When it comes to production of oil, emissions from burned Norwegian oil abroad have not been taken into account at all. In Canada, where emissions from production are high, consumption abroad can be used in the argument to show that relatively high emissions from production are still minor compared to consumption of the same oil. In contrast, we argue, along with other authors in this book, that what matters is total loadings—not the fuel, not production efficiency, not the nationality of the producer or consumer, not the supposed "greenness" of the players. This is yet another example of how boosters of more extraction, more production, more economic growth, and more return on investment use rhetoric (chapter 3).

### Norway: Part of the Solution?

The collapse of the Copenhagen 2009 meeting and following controversies was a blow to the global environmental movement and to the movement in Norway, which lost momentum to pressure for reduced emissions. At the same time, given that the Norwegian petroleum industry's intense pace of extraction had already reached its peak, one might conclude it was too late to once more raise the slogan of a moderate pace of extraction. If the sector is heading for a collapse anyway, Norway's part of world emissions will gradually decrease without a politically initiated transition strategy and a sustainable alternative. However, the history of Norwegian oil is far from over. The numbers of workers in the industry have nearly doubled since the production of oil reached its peak around 2000, and two large new discoveries in 2011 will probably again for a period increase production and mean even higher investments. This new oil boom following intensified exploration has led to a scarcity of engineers and skilled oil workers. To attract new students to the kind of education the industry needed, it was necessary for the industry to paint a rosy picture of an oil-powered future. The opening of Lo-Ve in the north was one part of such a future.

But from our perspective, the oil industry's new offensive illustrates how Norway, more than ever before, is locked into oil and gas dependency. The very size of the oil sector gives it huge lobbying power. Yet as a kind of dialectical response from an environmentally conscious public, increased knowledge of a potential climate crisis and the very feeling of being strangled by an ever more dominant oil industrial complex created a base for opposition. In the North, fishermen and their organizations are arguing against many of the large-scale planned developments, especially in Lo-Ve, which state environmental and fisheries agencies have warned against opening.[43] The consequences for the fisheries and environmental risks are the main basis for the resistance in the north, where there are strong interdependencies between the sustainability of the fisheries and societal development. Many polls show that a majority of Norwegians are now in favor of leaving the oil in the Lo-Ve area in the ground for the time being, this position due in part to the indirect threat of crowding out the fisheries.[44] These local arguments connect with national demands to leave the oil in the ground for the time being and the global arguments that relate to the climate crisis. Also, some voices inside the oil industry are arguing for a slower pace. The largest oil workers' union, Industry and Energy (IE), sent an official protest to the Ministry of Oil and Energy in June 2012 emphasizing this.[45]

In the spring 2013, a coalition of concerned groups launched a broad new campaign for an exit strategy from oil. The campaign is based largely on the trade union movement, the environmental movement, and the main organization representing the Norwegian Church. The campaign mostly stresses positive investment opportunities and green jobs. It is simultaneously confronting Norway's oil dependency by arguing that a large part of Norwegian oil and gas resources has to stay in the ground, at least until the climate crisis is under control. The alliance is linked to the historical current in Norwegian oil policy where a moderate pace was a central aim. For the first time since the 1970s, a demand to slow the pace of extraction and investment in the oil sector is being raised with significant force.

The aim of the alliance is to develop a concrete plan for how Norway as soon as possible can transform its economy from being fossil fuel based, to become not only carbon neutral but part of a global solution. In the church's yearly congress in April 2013, one of the main resolutions was a demand that "the term climate justice has to be given a concrete content, an increase in green jobs, reduced pace in extraction of oil … and a climate policy that takes solidarity between generations seriously."[46] The new mood was reflected in the yearly May Day parades organized by unions. In Oslo, an event where the prime minister, Jens Stoltenberg, was speaking, one of the dominant slogans in the large parade of thirty thousand that followed was, "No to drilling in Lofoten. For a slower pace of extraction of oil." The argument against Norway's oil dependency was surprisingly supported by a report from a large group of researchers at Norwegian Statistics. The report concluded that given a desire for domestic action and a concern for global emissions, "the majority of emission reductions should come through supply side measures, i.e., by downscaling Norwegian oil extraction."[47]

As expected, the "red-green" Stoltenberg-led government after eight years lost the election in September 2013. From an environmentalist point of view, the political situation might not have shifted substantially.

The government had given substantial support to rain forest protection abroad and taken a lead role for financing climate adaptation as part of the ongoing Kyoto II negotiations. However at home, the majority in Stoltenberg's Labour Party had a pro-oil position.

After the election, the Conservative Party had to rely on support from two center parties. These parties had put the protecion of Lo-Ve at the top of their agenda. As part of a deal the new government accepted these parties' position, thus keeping this important and sensitive area out of all rounds of oil concessions. Although this policy will likely last only as long as the coalition lasts, it is an important temporary victory for environmentalists and fishermen in the area. Still, because the potential oil reserves in the Lofoten-area is expected to be just a few percent compared to the total reserves on the Norwegian continental shelf, it is a more limited success seen from a climate perspective. However, the new Government soon discovered that giving in on the Lo-Ve issue was not enough to stem a more general, radical critique of Norway's oil policies and climate responsibilities.

The climate campaign also claimed some success in the sense that it was instrumental in making oil dependency one of the central issues in the election campaign. Norway's green party (not part of the red-green government) ran on a platform of phasing out Norwegian oil production over the next two decades and got their first representative in parliament in the election. In its first year in opposition the parties in the former government alliance have focused on a similar agenda. Most important was a development in the Labor Party, still Norway's largest, where, in a speech at the Social Democrat Youth's summer camp in July 2014, Jonas Gahr Støre, the labor party's new leader, and the opposition's prime minister candidate, introduced the prospect of leaving substantial amounts of Norwegian oil in the ground for climate reasons.[48]

Although the new climate alliance has yet to formalize a program, the aim is to show that Norway should use its comparative advantage when it comes to green energy, as extensively discussed in a book financed by and distributed among the participants.[49] This can uphold energy exports and employment, which in turn compensates for a necessary and substantial reduction in investment and production in the oil sector and instead intensifies investments and developments in the renewable industry where Norway is in a good position on both land and at sea with its long coastline. Another book, with a foreword by the leader of one of Norway's two public sector unions that was part of the alliance, presents a concrete strategy for how to reduce Norway's production of oil.[50] The strategy is based on the International Energy Agency's *World Energy Outlook for 2012* and other reports that state that if the global goal of maintaining no more than a two-degree rise in average temperatures is to be reached before 2050, only one-third of all proven petroleum reserves or less could be extracted.[51] This would mean that not only proven coal resources but also a large part of proven oil and gas reserves have to stay in the ground. A reduction will at one and the same time both open space for other activities and store resources for future production. Given the fact that Norway has already taken out so much more petroleum than it needs, the strategy assumes Norway has a moral responsibility to be an example for others; it can do that by deciding that proven reserves have to stay in the ground. At the same time, with the petroleum fund, Norway also has the means to make large investments in alternative sustainable energy forms globally.

As we saw in figure 10.1, the production of oil was halved between the peak in 2000 and 2012. However, if we look at the production of oil alone, it could be halved once more and still be greater than it was in the late 1980s, when Norway hit the limit of production set in the 1970s. In the mid-1980s, an even smaller production level gave Norway great incomes with oil prices around $20 per barrel. In 2012 the production could have been halved and more, and still contribute to large Norwegian export surplus on goods and services.

Given the climate dimension and the complexity of the oil sector, some of the demands from the campaign have to be complex. As of August 2014 there is no decision within the alliance on how this should be done. However, there is an agreement that new areas have to be temporally stopped. This includes Lo-Ve and areas farther north in the Norwegian Arctic, among them the island Jan Mayen, close to the eastern coast of Greenland.[52] In the book distributed by the campaign, a possible reduction in the pace of extraction and investment is combined with a plan to make sure that the consequences for workers are not too great. In this preliminary plan, temporary production of existing gas fields can be acceptable, but only if both contracts with buyers and income in Norway are linked to investment in a transition strategy (see chapter 4 for a related ethical argument). For example, the giant gas field Troll alone gives the Norwegian state an income of $5 billion every year. A large part of that income should be used for investment in offshore windmills and other important parts of a transition.

On May 5, 2013, the red-green government for the first time since the late 1980s pointed in the direction of cooling down the Norwegian oil sector by instituting new tax codes.[53] The oil companies' right to draw investment from income was reduced and its general company tax reduced 1 percent. Tax on oil was increased by 1 percent, leaving the oil industry with the same tax level as before but with other industries in a slightly better position. The changes might have been more symbolic than real, and their motive was purely economic, not based on climate concerns. Nevertheless, the symbolic value of these measures prompted an oil industry reaction. Statoil CEO Helge Lund and fourteen other oil CEOs strongly criticized the government's tax changes for oil companies in an article in the business daily newspaper *Dagens Næringsliv*.[54] Furthermore, Statoil postponed a prestigious oil development project in the Barents Sea (the Johan Castberg field) in June 2013. Several oil analysts read this as a "hit back" toward the state, and thus a politically motivated delay.[55] It had an effect: a vice minister at the Ministry of Finance assured the industry that the very modest tax raise would not be followed up with further rises in the future.[56] When the new Labor Party leader Gahr Støre one year later hinted that he was open to the argument of leaving some of Norway's oil in the ground based on climate concerns, he was criticized by both Lund and other key players in the industry.[57] These political contests are illustrative of how much power and money is involved in Norway's contemporary oil and gas struggles, as demonstrated in Ecuador's struggles, set out in chapters 5 and 11 in this book.

### Conclusion: The Return of the Tempo Debate

This chapter has analyzed how the Norwegian state departed from the pivotal objective of keeping a steady and low tempo on oil and gas extraction and offered some lessons for cooling down the petroleum state. The material outcomes that the Norwegian experience has produced are to a great extent what people feared in the early 1970s: that Norway would become a petroleum-dependent state. We have analyzed this as a politically uneven process where economic interests have prevailed over environmental interests, climate change concerns in particular. While environmentalism in the 1990s seemed to be partly co-opted by the state and partly made harmless by its focus on quotas in the Kyoto regime, we now see an emergent new third wave of environmentalism, an alliance that is willing to confront Norway's fossil fuel-based economy.

As we have emphasized in the chapter, the greatest challenge to keeping oil in the ground is the all-embracing oil sector, which has grown immensely over the past two decades. Leaving aside Norway's foresight, vision, and policies in the early 1970s, the past two decades of development suggest that indeed oil is a substance that is difficult for governments to handle (chapter 4). This means that the road to ending oil dependency is a long one (chapter 2), fraught with difficulties but also with hopeful signs. The main basis for change is a strong citizen movement, mobilizing from below, and developing a complex exit strategy. The challenge and the opportunity lie in combining the goal of ending oil dependency with concrete political struggles. Stopping expansion on a single site such as Lo-Ve, with its important ecosystem and fisheries is one step. To stay on target to fight climate change there is an obvious need to make sure that oil in large areas with both proven and potential unproven reserves is kept in the ground.

## Notes

1. Helge Ryggvik, *Til siste dråpe, om oljens politiske økonomi* [To the last drop, the political economy of oil] (Oslo: Aschehoug, 2009).

2. Berit Kristoffersen, "'Securing' Geography: Framings, Logics and Strategies in the Norwegian High North," in *Polar Geopolitics: Knowledges, Resources and Legal Regimes*, ed. Richard Powell and Klaus Dodds (Cheltenham, UK: Edward Elgar, 2014). The chapter emphasizes how this competing strategy toward mitigating the causes of climate change can be framed as opportunistic adaptation, where leading governmental figures and the petroleum industry argue in terms that it is not Norway's task to resolve this Arctic paradox but to manage it.

3. Ryggvik, *Til siste dråpe*.

4. Helge Ryggvik, *The Norwegian Oil Experience: A Toolbox for Managing Resources?* (Oslo: Centre for Technology, Innovation and Culture, 2010).

5. B. Kristoffersen and S. Young, "Geographies of Security and Statehood in Norway's 'Battle of the North,'" *Geoforum* 41 (2010): 577–84.

6. Damian Carrington, "Carbon Bubble Will Plunge the World into Another Financial Crisis: Report," *Guardian*, April 18, 2013, http://www.guardian.co.uk/environment/2013/apr/19/carbon-bubble-financial-crash-crisis.

7. Andreas Ytterstad, *100.000 klimajobber og grønne arbeidsplasser, For en klimaløsning nedenfra* [A 100.000 green and climate firendly jobs, working towards climate solutions from below ] (Oslo: Gyldendal, 2013). Helge Ryggvik, *Olje og Klima, En strategi for nedkjøling* [Oil and the climate: A strategy for cooling down] (Oslo: Gyldendal, 2013).

8. *Innstilling fra den forsterkede industrikomité om undersøkelse etter og utvinning av undersjøiske naturforekomster på den norske kontinentalsokkel m.m.* [Recommendation from the extended industrial committee about the exploration for and extraction of underwater natural deposits on the Norwegian Continental Shelf etc.]. *Innstillingen* 94 (1970–1971): 7–94.

9. Ibid.

10. Petroleumsvirksomhetens plass i det norske samfunnet [The role of petroleum activities in Norwegian society]. St. meld. [White paper] no. 25. (1973–1974)

11. Ibid., 6.

12. Ibid.

13. Norway most often uses the term *oil equivalent* when production level of oil and gas is measured. This is partly because it makes it possible to lump oil and gas together. In the first years, the term *ton* was used. Soon it became standard cubic meter (Sm³): 1 Sm³ = 0. 84 ton oil. 1 Sm³ oil = 6.29 barrels. Innst. S. 275 (1973–74), 4. (Parliament Resolution).

14. D. H. Meadows, D. Meadows, J. Randers, and W. Behrens, *Limits to Growth* (New York: Universe Books, 1972).

16. Ukontrollert utblåsning på Bravo 22 [Uncontrolled blowout on Bravo 22], NOU [Official Norwegian report] (1977), 47.

17. Petroleumsundersøkelser nord for 62. Breddegrad [Petroleum exploration north of the 62nd latitude]. St. meld. [White paper] 91 (1975–1976).

18. Petroleumsundersøkelser nord for 62. Breddegrad, St.meld. 91 (1975-1976), 42.

18. Helge Ryggvik and Marie Smith Solbakken, *Norsk oljehistorie Bind 3, Blod, svette og olje* [Norwegian oil history, vol. 3, Blood, sweat and oil] (Oslo: Ad Notam, 1997).

19. Ril styrking av oljevernberedskap, 4179 [Debates in the Parliament]. St. tid., sak 26, till. bev.

20. www.arkivverket.no/arkivverket/Arkivverket/Stavanger/Nettustsillingar/Gullfaks-A/Utbygging-og-drift.

21. *Petroleumsvirksomhetens fremtid (Tempoplanen)* [The future of oil activities [(The Tempo plan)]. NOU [Official Norwegian report] (1983), 23.

22. St. meld. [White Paper] no. 1 (1987–88).

23. See note 13.

24. Statistics Norway, *Historisk statistikk* [Historical statistics] (1994), 396; The Petroleum Directorate's *Facts* gives a figure of 57 billion.

25. *Utfordringer og perspektiver for petroleumsvirksomheten på kontinental-sokkelen* [Challenges and perspectives for petroleum activities on the continental shelf]. St. meld. [White paper] no. 26 (1993–1994), 54.

26. Ministry of Petroleum and Energy, *Fakta norsk petroleumsverksemd* (2009), 210. Available in English as *Facts—the Norwegian petroleum sector*, http://www.npd.no/en/Publications/Facts.

27. United Nations World Commission on Environment and Development, *Our Common Future: Report of the World Commission on Environment and Development* (New York: Oxford University Press, 1987).

28. Christian Hunold and John Dryzek, "Green Political Strategy and the State: Combining Political Theory and Comparative History," in *The State and The*

*Global Ecological Crisis,* ed. John Barry and Robyn Eckersley (Cambridge, MA: MIT Press, 2005).

29. Statistics Norway, *Emissions of Greenhouse Gases by Source, 1990–2009.*

30. Marit Reitan, "Interesser og institusjoner i miljøpolitikken" [Interests and institutions in environmental politics] (PhD diss., University of Oslo, 2008).

31. Hunold and Dryzek, "Green Political Strategy and the State," 82–84.

32. E. Hovden and G. Lindseth, "Discourses in Norwegian Climate Policy: National Action or Thinking Globally?" *Political Studies* 52 (2004): 63–81.

33. Ibid.

34. In 2005 Norway established the agency Oil for Development under the Ministry of Foreign Affairs for this purpose.

35. Norwegian Petroleum Directorate, *Fact Sheet 2012,* 135. Online at www .npd.no

36 Spår kraftige lønnskutt i Norge [Predicts sharp wage cuts in Norway], Dagens næringsliv [Business Daily], April 4, 2013, http://www.dn.no/forsiden/naringsliv/article2597142.ece.

37. Norwegian hydropower system generates an average of 122 terawatt-hours of electricity a year.

38. Global Wind Energy Council (http://www.gwec.net). At the end of 2012 Norway was producing 708 megawatts from wind. Sweden was producing 3,745 megawatts. Denmark was producing 4,162 megawatts.

39. "Myten om den reine, norske oljen" [The myth about the clean Norwegian oil], *Dag og Tid* [weekly newspaper Day and Time], January 13, 2012, http://dagogtid.no/nyhet.cfm?nyhetid=2192.

40. *KonKraft Rapport 5,* Konkraft, Petroleumsnæringen og klimaspørsmål [Konkraft, the petroleum industry and questions about climate change], s. 30 (2008), http://www.konkraft.no/default.asp?id=1005.

41. International Oil & Gas Producers Association, *Environmental Performance Indicators— 2011 Data.* (London: International Oil & Gas Producers Association, 2012, http://www.ogp.org.uk/pubs/2011e.pdf

42. Ibid.

43. Kristoffersen and Young, "Geographies of Security and Statehood in Norway's 'Battle of the North'"; O. A. Misund and E. Olsen, Lofoten–Vesterålen: For Cod and Cod Fisheries, But Not for Oil?" *Journal of Marine Science: Journal du Conseil* 70 (2013): 722–25.

44. Leif Christian Jensen and Berit Kristoffersen, "Nord-Norge som ressursprovins: Storpolitikk, risiko og virkelighetskamp" [Northern Norway as a resource province: High politics, risks, and the negotiation of reality], in *Hvor går Nord-Norge? Politiske tidslinjer* [Where is Northern Norway headed? Political timelines] ed. Svein Jentoft, Kjell Arne Røvik, and Jens-Ivar Nergård (Stamsund: Orkana Forlag, 2013), 67–80.

45. "Ber regjeringen dempe aktiviteten på norsk sokkel" [Asks the government to lower the activity level on the Norwegian Shelf], Dagens Næringsliv [Business Daily], June 29, 2012.

46. The Church Conference statement on April 12th 2013: "Forsvarlig etisk forvalting av norske petroleumsressurser og bruk av oljefondet" [An ethically responsible management of the Norwegian petroleum resources and use of the oil fund].

47. Taran Fæhn, Cathrine Hagem, Lars Lindholdt, Ståle Mæland, and Knut Einar Rosendahl, "Climate Policies in a Fossil Fuel Producing Country: Demand versus Supply Side Policies" (Statistics Norway, discussion paper 747, June 2013), http://www.ssb.no/en/forskning/discussion-papers/_attachment/123895?_ts=13f51e5e7c8.

48. Støres tale till AUF [Støre's speech to the Labour Youth party], at Norwegian Broadcasting, July 12, 2014. http://www.nrk.no/video/stores_tale_til_auf/E9D630A8D43BA638.

49. Ytterstad, *100.000 klimajobber og grønne arbeidsplasser*, 29.

50. Ryggvik, *Olje og klima.*

51. International Energy Agency, *World Energy Outlook for 2012* (Paris: International Energy Agency, 2012), 3.

52. Ryggvik, *Olje og klima.*

53. Ministry of Finance, "Redusert friinntekt i petroleumsskatten" [Reducing free income in the petroleum taxation], press release, May 5, 2013, http://www.regjeringen.no/nb/dep/fin/pressesenter/pressemeldinger/2013/redusert-friinntekt-i-petroleumsskatten.html?id=725999.

54. "Rokker ved forutsigbarheten" [Undermines the predictability], *Dagens Næringsliv* [Business Daily], May 22, 2013, 3.

55. "Statoil Johan Castberg Postponement a Political Stance," Say Oil Analysts," Stavanger Aftenblad [Stavanger evening post], June 5, 2013, http://www.aftenbladet.no/energi/aenergy/Statoil-Johan-Castberg-postponement-a-political-stance_-say-oil-analysts-3191049.html#.UbC4suuHfIp.

56. "Uenighet om fri-inntekt" [Disagreement on taxation write-offs], Aftenposten [The Evening Post], June 6, 2013, see also commentary by the government's State Secretary Kjetil Lund http://www.regjeringen.no/nb/dep/fin/aktuelt/taler_artikler/taler_og_artikler_av_ovrig_politisk_lede/taler-og-artikler-av-statssekretar-kjeti/2013/uenighet-om-friinntekt.html?id=729218.

57. "Statoil-sjefen advarer mot ny oljekurs Dagens Næringsliv [Statoil CIO warns against new oil policy], Dagens Næringsliv Business Daily, August 21, 2014 http://www.dn.no/nyheter/energi/2014/08/20/2157/Olje/statoilsjefen-advarer-mot-ny-oljekurs.

# Part 3

## The Politics of Delegitimization

# 11

## The Good Life (*Sumak Kawsay*) and the Good Mind (*Ganigonhi:oh*): Indigenous Values and Keeping Fossil Fuels in the Ground

Jack P. Manno and Pamela L. Martin

This book has evolved out of what we three coeditors have come to call the Keep Them (fossil fuels) in the Ground (KTIG) Project. We were drawn together by similar research interests and an emerging theoretical stance on why it's important to focus our analyses of fossil fuels on the subterranean, on that which happens before fossil fuels enter the accounting books, become part of the supply chain, and get registered as available energy. In these discussions we (Martin and Manno) noticed that we had both been deeply influenced by our work with Indigenous[1] communities: Pam by the Indigenous peoples of the Ecuadorian Amazon and Jack by the Onondaga people, the "People of the Hills," the "keepers of the fire," or the capital of the Haudensoaunee (Iroquois) Confederacy in what is now central New York State.

These communities are very different, but each has been embattled in efforts to keep fossil fuels in the ground. The Indigenous of the Amazon are a forest people, some living in voluntary isolation in forests targeted for oil development. While some have been spokespeople of the opposition to oil development in their territory, many remain largely out of the limelight of Ecuadorian politics, trying to sustain their lives as forest people. The people of the Onondaga Nation live on unceded, sovereign territory close to a metropolitan area. For the Onondaga, the threat is gas embedded in vast shale formations suddenly being promoted as available and therefore, as energy politics have it, necessarily and inevitably to be mined. The shale gas boosterism has followed from new industrial-scale techniques of blasting shale rock with chemical-laden water designed to free microseams of gas from deep beneath the surface. This is land the Onondaga consider part of their aboriginal territory, subject to a 2006 legal action in US federal court in which the Onondaga sought a declaration that their land had been taken illegally and their aboriginal title was legally intact. In both cases, oil in the Amazon and fracking in New

York, arguments have been made that draw on an Indigenous philosophy of "right relation" between humans and the land.

In this chapter, we focus on two key concepts: one known as the good life—*sumak kawsay*, in the Quichua language and *buen vivir* in Spanish—and the other the good mind (*ganigonhi:yoh*), in the language of the Onondaga, the central keepers of the fire, or capital, of the Haudenosaunee confederacy.[2] In these cases, we were interested to see how these two similar concepts influence and inspire social movements that are beginning to imagine a path beyond fossil fuels. In many ways, the conflicts over energy are over what constitutes a good life and how reasoned minds think about their world and their place in it. After all, fossil fuels, especially as refined into liquid gasoline, have become the lifeblood[3] of the American way of life, a particular model of the good life defined by ownership of a single-family home, a thick grass lawn, and one's own gas-powered automobile rolling down highways; by a life detached from a particular place and one inherently dependent on inexpensive fossil fuels; by a way of life vigorously exported far beyond US borders. In opposing fossil fuel mining, the people of the Ecuadorian Amazon and their allies and the Haudenosaunee (Iroquois) people and their allies in upstate New York have drawn on their traditional values and philosophical perspectives that are complex and highly developed, having evolved over centuries, even millennia.

While we focus on Amazonian and Haudenosaunee peoples, many other Indigenous peoples and nations have been in the forefront opposing extreme energy and mineral developments in many parts of the world: Nigeria, Philippines, Canada, Peru, Colombia, Bolivia, and others.[4] Indigenous voices often provide a particular kind of ethical rationale for keeping fossil fuels underground. In addition to defending their lands and livelihoods, their opposition is often based on a critique of the American way of life, which they see lacking an appreciation of the duties and responsibilities that human beings have to the land and the environment, a duty that comes as a condition for being fully human, fully alive, and in good mind.

We would be exaggerating if we asserted that all Indigenous peoples share common philosophies of environmental stewardship or that no Indigenous peoples have enthusiastically embraced extreme forms of fossil fuel extraction. Yet despite often bitter and sometimes violent differences in opinion among Indigenous peoples about specific energy developments, it is also true that Indigenous peoples around the world have at their core a sense of gratitude for the riches of the natural world, a sense inherently violated by fossil fuel extraction. As Potawatami scientist and author Robin Kimmerer has written. "Many Native peoples across the world, despite myriad cultural differences, have this in common—we are rooted in cultures of gratitude."[5] As Oren Lyons, Onondaga Nation faithkeeper and activist for the rights of Indigenous peoples, has written, "Industrial societies lack the understanding of the interrelationships that bind all living things."[6] The interrelationship is one of mutual dependence, or interdependence and mutual gratitude for the care taken and the work done in carrying out one's duties, also sometimes referred to as one's "original instructions."[7] Kimmerer has also written about a "Code of the Honorable Harvest" that is similar among the varied peoples of the Great Lakes of North America:

The guidelines for the Honorable Harvest are not written down, or even consistently spoken of as a whole—they are reinforced in small acts of daily life. But if you were to list them, they might look something like this:
Know the ways of the ones who take care of you, so that you may take care of them.
Introduce yourself. Be accountable as the one who comes asking for life.
Ask permission before taking. Abide by the answer.
Never take the first. Never take the last.
Take only what you need.
Take only that which is given.
Never take more than half. Leave some for others.
Harvest in a way that minimizes harm.
Use it respectfully. Never waste what you have taken.
Share.
Give thanks for what you have been given. Give a gift, in reciprocity for what you have taken.
*Sustain the ones who sustain you and the earth will last forever.*[8]

Kimmerer concludes, "Deeply rooted in cultures of gratitude, this ancient rule is not just to take only what you need, but to take only what is given. Taking coal buried deep in the earth, for which we must inflict irreparable damage, violates every precept of the code. By no stretch of the imagination is coal *given* to us. We have to wound the land and water to gouge it out of Mother Earth."[9]

In this chapter, we focus on two examples of how Indigenous knowledge and Indigenous values have become significant features of movements to keep fossil fuels in the ground. Indigenous peoples in the Americas and their allies have drawn on these values in articulating reasons for keeping fossil fuels in the ground and bolstering society-wide efforts to oppose expanded oil development in the Ecuadorian Amazon and maintain a moratorium on hydrofracking in New York State.[10] We argue that the Indigenous perspectives outlined here provide an alternative understanding, ancient yet new, of contemporary politics. These perspectives are pregnant with the potential for deciding to keep fossil fuels in the ground, with accepting the responsibilities that accompany using the gifts of nature, with articulating meaningful goals such as the good life in exercising such power.

The Haudenosaunee environmental task force has stated, "As more time passes, western society has begun to feel the limit of our resources and the message of the Haudenosaunee has begun to be heard."[11] In South America, that message has been heard at the highest levels of government. Bolivia and Ecuador have incorporated Indigenous values in their recent constitutions (2008) and plans for development. The concept of the good life (*buen vivir*) is a pillar. In essence, these movements seek paradigm change, locally, nationally, and even globally.[12] As the world economy expands and seeks new reservoirs of natural resources and as the low-hanging fruit of raw materials and high-quality fuels becomes scarce, transnational and national companies are penetrating farther into the lands of traditional peoples. In fact, worldwide a growing number of natural resource extractive industries and countermovements are located in Indigenous territories, which explains the great need to understand the impacts on these ancient peoples and their societies, as well as their proposed solutions for a sustainable future. In the remainder of this chapter, we discuss the international Indigenous rights movement more broadly and then interweave the two cases, the Yasuní-ITT Initiative and new oil concessions into the Amazon, the good life and oil, and the Onondaga, the good mind, and hydrofracking for shale gas.

## Indigenous Resurgence, Indigenism, and the UN Declaration on the Rights of Indigenous Peoples

Indigenous peoples who have survived politically and culturally, now numbering around 400 million people worldwide, are those whom Martin and Wilmer refer to as peoples descended from (and who have maintained cultural ties to) the original inhabitants of a place where state institutions not of their own making assert jurisdiction and, as a consequence, do not now control their political destinies."[13] We think it's important to emphasize that such a definition is not just about knowledge or place; it is political. The growing movement for the rights of Indigenous peoples to regain control of their destinies and associated movements for the rights of nature and the rights of Mother Earth are efforts to reassert a degree of Indigenous sovereignty along with a recognition of Indigenous sense of responsibility for Mother Earth and the nonhuman beings in which they once cohabited in relations of interdependence and coevolution.[14]

Ronald Niezen, an anthropologist of law and social movements at McGill University, describes how many Indigenous groups have turned to the United Nations (UN) and the international movement of human rights after experiencing a common sense of futility when appealing to the courts and national legislatures, all dominated by the worldviews of colonial and settler societies.[15] The strength of this movement is reflected in the 2007 adoption by the UN General Assembly of the UN Declaration on the Rights of Indigenous Peoples (UNDRIP). Its preamble states that "Indigenous peoples have suffered from historic injustices as a result of, inter alia, their colonization and dispossession of their lands, territories and resources, thus preventing them from exercising, in particular, their right to development in accordance with their own needs and interests," and article 32 that "Indigenous peoples have the right to determine and develop priorities and strategies for the development or use of their lands or territories and other resources."[16]

The goal of the movement Niezen terms *Indigenism* is to gain international recognition of the distinct collective rights of self-determination of Indigenous peoples. These peoples insist that UN member states recognize and honor Indigenous history, Indigenous culture, the harms done by their past dispossession, their right to pursue their own forms of development, and their right to just compensation for past wrongs. Niezen defines Indigenous peoples in this movement as descendants of original inhabitants who were displaced or outnumbered by the arrival of newcomers (settlers), maintain significant aspects of the their distinct culture, select or are forced to remain on the margins of the dominant society's political and economic life, and whose land and resources are subject to or threatened by unwanted development.

While the General Assembly's adoption of UNDRIP has arguably had little effect on member states' energy development policies, the declaration has created opportunities for Indigenous understandings, expectations, legal rights, and spiritual perspectives to be broadly communicated. It has empowered Indigenous peoples to join as equals, or even as elder relatives, in the debate over international environment and development norms, nudging the discourse toward the conclusion that the rights of Indigenous peoples necessarily imply that the nation-states also have duties to protect land, air, water, and nonhuman beings who cannot speak themselves. In this book, we suggest this may sometimes mean keeping fossil fuels in the ground.

The internationalization of Indigenous voices is often traced back to the early efforts of the International Indian Treaty Council (IITC) formed during a 1974 gathering organized by the American Indian Movement and attended by representatives of ninety-eight Indigenous nations. The IITC became a voice of Indigenous peoples of the Western Hemisphere. They recognized "common bonds of spirituality, ties to the land and respect for traditional cultures."[17] In 1977, Indigenous peoples were invited for the first time to a meeting of nongovernmental organizations (NGOs) recognized by the United Nations to be held in Geneva to discuss a way forward toward international recognition of Indigenous rights.

From the beginning of their interventions into international law and policy, Haudenosaunee intellectuals focused on energy and extraction in explaining the oppression of Indigenous peoples and the dispossession of their lands. As the Haudenosaunee prepared for the NGO meeting in Geneva, the chiefs of the Grand Council of the Haudenosaunee Confederacy asked John Mohawk (Seneca), who would later become a professor at the State University of New York, to draft three papers presenting the history and current situation facing the Haudenosaunee.[18] These papers were presented in Geneva and later published in a book, *Basic Call to Consciousness: The Haudenosaunee Address to the Western World*.[19] In it, the Haudenosaunee assert that the reason European settlers attempted to destroy Indigenous peoples and nations was that the original peoples stood in the way of the total extractive exploitation of the forests and the lands. It was all about energy: "The hardwood forests of the Northeast were not cleared for the purpose of providing farmlands. Those forests were destroyed to create charcoal for the forges of the iron smelters and blacksmiths. By the 1890s, the West had turned to coal, a fossil fuel, to provide the energy necessary for the many new forms of machinery that had been developed. During the first half of the Twentieth Century, oil had replaced coal as a source of energy."[20] *Basic Call* asserted that the Western culture with its the leveling of the forests, polluted waters, and the near extinction of the buffalo "has been horribly exploitative and destructive of the Natural World." The authors pointed out the many species that had been eliminated "because they were unusable in the eyes of the invaders." Throughout their message runs a theme of overexploitation of resources. Someday, they asserted, the world will inevitably come up against limits: "Like the hardwood forests, the fossil fuels are also finite resources."[21] With great prescience, the Haudenosaunee wrote in 1977, "We think even the systems of weather are changing. Our ancient teaching warned us that if Man interfered with the Natural Laws, these things would come to be. When the last of the Natural Way of Life is gone, all hope for human survival will be gone with it. And our Way of Life is fast disappearing, a victim of the destructive processes."[22] Such is the message that has been long ignored in mainstream policymaking in the United States and elsewhere. But as the Indigenous predictions are bearing out, some actors are taking notice. One set comprises leaders and civil society members in Ecuador and New York State.

### Resistance to Oil and Gas Extraction

The Ecuadorian and Haudenosaunee peoples have created political movements to resist further destruction of their lands. Within an emerging system of global environmental governance, the Yasuní-ITT Initiative of Ecuador and its previously associated United Nations Development Programme (UNDP) Trust Fund have established a global precedent and compensation mechanism for leaving oil underground in the ITT block of the Ecuadorian Amazon (chapter 5). It is a "concrete demonstration of the good life" for the peoples of Ecuador and around the globe, says Daniel Ortega, director of the environment and climate change in Ecuador's Ministry of Foreign Affairs.[23] Haudenosaunee (Iroquois) environmental leaders are educating and organizing to prevent hydrofracking in shale rock throughout their aboriginal territory in much of New York State and southern Ontario.

In this section, we survey first the resistance to new oil concessions in the southern Amazon, particularly the Yasuní-ITT Initiative designed to preclude oil development in the Amazonian lands of, among other Indigenous groups, the Tagaeri and Taromenane peoples who live in voluntary isolation within the boundaries of Ecuador's Yasuní National Park and, second, the resistance of industrial-scale hydraulic fracturing of shale to access natural gas in upstate New York in traditional Haudenosaunee territories. The cases feature very different peoples who nonetheless bring traditional knowledge and Indigenous norms of sustainable development to bear on modern problems. Both have turned to international institutions for intervention and redress.

In Ecuador, Indigenous peoples are struggling to enforce the Yasuní-ITT Initiative to keep oil in the ground through payments to compensate

Ecuador for forgoing the opportunity to develop a significant source of oil in the Ishpingo Tambococha Tiputini (ITT) oil block of Yasuní National Park. And to the south of Yasuní, Indigenous groups are fighting against a new round of oil concessions in their territories, which have never before been touched for extractive processes. In the United States, the Haudenosaunee have looked to a newly articulated body of human rights particular to Indigenous peoples as expressed in the United Nations Declaration on the Right of Indigenous Peoples and have asked the UN special rapporteur on the rights of Indigenous peoples, James Anaya, to address rights issues associated with hydrofracking for natural gas near and beneath Haudenosaunee territory. The Haudenosaunee and other Indigenous groups first took their grievances to the United Nations in 1977 when they began the drafting process that eventually became the UN Declaration on the Rights of Indigenous Peoples, adopted by the General Assembly in 2007. In May 2012 at the North American Indigenous Caucus of the UN Permanent Forum on Indigenous Issues, they called for "a complete ban on the method of natural gas drilling known as "hydraulic fracturing," or "hydrofracking," within the traditional territory and treaty lands of the Haudenosaunee Confederacy, because everywhere hydrofracking will contaminate land, air, and water.

The Yasuní-ITT Initiative is designed as an innovative global environmental governance structure whereby Ecuador would draw funds from a UNDP trust fund to compensate the Ecuadorian government for forgoing opportunities to mine for oil in the ITT block of the Amazon. As of this writing, Indigenous and non-Indigenous peoples of Ecuador are united in protest against President Rafael Correa's withdrawal of the first global plan to keep oil underground. The Haudenonsaunee efforts to ban industrial-scale hydrofracking require it to respond to the New York Department of Environmental Conservation procedures for gathering comments on a proposed approach to regulating industrial-scale natural gas mining from shale rock. In both cases, Indigenous peoples are allied with non-Indigenous groups while navigating a dominant institutional landscape that has historically been indifferent to Indigenous knowledge and hostile to the rights of Indigenous peoples. In both cases, Indigenous peoples and their allies are introducing a new logic of resistance to fossil fuel mining—in the Amazonian case, in a remote or hard-to-get-at area that is highly vulnerable to the environmental disruption, such as drilling for oil entails, and in the Haudenosaunee case, on or near ancestral land actively in adjudication by US federal courts. In both cases, the keep it in the ground (KIIG) proponents are drawing on recent global initiatives for the rights of Indigenous peoples and the rights of nature (and for the rights of all components of the natural world, including humans, to be free and able to carry out their duties and responsibilities) to bolster and support their goals. Key themes from these nascent movements are both influenced by and influence two very different peoples, each publicly expressing Indigenous perspectives to make a case for keeping valuable but harmful resources in the ground.

### Yasuní-ITT and the Southern Amazon

The Yasuní-ITT proposal calls for coresponsibility with the rest of the world (common, but differentiated) in avoiding emissions that the nearly 900 million barrels of oil in the ITT block could produce. The world would pay for avoided carbon emissions in order to protect one of the most biodiverse plots on earth. The $350 million per year (50 percent of its expected earnings from oil extraction) that Ecuador seeks each year for ten years would be placed in a UNDP trust fund with a board of directors that includes Ecuadorians, as well as members of the global community. If successful, it would be one of the largest global environmental trust funds of its kind. The funds would be directed to two funding windows: a capital fund and a revenue fund. As of August 2013, President Correa closed the funding mechanisms, with approval from the National Assembly, in part of the Yasuní Park block 43 while Indigenous movements and supporters protest in the streets of Quito against any oil extraction in this area and are seeking international legal mechanisms to challenge this decision (Inter-American Commission on Human Rights, for example) (see chapter 5).

While scientists globally have noted the high levels of biodiversity in the Yasuní National Park,[24] there is also a significant human survival element to this story: people who live in voluntary isolation, the Tagaeri and Taromenane Indigenous peoples. Their designated protected zone (*zona intangible*) is within the park and to the southern portion of the ITT block. The combination of nonemitted carbon, or leaving oil underground, plus the high levels of biodiversity and social components of this proposal illustrate its multifactor approach centered on Ecuador's 2008 Constitution, which guarantees rights to a diverse citizenry, as well as rights to nature. This constitution is based on the indigenous norm of *buen vivir*, or *sumak kawsay*—well-being or the good life. Thus, the Yasuní-ITT Initiative is not one born of transnational environmental networks from the North or goals of conservation from a Western perspective, but rather is deeply rooted in building a local and global community infused with the indigenous norm of well-being. Significantly, the funds from the initiative were to be invested in renewable-energy projects with the goal of transitioning Ecuador to a postpetroleum state—an affirmative indicator of KIIG policy that now hangs in the balance as Ecuador readies for extraction of oil from 0.01 percent of the ITT block.

Still, the boosterist oil economy (chapter 3) of the Ecuadorian state and growing debt from China have prompted new plans for sixteen oil blocks in the untouched southern Amazonian provinces of Pastaza and Morona Santiago where seven indigenous communities—the Zapara, Quichua, Huoarani, Achuar, Shuar, Taromenane, and Tagaeri—live. The Quichua community of Sarayacu, which lies within the oil round area proposed by the Ecuadorian government in 2013, has a long history of resistance to oil development and took its case to the Inter-American Commission on Human Rights in July 2012. The commission ruled in favor of the Quichua community of Sarayaku and recommended the Ecuadorian government enforce its policy of prior consent and respect for the community and its territory.[25] And so the fight against oil extraction continues in the southern Amazon. The underpinning of resistance to what the Ecuadorian government terms "development" is *sumak kawsay*, the good life.

Manari Ushigua, president of the National Indigenous Confederation of Ecuador (CONAIE), explains that for Indigenous peoples of the Amazon, oil extraction is not "development; for us it is a project that will end our culture, our lives, our history, our political ideology, and our form of living with the natural world."[26] For this reason, CONAIE has mobilized against the new round of oil concessions to continue the impetus of the Yasuní-ITT Initiative and keep the oil under the ground. While the Yasuní-ITT Initiative and Ecuador's constitution provide a positive road map for change, the case of southern Ecuadorian resistance to new oil extraction also highlights the challenges of implementing the good life within the modern world system of natural resource extraction and the integration of Indigenous views through Ecuador's constitution to the broader Ecuadorian society and global economy.

### Onondaga and Haudenosaunee Opposition to Industrial Scale Hydrofracking

The Onondaga Nation to the north faces similar constraints and opportunities. *Hydrofracking* is a shorthand term referring to innovations that have made it technically feasible and deceptively profitable to extract natural gas from shale formations deep underground.[27] The process injects high volumes of a potentially toxic mixture of water, sand, and "fracking" chemicals under high pressure to create cracks in the shale through which gas is released to be brought to the surface. To produce enough natural gas at an economically viable price, the process must be done at an industrial scale through the innovation of horizontal drilling, where multiple drill heads fan out horizontally in several directions from a single shaft. Even before fracking began in the Pennsylvania portions of the Marcellus Shale, "landsmen," or door-to-door purveyors of leasing agreements, were approaching New York landowners, and applications for drilling permits began to appear at the offices of the New York State Department of Environmental Conservation (NYSDEC). Because of the size and variety of potential environmental effects, NYSDEC determined that New York's normal procedures for regulating gas drilling would not suffice to regulate the new technology.[28] The usual process for conventional gas drilling involved the gas company submitting what is known as a general environmental impact statement (GEIS) that would be reviewed by NYSDEC. The department determined that the standard GEIS would not suffice to assess the potential environmental consequences of the new fracking technology and the potential hazards that accompanied it. It then drafted a supplemental GEIS and asked for public comment.

What ensued was a flood of commentary. More than 15,000 comments were received. As of this writing, the natural gas in the Marcellus Shale remains deep in the ground. From the beginning, the Haudenosaunee, through the traditional leadership of the Onondaga Nation, the capital of the confederacy, and through the Haudenosaunee Environmental Task Force, has actively opposed hydrofracking in the ancestral territory. Their arguments are based on their Indigenous perspective on the relation between humans and the natural world. Their philosophy of the good mind echoes and perhaps reinforces the Indigenous norm of *buen vivir*, or *sumak kawsay*—well-being or the good life.

## The Politics of the Good Life (*Buen Vivir*) and the Good Mind (*Ganigonhi:oh*)

The concept of well-being/the good life, or *sumak kawsay*, refers to living in harmony with nature rather than dominating nature or conserving it like a museum to observe and preserve but not actively live in it. The "good life" refers to Indigenous recognition of the inherent human connection to the earth and the need to live within the natural environment as a balanced, or sustainable, process. This norm is codified in the 2008 Ecuadorian Constitution, which states in the preamble, "We decide to construct a new form of civil society, in diversity and harmony with nature to achieve el *buen vivir*, el *sumak kawsay*."[29] This norm has echoed throughout the Andes in Venezuela, Bolivia, and Colombia, including in the Bolivian Constitution and more broadly in the international declarations of indigenous rights movements.[30]

The words *buen vivir* and g*anigonhi:oh* express a particular understanding of fundamental relations between humans and nature and the way that Indigenous peoples weave their lives within and around their environments. In our work with these Indigenous peoples, we have found that one of the key forces driving the antiextraction movements to keep fossil fuels in the ground is a new politics formulated in the context of an Indigenous understanding of the good life and the good mind. These concepts are spreading to the descendants of the settlers, nonnative people who are aligning themselves with the Indigenous rights movement not only as a human rights issue but also as a way to promote a responsibilities-based legal framework based on a rich conceptualization for humanity's responsibility to care for the ecosystems on which all life depends. Ecuadorians, both Indigenous and not, believe so strongly in the goal of living the good life that they have included it in their new constitution, passed in Montecristi in 2008. The term *sumak kawsay* is the backbone of the policies that outline the rest of the constitution and the country's National Development Plan (Senplades). In their constitution, Ecuadorians have granted rights to nature—a first of its kind. Overall, the politics of the good life are driving elements in keeping oil underground in one part of the Amazon (for now) and in pursuing alternative energy policies in this resource-rich country.

The Haudenosaunee concept of the good mind similarly refers to a state of recognition of one's dependence on all aspects of the natural world and the resulting responsibility that humans have to maintain and sustain conditions conducive to the flourishing of life. It is captured most fully in what is known as the Thanksgiving Address. Before and following every important occasion, council meeting, or ceremony, the people are asked to bring their minds together to acknowledge with greetings and thanks, love, and respect for the special gifts provided by each of the elements of Creation. The ceremonial oration is known as The Words That Come before All Else. It is not a prescribed speech that is always the same. It is shorter or longer, more florid or direct, depending on the occasion. But the elements always remain the same. It begins by thanking the people who are present and giving thanks that they are well and could make it to the gathering. From there the earth is thanked; the waters come next, then the green plants, the trees, the forest animals and birds, the winds and thunders, the sun, moon, and stars, the teachers who remind the people and help them sustain the good mind, and finally the Creator. The speaker always mentions that surely he left things out and asks those present to give thanks to the forgotten ones. The address is meant to remind everyone of the web of life made up of beings that are intrinsically and perpetually dependent on each other. Life is fragile. In the Haudenosaunee Creation story, the continuance of life is the result of a game of chance, with the odds heavily against it. Each day that the world goes on is a day for gratitude for each element of Creation, each fellow being who continues to carry out its particular way of life, its intrinsic duties and powers that are considered by the Haudenosaunee to be original instructions from the Creator. Among the obligations that human beings have is to give thanks together, and that, in turn, encourages life to continue.

Onondaga Turtle Clan Mother, Freida Jacques, explains: "As Haudenosaunee, we give thanks to all the parts of Creation that make life possible here on Earth. This keeps us connected with the very vital purpose of all living things. So our respect and love includes all parts of Creation. This understanding helps us use the Good Mind in our interactions with the natural world around us."[31] The good mind is central to how the Haudenosaunee consider gas drilling in their ancestral territory. As the Tadodaho Sid Hill, a political and spiritual leader, put it, "You do one thing over here, hydrofracking, now what is that going to do to the animals, what's it going to do to the water, what's it going to do to the air. No, we can't be a part of that (hydrofracking), we can't promote it. .... Everything is related ... we're talking about the animals and who their leaders are and the plants and who their leaders are. They are not just things. They have duties. They're related to us. We understand this. We do take from Mother Earth but we don't take more than what we need." According to the late author and Haudenosaunee philosopher John Mohawk, three main principles—peace, justice, and the power of good minds coming together as one to make decision by consensus—have guided Haudenosaunee thought since its founding constitution, known as the Great Law of Peace, ended a long period of warfare and brought the five nations together a thousand or more years ago and established a confederacy.

The Onondaga Nation responded to the NYSDEC's suggestion to ban hydrofracking in the watersheds from which New York City and Syracuse draw their drinking water:

The Onondaga Nation respectfully disagrees with the criteria used above to protect some areas from hydrofracking and not others. All waters are used by someone, be they human or animal, bird or fish. Also, outside of New York's cities, everyone relies on private wells that draw drinking water from mostly shallow aquifers. Therefore, all waters deserve protection. The Onondaga further stated that they are guided by Original Teachings, so that they may have a good life on this earth and in the future. These teachings include not digging deep into Mother Earth below the surface that people live on. These teachings are a form of Traditional Ecological Knowledge, and are supported by the scientific realities of methane migration, radium in shale, and earthquakes caused by injection wells.[32]

Like the Haudenosaunee in New York, there are signs that abound in Ecuador saying, "*Agua es vida*" ("Water is life") as part of the government's *buen vivir* plan to encourage people to protect and care for their water supplies. Extractive industries have a direct impact on this link to water and human/ecological survival. Making this link, Alberto Acosta, former president of Ecuador's Constituent Assembly, says:

In [this new economic paradigm of the good life] life itself is at stake. Following this holistic concept, from the diversity of elements that condition human actions that bring the Good Life, material goods are not the only determining factors. There are other values at play, such as: knowledge, social and cultural recognition, ethical and even spiritual codes of conduct in relation to society and Nature, human values, a vision of the future, among others. The Good Life constitutes a central category in the philosophy of life for Indigenous societies.[33]

Fander Falconí, former foreign minister of Ecuador and former director of Senplades, the National Development Plan, argues that the Yasuní-ITT Initiative is a marked departure from the current model of sustainable development in which the concentration of carbon dioxide in the atmosphere has not declined and biodiversity loss continues. In this sense, Falconí argues that the international system currently has applied the logic of economic adjustments to the environmental realm, or simply a "Green Bretton Woods, as if the poorest economies should adjust to arrive at the level of global environmental sustainability."[34] For Acosta, Falconí, and others from Ecuador and around the rest of the world fighting to move the international system toward a different path, a sustainable one (in this case based on *sumak kawsay*), the Yasuní-ITT Initiative is more than one international campaign. It is an attempt to restructure the normative underpinnings of local, national, and global environmental governance beyond mere economic calculations, adding environmental and social factors to the equation, and creating a language of the good life, *buen vivir*.[35] And so this struggle continues into the Ecuadorian government's 2011 announcement of the eleventh round of oil concessions in its southern Amazonian provinces of Pastaza and Morona Santiago, totaling nearly 495,000 acres of pristine rain forest and indigenous communities.

According to Jose Maria Tortosa, *buen vivir*

comes from the social periphery of the world and does not contain the deceptive elements of conventional development. There will not be a question of the "right to development" or the "principle of development" as a guide to State action. Now, it is about the buen vivir of concrete people in concrete situations analyzed in a concrete manner and with a vocabulary that comes from these marginalized peoples who have been excluded from societal institutions and whose language was considered inferior and incapable of expressing the abstract.[36]

In the words of CONAIE president Manari Ushigua,

Buen vivir is an economic system built from various systems that include the 18 indigenous communities, the 14 indigenous nationalities, mestizos, whites, Africans, and coastal peoples. We offer this system as alternative because the capitalist system is in crisis today. What other alternative do you see today? There isn't one. This is our idea for equitable change.[37]

While scholars, activists, and policymakers who live within communities and write on *buen vivir* as a form of life and an alternative development path emphasize the indigenous underpinnings of the concept, they also acknowledge that its critique of the Northern notion of development is not a complete rejection of Western ideas and institutions or of scientific knowledge. They see pluricultural diversity in the concept within Andean and Amazonian communities, a diversity that is constantly evolving. Atawallpa Oviero Freire, Ecuadorian Quichua author and activist, adds that *buen vivir* has ancient roots, but its interpretation today is quite different from the precolonial times. It is a living concept applied to community realities and aimed toward an equitable system of living within nature for all of humanity, not for certain populations of humanity. In addition, he asserts that *buen vivir* in this fashion is a form of global emancipation for the millions of poor on the planet who have been subjected to so-called development projects that have resulted in the destruction of their lands, livelihoods, sacred places, and even human life. For Oviero Freire and others, including representatives of Achuar, Shuar, and Kichwa communities in Ecuador, the Yasuni-ITT Initiative and other means of resistance to further oil extraction in southern Ecuador are simultaneously about leaving oil underground and about the ethical and moral imperative of living in peace on this planet.[38]

**Box 11.1**
Against the Expansion of the Oil Frontier in Nigeria

Nigeria would be one of the least likely places to keep oil in the ground. Production started with the Royal Dutch Shell Oil of Britain and the Netherlands in 1958.[a] Now Nigeria is the largest producer of oil in sub-Saharan Africa[b] and ranks seventh on the list of countries from which the United States imports petroleum.[c] In 2009, the Nigerian advocacy group Environmental Rights Action/Friends of the Earth (ERA) effectively said, "Enough!" It proposed leaving oil in the soil, citing the inevitable exhaustion of fossil fuels in Nigeria, as well as the ongoing ecological degradation, toxic discharges, oil spills, and continued poverty as a consequence of oil exploration.

In 2011, the United Nations Environment Program assessed the environmental impact on the Ogoni people of Ogoniland, a region in the southeast Niger Delta basin. It found that levels of benzene, a cancer-causing chemical, were over 900 times above accepted World Health Organization standards, which would take an estimated thirty years without further oil activity to clean.[d] The ERA proposal asks the Nigerian government not to increase oil production from 2.3 million to 5 million barrels of oil per day per government plans, but rather to leave all newly discovered oil in the ground. It asks every Nigerian citizen to contribute $156 per year (compared to their per capita gross national income of $1,430) into a "crude solidarity fund." ERA estimates that about 100 million out of Nigeria's 140 million citizens would be able to make the payments. International aid agencies, philanthropists, and other countries could pledge to symbolically buy the remainder of the barrels.[e] This proposal differs from the Yasuní-ITT Initiative in that the movement does not emphasize coresponsibility with the rest of the planet, but rather coresponsibility with its own peoples and government.

As of July 2013, Nigeria's minister of petroleum resources, Diezani Alison Madueke, has, not surprisingly, ignored the ERA's call and gone ahead with seismic and exploratory work in nine inland basins. Preliminary results suggest that hydrocarbons are present in the Anambra, Chad, and Bida Basins,[f] where the deposits of the latter appear to contain 30 percent oil and 70 percent gas, compared to 25 percent oil and 75 percent gas in the Niger Delta region.[g] Nnimmo Bassey, chair of Friends of the Earth International and a Nigerian environmental activist, says, "We need to be able to look power in the face and tell the truth—that our political structures have been colonized by corporations like Shell, Monsanto, and the rest of them. It's time they get their dirty hands off our lives, and to regain our sovereignty over our political structures."[h] In other words, it's time to keep them in the ground.

**Box 11.1**
(continued)

---

**Sources**

a. Environmental Rights Action/Friends of the Earth, "Nigeria. Building a Post Petroleum Nigeria," November 2009, http://eraction.org/publications/presentations/leaveoilproposal.pdf.

b. Ibid.

c. U.S. Energy Information Administration, "Petroleum and Other Liquids," July 2013.

d. United Nations Environment Programme, "Environmental Assessment of Ogoniland Report (Nairobi, Kenya, 2011)," http://postconflict.unep.ch/publications/OEA/UNEP_OEA.pdf.

e. Environmental Rights Action/Friends of the Earth, "Nigeria: Building a Post Petroleum Nigeria," November 2009, http://eraction.org/publications/presentations/leaveoilproposal.pdf.

f. Juliet Alohan, "Alison-Madueke Insists on Deregulation of Petroleum Sector," *Leadership*, July 24, 2013, http://leadership.ng/news/240713/alison-madueke-insists-deregulation-petroleum-sector.

g. "Nigeria: Bida Basin Hydrocarbon is of High Oil Content–Report," *Oil Spill Witness*, July 22, 2013, http://www.oilspillwatch.org/news/2013/07/23/nigeria-bida-basin-hydrocarbon-high-oil-content-report-daily-trust-22-july-2013.

h. Nnimmo Bassey, "Sustainability Resides in Solidarity," *New Internationalist Blog*, June 2012, http://newint.org/blog/2012/06/21/sustainability-solidarity-rio/#sthash.GxMiFR2f.dpuf.

---

At the international level, Uruguayan scholar Eduardo Gudynas notes that both the Ecuadorian concept of *sumak kawsay* and the Bolivian concept of *suma qamaña* (an Aymara term) include ethical and spiritual perspectives on nature and are critical of the conventional utilitarian approach to development. Gudynas argues that the good life in both conceptions is not necessarily a critique of capitalism or socialism, or even postmodernism. Rather, *buen vivir* "is a platform to build alternatives beyond European modernity, it is moving away from Euro-centric political thought. But, buen vivir did not imply a complete rupture with those traditions ... rather a bridge to ... social justice positions."[39]

According to Indigenous leader and former minister of foreign affairs Nina Pacari, the indigenous *cosmovisión* that surrounds the concept of *buen vivir* and rights of nature in the Ecuadorian and Bolivian constitutions is a natural outgrowth of the relationship of humans to Mother Earth. She explains that when babies are born in Indigenous communities, the placenta is planted next to a tree to symbolize the life connection between *allpa-mama* (Mother Earth) and all living things. For Pacari, there is no disconnect between humans and Earth: "we are all part of everything."[40] Granting rights to nature in Ecuador developed from this Indigenous conception of life, including inanimate objects (such as rocks), as part of Mother Earth and should be protected in equilibrium. It is this strong Indigenous belief that informs not only *buen vivir* in Ecuador and the Andes at the scale of each society but also the larger global movement of rights of nature.[41]

In examining the Yasuní-ITT Initiative and its role in catapulting alternative approaches to the green economy as proposed by the Rio+20 dialogue of June 2012, anthropologists Laura Rival and Alex Barnard also found a local-global connection in the concepts of *buen vivir* and the rights of nature. Both have found that the Yasuní-ITT Initiative reflects an "environmentalism of the people" rather than a radical perspective as some would claim.[42] Barnard describes an encounter with a cab driver in Coca, an Amazonian town near the Yasuní National Park and the ITT block, where the driver is concerned about the lack of rainfall, more sun, and impacts on cancer among residents, all adding up to a need to protect the park. Barnard calls such an understanding "multiscaling impacts," or the ability of residents around the park to attribute multiple factors to the initiative. Barnard argues that "leaving oil underground in Yasuní represented not so much a radical change to their own livelihoods as a responsible step towards acknowledging what was—and was not—a sensible developmental path for Ecuador."[43] Thus, the indigenous norm of *buen vivir* is not only codified in the constitution, but is embedded in non-Indigenous Ecuadorian society as a civic goal.

Rival's research in and around Yasuní National Park also confirms the local support of the initiative and the deep complexities surrounding it, which include illegal logging, endangered species trade, and narcotics trafficking, in addition to oil extraction. While uncontacted Indigenous groups and world-renowned levels of biodiversity are what proponents of the initiative emphasize, Rival and Barnard note the myths surrounding Yasuní. Despite such challenges to *buen vivir*, both found local residents supportive of the initiative and having a deep respect for protecting what Northerners call biodiversity. For Rival, the Yasuní-ITT Initiative challenges long-standing notions of conservationism and developmentalism "as a Western myth imposed on the rest."[44] She also suggests that the initiative should be used as a tool in regional negotiations, such as the Alternativa Bolivariana para las Americas (a regional trade organization) and the Iniciativa para la Integración de la Infraestructura Regional (the regional infrastructure initiative). The Indigenous concept of *buen vivir* includes change and adaption to new environments—terms not unfamiliar to the Kyoto Protocol and, in fact, all ecologically grounded environmental measures. In the Andes, these changes are part of a collective process within the environments of living communities. Being careful about importing Northern ideas that connote abstract ideas like biodiversity and rather than call the implementation of such strategies "sustainable," Rival uses an increasingly popular term among ecologists and even engineers: *resiliency*. *Resilience* is thus a modern term that captures the indigenous contribution of *buen vivir*, the ideals of the Yasuní-ITT Initiative, and the spirit of those fighting to keep oil in the ground.[45]

## Yasunízate!

The Yasuní-ITT movement is now part of Yasunízate—a larger global campaign for the rights of nature, regardless of its on-again, off-again implementation in Ecuador. In September 2010, activists from South Africa, the United States, Australia, Bolivia, Peru, and Ecuador met in Patate, Ecuador, to form the Global Alliance for Rights of Nature. According to the organizers, "Their intention was to explore ways to expand the concept of Rights of Nature as an idea whose time has come."[46] The purpose of this meeting, according to its organizers from Fundación Pachamama and Pachamama Alliance of Ecuador and the United States, Bolivian activists, and the US-based Community Environmental Legal Defense Fund and Indian activist Vandana Shiva, was "to further galvanize the momentum of Ecuador's recent adoption of Rights of Nature in its Constitution, the Universal Declaration of Rights of Mother Earth from the People's Conference in Cochabamba, Bolivia, and growing community developments in the United States."[47] The Global Alliance was launched in Quito following this meeting.

Although it is only a few years old, the alliance has worked on issues such as river pollution from mining activities in Vilcabamba in southern Ecuador, a Sustainability Bill of Rights in Santa Monica, California, and Pittsburgh, Pennsylvania; and shark fin hunting in the Galapagos Islands, Ecuador. They have presented at the COP 17 talks in Durban, South Africa, and at the Rio+20 Summit in June 2012.

In April 2009, the UN General Assembly proclaimed April 22 to be International Mother Earth Day. Bolivian president Evo Morales Ayma and the Bolivian delegation proposed the resolution to the General

Assembly. Bolivia and Ecuador, in conjunction with the Global Alliance of Rights of Nature, are working to present a Universal Declaration of the Rights of Nature to the General Assembly for adoption. Environmental legal scholar Cormac Cullinan, also a founding member of the Global Alliance and author of *Wild Law*, argues that property rights centered solely on humans need to be expanded to rights to all living things and their right to live in a healthy and sustainable environment.[48]

Other Indigenous groups in the Amazon have a long history of resisting oil extraction and are joining the Yasuní call to leave the oil in the ground. The Shuar, Kichwa, and Achuar peoples to the south are arguing that new oil concessions and mining programs are not part of *buen vivir*, but rather are development as usual. Domingo Ankuash, a Shuar leader from Bomboiza in the southern Ecuadorian Amazon, decries the renewed emphasis on state oil extraction: "There are threats and you cannot live in peace. This is not *buen vivir*!" Franco Viteri, another Shuar leader, added, "Here we are intensely rich. The fact that we do not have cars, motorcycles or light does not mean we are poor."[49] These groups, albeit of different Indigenous communities, agree that increased extraction is not *buen vivir*. The Indigenous communities' view of development is distinct from the Northern world's view. Through the international NGO Pachamama Alliance, they are supporting rights to nature and the post-extractive goal of the Yasuní-ITT Initiative, including protests in Calgary, Canada, Paris, France, and Beijing, China, against new oil round concessions in the southern Ecuadorian Amazon during oil round talks with Ecuadorian government officials and potential investors.

### Gayanashagowa, the Great Law of Peace

The Marcellus Shale is believed to contain immense volumes of natural gas and lies beneath what is now known as New York's Southern Tier and its Finger Lakes region, as well as large sections of northern and western Pennsylvania, eastern Ohio, western Maryland, and most of West Virginia. The Marcellus Shale name comes from Marcellus, New York, a small community of around 6,000 in Onondaga County near outcroppings of the vast shale formation. Onondaga County is named for the Onondaga Nation, the Indian Nation and Indigenous community whose currently recognized sovereign territory lies five miles south of the midsize city of Syracuse, New York. Onondaga Nation has been the capital, known as the keepers of the central fire, where the chiefs of the Haudenosaunee Nations continue to meet in council to make decisions guiding their confederacy as they have for more than a thousand years.

In March 2009, the Haudenosaunee Environmental Task Force (HETF) released a statement on hydrofracking: "Every part of the natural world is important and interrelated; when humans tinker more and more with the natural balance, we do so at the peril of our grandchildren." The statement closes by asking NYSDEC to partner with the HETF to find alternative energy sources that "do not destroy our grandchildren's ability to live long and healthy lives."[50] In November, the traditional chiefs of the Haudenosaunee Council banned hydrofracking on their lands:

The Haudenosaunee will not allow hydrofracking on or near their aboriginal territory, and call on the Government of New York State to similarly ban hydrofracking and other unconventional gas drilling methods within New York State. … We do so for the future of all our relations.[51]

The Onondaga Nation posted an analysis of hydrofracking on its official website titled, "You Can't Drink Gas!! The Onondaga Nation & H.E.T.F. Oppose Gas Drilling By Hydro-Fracking." It begins with the same language that opens their legal complaint in the Onondaga Nation land rights action:

The Nation and its people have a unique spiritual, cultural and historic relationship with the land, which is embodied in the Gayanashagowa, the Great Law of Peace. … The people are one with the land and consider themselves stewards of it. It is the duty of the Nation's leaders to work for a healing of this land, to protect it and to pass it on to future generations.

This obligation to act as stewards of the land and the waters is being fulfilled in the Nation's work to stop the dangerous natural gas drilling method known as hydrofracking. Around 1600 leases have already been signed by landowners in Onondaga County to gas drilling companies and some are just outside the Nation's currently recognized territory. The Nation is very concerned with the dangers that will be created to the waters in general and to the Nation's drinking water system in particular.[52]

The statement expresses several components of the Haudenosaunee worldview relevant to the controversies surrounding hydrofracking that can be summarized:

• Native sovereignty and treaty rights require nation-to-nation consultation and agreement.

• The territory under consideration is peopled by many nonhuman beings that are considered relatives.

• Water is sacred, "the first law of life," and humans have special duties given by the Creator to protect water.

• Environmental decisions are moral decisions that require deliberation over a long period of time to consider the unintended consequences of similar past decisions and the potential consequences for future descendants at least to the seventh generation.

• While New York State officials believe that gas companies have a legal right to obtain natural gas and the NYSDEC has a legal duty to permit and regulate the practices, the Haudenosaunee have a legal right though treaties to carry out their sacred duties to care for and protect the land and waters.

In addition to these broad principles, the Onondaga Nation press release states that "two major concerns were raised [by the Haudenosaunee Environmental Task Force] that are unique to the Indian Nations: (a) horizontal drilling under Indian territories will be a violation of treaty protected mineral rights; and (b) the proposed regulatory system will not have any mechanism for the protection of cultural resources or the Nation's sacred sites, unmarked burial sites and former village and other archeological sites." The HETF concluded that despite the best efforts of the state environmental agency to regulate the gas industry's practices, unintended consequences from the hydrofracking process will inevitably damage the environment long into the future:

The DEC lacks the authority to protect the environment or its people by identifying an industry as inherently detrimental and banning it. The best they can do is regulate and fine polluters when the inevitable accidents occur; their hands are tied by policy set higher up in the government.[53]

According to HETF members present, the commissioner told them that "the Creator put riches above the earth, and he put them below the earth for the benefit of the people" (personal communication to the authors). While clearly the commissioner was attempting to communicate in a language he believed he shared with the Haudenosaunee, the idea that the riches of the earth were placed here to benefit people contrasts sharply with a responsibility-based worldview common to Indigenous perspectives.[54] Aboriginal studies scholar Deborah McGregor recalls a conversation with an Ojibwa man, Mahgee Binehns, an Anishnabe speaker. She asked Binehns whether there was an Ojibwa concept that corresponded to "sustainable development." She reported:

Indigenous views of development are based not on taking but on giving. Indigenous people ask themselves what they can give to the environment and their relationship with it. The idea of sustaining, maintaining and enhancing relations with all of Creation is of utmost importance from an Indigenous point of view.

Indigenous ways of life focus on this type of relationship with Creation. Indigenous people understand that with this special personal relationship with Creation comes tremendous responsibility; it is not something to be taken lightly.[55]

In this framework, Indigenous self-determination is understood as the capability of performing one's responsibility to assist the various elements of Creation in carrying out their duties. Thus, self-determination includes the political and legal rights of a people to make its own choices. Beyond that, Indigenous self-determination is understood as necessary for a healthy environment, because without it, those with the duties of ecosystem stewardship are rendered powerless. At the core of this Indigenous sense of self-determination are spiritual and relational responsibilities that are continuously renewed and conceptualized as privileges. One's relationship to the land on which one lives is not one of rights of ownership toward which the proper relation is one of development and exploitation, but rather one of privileges granted one by the land to the resident for which the proper relationship is stewardship and gratitude. Robin Kimmerer describes the Haudenosaunee perspective as a "radical new vision of property rights" that has everything to do with the right to continue to fulfill one's sacred responsibilities.[56] To attempt to articulate this perspective within a decision process designed to balance economic and political claims and rights of ownership or entitlement invariably leaves Native representatives experiencing their worldview marginalized, unheard, dismissed, and disrespected. This can happen even when those facilitating these processes are gracious, welcoming, polite, and eager to receive the input from Native constituents.

In this case, the HETF and Onondaga leadership articulate a holistic, spiritual perspective based in reciprocal models of collective responsibilities. In contrast, the NYSDEC and similar modern Western perspectives are abstract, economically efficient, and based on linear models of causality, risk, and individual rights (with corporations considered to be collective individuals and communities, collections of individuals). In many ways, these differences in emphasis on rights and responsibilities most clearly define the cultural, and thus political, differences between the Indigenous and Western ways of understanding issues of extraction.

"The ice is melting in the North," Oren Lyons, Onondaga Faith Keeper, told the United Nations World Peace Summit in 2000. Lyons had made many trips to the Arctic and visited with northern Indigenous leaders as well as Arctic scientists. He frequently warns of urgent dangers from climate change. According to Lyons, many Indigenous cultures have stories and prophecies about environmental catastrophes brought on by humans' forgetting their original instructions and no longer using the good mind to guide their actions. Climate change has accelerated not only the melting of the North, but also, we have argued here, the influence of Indigenous peoples and their teachings. This influence can be seen in the adoption of the UN Declaration of the Rights of Indigenous Peoples and the emergence of new social movements for the rights of Mother Earth.

Several Onondaga and Haudenosaunee leaders have played major roles in the creation of the UN Permanent Forum on Indigenous Issues, in the drafting of the UN Declaration on the Rights of Indigenous Peoples and its eventual adoption, twenty years later, by the General Assembly. Recently Haudenosaunee representatives formally requested that the permanent forum take up the issue of unconventional natural gas drilling on Indigenous territories. They believe it is a global issue. Shale gas drilling leads to multiple environmental impacts and further perpetuates humanity's dependence on nonrenewable fossil fuels. Methane, the gas being released from where it is now trapped deep in the earth, is a greenhouse gas, substantially more powerful than carbon dioxide. Leaving the gas underground where it has been for millions of years is consistent with the rights of Indigenous peoples and the rights of Mother Earth. This coming together of the discourse of environment and indigeneity appears to be growing in terms of both activity and potential for significant social change. Both of our cases tell a story not only of Indigenous peoples as activists but also of an emerging alliance of organizations and individuals who align themselves with the Indigenous movements to promote specific human rights as peoples. They understand Indigenous worldviews to offer an alternative way of understanding human rights as the right to take care of and protect the environment, Mother Earth, the source of life.

## Toward a New Politics of Keeping It in the Ground

Such movements from the remote Ecuadorian Amazon to upstate New York are tied together by the common infusion of Indigenous norms of well-being—the good life or the good mind—in public policies and codified in Ecuador into law. While some analysts and scholars have examined the policy outcomes and options around keeping oil and natural resources underground, we find that such analysis lacks the deeper lessons derived from an understanding of the Indigenous norms behind these policies and movements—norms that may help point the planet toward a better life for all.[57]

This complex array of connections and multiple directions of influence (North-South, South-North, and South-South) belie the traditional view of transnational networks and global governance, which portrays global norms and policies as being developed at the global level and then transferred to the domestic level, moving principally from the global North to the global South. As Carlos Espinoza, director of research at La Facultad Latinoamericana de Ciencias Sociales (FLACS) Quito, Ecuador, commented, "This is not the globalization of the 1990s anymore."

The redirection of norms from local, Indigenous actors to global partners in the form of networks with NGOs and international governmental organizations, such as the UN Working Group on Indigenous Peoples, outlines the dynamic process of normative development and compliance that combines the rational bargaining for maximized utility with a serious emphasis on social interaction and learning on domestic and international levels simultaneously.[58] In other words, local peoples want to earn a living, but not at the expense of their natural environments, and they want a plan for resource management that will foster a new economic paradigm of well-being, not unlike the case made in this book for livelihood (chapter 4) and for well-being within a healthy economy (chapter 2). In both cases, the Indigenous actors have collective memories and experiences of ecological degradation of their lands. This learning process has informed their current mobilization, movements, and policy outcomes in the form of the Yasuní-ITT Initiative, resistance to the eleventh round of oil concessions in the southern Amazon, and lawsuit against New York State.

When searching for solutions to these issues, actors organize at local levels with local concerns, norms, and ideas yet work in tandem with global partners for goals at both levels.[59] This process also denotes the ethical premise of this book that keeping fossil fuels in the ground based solely on neoliberal economic data or plans to "marketize" the value of them underground is not enough. Building a postpetroleum plan on ethical principles such as the good life and good mind that Indigenous peoples have taught for centuries as developed in public policies at local, place-based levels and as a global dialogue and mechanism is not only needed; it is urgent. Alberto Acosta uses the term *glocalization* to highlight the directional focus of analysis from the local needs to global demands and solutions.[60] The Indigenous peoples of Ecuador's Yasuní

National Park, the southern Amazon, and the Onondaga Nation may provide new optics and solutions for pathways to the good life toward keeping fossil fuels in the ground.

The discontent in Ecuador and places like Nigeria and Sudan may suggest that resistance to fossil fuels is a phenomenon of the Global South. But Indigenous groups in North America are increasingly voicing their opposition to tar sands and pipelines. In Canada ten indigenous groups whose territories are either rich in tar sands or are crossed by proposed pipelines are claiming that Ottawa and Washington are ignoring their people's rights as well as neglecting eighteenth- and nineteenth-century agreements that give aboriginal groups decision authority, however contested, in their territories.

Canada has proposed building three multibillion-dollar pipelines. But native leaders, such as hereditary chief and global indigenous leader/activist Phil Lane Jr. from the Ihanktonwan Dakota in South Dakota, are fighting it. Lane says, "Along with every single legal thing that can be done, there is direct action going on now to plan how to physically stop the pipelines."[61] This resistance comes from a country where 33.23 percent of Canadian GDP is derived from mining, oil, and gas extraction.[62] What's more, the Canadian government expects oil production to increase from 1.31 million barrels per day in 2008 to 3 million barrels per day in 2018.[63]

There is a larger political context here, though. Indigenous groups in Canada are fighting for their traditional rights in a country where environmental attitudes among all Canadians have been strong. As a result, overall Canadian sentiment is aligning with Indigenous resistance that fears impacts on their lands. Studies show, for example, that 20 percent of Canadians oppose Enbridge's Northern Gateway, a $6 billion proposed pipeline intended to transport 525,000 barrels of diluted bitumen from Bruderheim Alberta Kitimat in British Columbia.[64] British Columbia's government rejected the pipeline in June 2013.[65] And six out of ten Canadians, according to recent polls, are unhappy with the government handling of environmental issues overall.

Environmental concerns may be most intense regarding tar sands. The United Nations Environment Programme has listed Alberta's tar sands mines among the top 100 "hot spots" of environmental degradation.[66] The processing of tar sands oil results in roughly 14 percent more greenhouse gas emissions than the average oil used in the United States,[67] and greenhouse gas pollution in Canada from tar sands is predicted to triple by 2020.[68] Land and water degradation is massive, much of it effectively permanent. These impacts combine with the British Columbia government's rejection of the pipeline and local Indigenous groups' resistance to suggest that pressure from local and global movements is mounting to keep fossil fuels in the ground.

## Notes

1. The Haudenosaunee Confederacy, believed to have been formed through the influence of a Huron man who was known as the Peacemaker, was founded on the shores of Onondaga Lake where the five original nations, from east to west— Mohawk, Oneida, Onondaga, Cayuga and Seneca Nations—agreed to bury their weapons of war beneath a tree of peace and to henceforth live under a constitution known as the Great Law of Peace.

2. See Matthew Huber, *Lifeblood: Oil, Freedom, and the Forces of Capital* (Minneapolis: University of Minnesota Press, 2013).

3. Eghosa E. Osaghae, "The Ogoni Uprising: Oil Politics, Minority Agitation and the Future of the Nigerian State," *African Affairs* 94 (1995): 325–44; Theodore E. Downing, Jerry Moles, Ian McIntosh, Carmen Garcia-Downing , "Indigenous Peoples and Mining Encounters: Strategies and Tactics," *Mining, Minerals and Sustainable Development* 57 (2002): 1-41; Ian Urquhart, "Between the Sands and a Hard Place? Aboriginal Peoples and the Oil Sands" (Buffett Center for International and Comparative Studies, Working Paper 10–005, 2010), 18–19; Martí Orta-Martínez and Matt Finer, "Oil Frontiers and Indigenous Resistance in the Peruvian Amazon," *Ecological Economics* 70 (2010): 207–18; Cesar A. Rodrıguez-Garavito and Luis Carlos Arenas, "Indigenous Rights, Transnational Activism, and Legal Mobilization: The Struggle of the U'wa People in Colombia," in *Law and Globalization from Below: Towards a Cosmopolitan Legality* eds. Boaventura Sousa Santos and Cesar A. Rodriquez Garavito (Cambridge: Cambridge University Press, 2005), 241–266; Andrew Canessa, "Conflict, Claim and Contradiction in the New Indigenous State of Bolivia" (desiguALdades.net, Working Paper 22, 2012).

4. Robin Wall Kimmerer, *Braiding Sweetgrass: Indigenous Wisdom, Scientific Knowledge and the Teaching of Plants* (Minneapolis, MN: Milkweed Editions, 2013), 106.

5. Oren Lyons, in ibid., dust jacket.

6. Melissa Nelson, ed., *Original Instructions: Indigenous Teachings for a Sustainable Future* (Rochester, VT: Bear & Company, 2008).

7. Kimmerer, *Braiding Sweetgrass*, 183.

8. Ibid., 187.

9. Jack P. Manno, Andrea Parker, and Paul Hirsch, "Introduction by the Onondaga Nation and Activist Neighbors of an Indigenous Perspective on Issues Surrounding Hydrofracking in the Marcellus Shale," *Journal of Environmental Studies and Sciences* 4, no. 1 (2014): 47–55.

10. Haudenosaunee Environmental Task Force, "Haudenosaunee Position on the Great Lakes," November 12, 2010, section 4.2, http://hetf.org/index.php/haudenosaunee-position-on-the-great-lakes.

11. Renee Ramirez, "La transición ecuatoriana hacia el buen vivir," in *Sumak Kawsay/buen vivir y Cambios Civilizatorios*, ed. Irene Leo (Quito: FEDAEPS, 2010).

12. Pamela L. Martin and Franke Wilmer, "Transnational Normative Struggles and Globalization: The Case of Indigenous Peoples in Bolivia and Ecuador," *Globalizations* 5 (2008): 583–98.

13. Global Exchange, "The Stillheart Declaration," http://www.globalexchange.org/communityrights/rightsofnature/stillheart.

14. Ronald Niezen, *The Origins of Indigenism: Human Rights and the Politics of Identity* (Berkeley, CA: University of California Press, 2003).

15. United Nations, Declaration on the Rights of Indigenous Peoples, March 2008, http://www.un.org/esa/socdev/unpfii/documents/DRIPS_en.pdf.

16. International Indian Treaty Council, accessed September 20, 2013, http://www.treatycouncil.org/about.htm.

17. The Haudenosaunee, or "People of the Longhouse," is made up of the Mohawk, Oneida, Onondaga, Cayuga, Seneca, and Tuscarora Nations and is sometimes referred to as the Six Nations and sometimes as the Iroquois.

18. Akwesasne Notes, ed., *Basic Call to Consciousness* (Summertown, TN: Native Voices, 2005 [1978/1981].

19. Ibid., 88.

20. Ibid., 90.

21. Ibid..

22. Daniel Ortega, director of the environment and climate change in the ministry of foreign affairs, Ecuador, interview with Pamela Martin, Quito, Ecuador, December 13, 2012.

23. Margot S. Bass, Matt Finer, Clinton N. Jenkins, Holger Kreft, Diego F. Cisneros-Heredia, et al., "Global Conservation Significance of Ecuador's Yasuní National Park," *PLoSONE* 5, no. 1 (2010): e8767, doi:10.1371/journal.pone.0008767.

24. Organization of American States, Interamerican Commission on Human Rights, "Inter-American Commission on Human Rights Application to the Inter-American Court of Human Rights in the case of Kichwa People of Sarayaku and Its Members (Case 12.465) against Ecuador," April 26, 2010, http://www.cidh.oas.org/demandas/12.465%20Sarayaku%20Ecuador%2026abr2010%20ENG.pdf.

25. Manari Ushigua, interview by Andrea Lizarzaburu, Quito, Ecuador, January 28, 2013.

26. For more about the nature of the deception, see chapter 2.

27. New York Department of Environmental Conservation, "Draft Supplemental Generic Environmental Impact Statement (DSGEIS) Relating to Drilling for

Natural Gas in New York State Using Horizontal and Hydraulic Fracturing," accessed January 15, 2010, http://www.dec.ny.gov/energy/58440.html; Dusty Horwitt, "Drilling around the Law," *Environmental Working Group Drilling Around the Law Report* (2010), http://www.ewg.org/sites/default/files/report/EWG-2009drillingaroundthelaw.pdf.

28. Republic of Ecuador, Political Constitution of Ecuador 2008, http://biblioteca.espe.edu.ec/upload/2008.pdf.

29. Universal Declaration of Rights of Nature, April 22, 2010, accessed September 20, 2013, http://therightsofnature.org/universal-declaration/; Constitution of Bolivia, 2009, http://pdba.georgetown.edu/Constitutions/Bolivia/bolivia09.html.

30. Freida Jacques, "Use the Good Mind," accessed April 11, 2013, http://www.whitebison.org/magazine/2001/jan-feb/teaching3.pdf.

31. Onondaga Nation, "Onondaga Nation's Statement to NYSDEC on 'Hydro-fracking,'" August 12, 2014, http://www.onondaganation.org/news/2012/onondaga-nations-statement-to-nysdec-on-hydro-fracking/.

32. Alberto Acosta, "El buen vivir, una oportunidad por construer," *Portal de Economía Solidaria*, February 17, 2009.

33. National Secretary for Planning and Development (Senplades), Ecuador, "The Yasuni-ITT Initiative Analyzed by Different Perspectives," February 8, 2012, http://www.planificacion.gob.ec/la-iniciativa-yasuni-itt-se-analizo-desde-diferentes-opticas.

34. Ibid.

35. Jose Maria Tortosa, *Maldesarrollo y Mal vivir: Pobreza y Violencia a Escala Mundial* (Quito: Ayba Yala, 2011).

36. Manari Ushigua, interview by Andrea Lizarzaburu, Quito, Ecuador, January 28, 2013.

37. Atawallpa Oviero Freire, *Que es Sumak Kawsay: Mas alla del Socialismo y Capitalismo* (Quito: Abya Yala, 2011).

38. Gudynas, Eduardo, "Buen Vivir: Today's Tomorrow," *Development* 54, no. 4 (2011): 445.

39. Nina Pacari, "Naturaleza y territorio desdela mirada de los pueblos indigenas," in *Derechos de la Naturaleza: El Futuro es Ahora*, ed. Alberto Acosta and Esperanza Martinez (Quito: Abya Yala, 2009), 35.

40. Thomas Fatheuer, *Buen Vivir: A Brief Introduction to Latin America's New Concepts for the Good Life and the Rights of Nature* (Berlin: Heinrich Boll Foundation).

41. Augusto Tandazo, an oil analyst, argues that there is no market for unemitted carbon and that the Yasuní-ITT Initiative will simply encourage leakage or drilling for oil somewhere else on the planet. See Augusto Tandazo, "Audio: Agusut Tandazo Habla de Yasuni ITT)," *El Hoy*, September 26, 2011, http://www.hoy.com.ec/noticias-ecuador/audio-augusto-tandazo-habla-del-yasuni-itt-503125.html.

42. Alexander Voscick Barnard, "We Are the Lungs of the World" (master's thesis, Oxford University, 2010), 59.

43. Laura Rival, "Anthropological Encounters with Economic Development and Biodiversity Conservation" (Queen Elizabeth House working paper, January 2011): 287, http://www3.qeh.ox.ac.uk/pdf/qehwp/qehwps186.pdf.

44. Laura Rival, "Planning, Development Futures in the Ecuadorian Amazon: The Expanding Oil Frontier and the Yasuní-ITT Initiative," in *Extractive Economies, Socio-Environmental Conflicts and Territorial Transformations in the Andean Region,* ed. Anthony Bebbington (forthcoming); Laura Rival, "the Resilience of Indigenous Intelligence," in *The Question of Resilience: Social Responses to Climate Change,* ed. Kirsten Hastrup (Copenhagen: Royal Danish Academy of Sciences and Letters, 2009), 293–314.

45. Global Alliance for the Rights of Nature, "Founding the Alliance," accessed September 20, 2013, http://therightsofnature.org/founding-meeting/.

46. Ibid.

47. Cormac Cullinan, *Wild Law: A Manifesto for Earth Justice* (White River, Vermont: Chelsea Green Publishing, 2011).

48. "Shuar en resistencia contra la explotación petrolera y minera en sus territorios," *Fundacion Pachamama,* September 24, 2012, 2013, http://pachamama.org .ec/?p=4320.

49. Haudenosaunee Environmental Task Force, "Haudenosaunee Statement on Hydrofracking," November 4, 2009, http://www.onondaganation.org/news/2010/ hetf_position.pdf.

50. Onondaga Nation, "Traditional Native leaders: Hydrofracking Must Be Banned," news release, November 5, 2009, http://www.onondaganation.org/ news/2009/news-release-traditional-native-leaders-hydrofracking-must-be -banned.

51. Onondaga Nation, "You Can't Drink Gas!! The Onondaga Nation & H.E.T.F. Oppose Gas Drilling by Hydro-Fracking," accessed May 8, 2010, http:// www.onondaganation.org/news/2010/2010_0122.html

52. Haudenosaunee Environmental Task Force, "Statement on Hydraulic-Fracturing," March 4, 2009, http://hetf.org/index.php/hydrofracking/23-statement -on-hydraulic-fracturing-march-4-2009.

53. Jeff Corntassel, "Toward Sustainable Self-Determination: Rethinking the Contemporary Indigenous-Rights Discourse," *Alternatives* 33 (2008): 105–32; Jack Manno and Irving Powless Jr., "Brightening the Covenant Chain: The Onondaga Land Rights Action and Neighbors of Onondaga Nation," in *Governance for Sustainability: Issues, Challenges, Successes,* ed. K. Bosselman, R. Engel, and P. Taylor (Gland, Switzerland: IUCN, 2008), 149–58.

54. Deborah McGregor, "Traditional Ecological Knowledge and Sustainable Development: Towards Coexistence," in *In the Way of Development: Indigenous Peoples, Life Projects and Globalization,* ed. Mario Blaser, Harvey A. Feit, and Glenn McRae (London: Zed Books, 2004).

55. Robin Kimmerer, "The Rights of the Land: The Onondaga Nation of Central New York Proposes a Radical New Vision of Property Rights," *Orion* (November/December 2008).

56. Margot S. Bass, Matt Finer, Clinton N. Jenkins, Holger Kreft,,Diego F. Cisneros-Heredia, Shawn F. McCracken, Nigel C. A. Pitman, Peter H. English, Kelly Swing, Gorky Villa, Anthony Di Fiore, Christian C. Voigt, Thomas H. Kunz, eds. "Global Conservation Significance of Ecuador's Yasuní National Park," *PLoS ONE 5*, no. 1, doi:10.1371/journal.pone.0008767; Eric Bettelheim, "Yasuni's Means Won't Achieve Its Ends," Katoomba Group's Ecosystem Marketplace, August 27, 2009; Herman Daly, "International Policies to Accelerate Sustainable Development in Developing Countries and Related Domestic Policies," *Population and Environment* 15, no. 1 (September 1993): 66–69; Matt Finer, Varsha Vijay, Fernando Ponce, Clinton Jenkins, and Ted Kahn. "Ecuador's Yasuní Biosphere Reserve: A Brief Modern History and Conservation Challenges," IOP Publishing Environmental Research (July–September 2009), doi:10.1088/1748-9326/4/3/034005.; Matt Finer, Clinton N. Jenkins, Stuart L. Pimm, Brian Keane, and Carl Ross, "Oil and Gas Projects in the Western Amazon: Threats to Wilderness, Biodiversity, and Indigenous Peoples," *PLoS ONE* 2008, 3, no. 8, doi:10.1371/journal.pone.0002932; Joseph Henry Vogel, *The Economics of the Yasuní Initiative: Climate Change as If Thermodynamics Mattered* (London: Anthem Press, 2009); Sven Wunder, *Oil Wealth and the Fate of the Forest: A Comparative Study of Eight Tropical Countries* (New York: Routledge, 2003).

57. Jeffrey T. Checkel, "Why Comply? Social Learning and European Identity Change." *International Organization* 55 (Summer 2001): 553–88.

58. Ken Conca, *Governing Water: Contentious Transnational Politics and Global Institution Building* (Cambridge, MA: MIT Press, 2006).

59. Alberto Acosta, *Desarrollo Glocal: Con la Amazonia en la Mira* (Quito: Corporación Editora Nacional, 2005).

60. David Ljunggren, "Canadian and U.S. Native Bands Vow to Block Oil Pipelines," Reuters, March 20, 2013, http://uk.reuters.com/article/2013/03/20/canada-pipelines-aboriginals-idUKL1N0CC57120130320.

61. Government of Canada, Statistics Canada, "Gross Domestic Product (GDP) at Basic Prices, by North American Classification System (NAICS), Provinces and Territories," April 2013, http://www5.statcan.gc.ca/cansim/a46?lang=eng&childI d=3790028&CORId=3764&viewId=1.

62. Alberta Energy, "Oil Sands," n.d., http://www.energy.gov.ab.ca/OurBusiness/oilsands.asp.

63. Darren Campbell, "B.C. Just Says No to Enbridge Inc's Northern Gateway Pipeline," *Alberta Oil Magazine* (March 2013), http://www.albertaoil magazine.com/2013/05/b-c-just-says-no-to-enbridge-inc-s-northern-gateway-pipeline.

64. Public Radio International, "British Columbia Rejects Oil Pipeline, Casting Keystone in New Light," June 2013, http://www.pri.org/stories/science/environment/

british-columbia-rejects-oil-pipeline-casting-keystone-in-new-light-14056
.html.

65. Dan Woynillowicz, "Tar Sands Fever!" *World Watch Magazine* 20, no. 5 (September–October 2007), http://www.worldwatch.org/node/5287.

66. David Biello, "How Much Will Tar Sands Oil Add to Global Warming?" *Scientific American* (January 2013), http://www.scientificamerican.com/article.cfm?id=tar-sands-and-keystone-xl-pipeline-impact-on-global-warming.

67. Woynillowicz, "Tar Sands Fever!"

# 12

## Exit Strategies

Thomas Princen and Adele Santana

One way or another, the oil, gas, and coal industries will close up shop, or at least drastically reduce operations. The geology, the energy return on energy invested, the financial return on capital invested, and the desperate demand for alternative uses of capital (e.g., for food and sanitation) all ensure industry exit. But will it be soon enough to avert catastrophic biophysical and social outcomes? In this book, the coeditors have assumed not. Might, then, the strategies of resistance politics we have seen in Ecuador, Norway, and elsewhere be enough, and soon enough? In this chapter, we assume not. We assume that for an overall societal exit from the fossil fuel era, industrial inertia must be overcome in part by industry itself. And for that, individual firms must lead the industry in a direction heretofore unimaginable—out of business, or at least out of a pattern of extraction rates incompatible with sustainable and just living on this planet.

In fact, in this chapter, we assume that farsighted industrialists would want it that way. They know that no business is permanent—that for all the desire to perpetuate an industry, provide jobs, and generate wealth, there comes a time to move on, perhaps declaring victory ("We created the modern world!"), and call it a day (or a century). They know that a higher calling is possible and that in a biophysically constrained and highly unjust world, they too can rise to the occasion. And that occasion is figuring out how to live well by living well within our means. In a world full of human activity, exceeding regenerative capacities by just about every biophysical measure (ecological, water and carbon footprints, ecosystem services, net primary photosynthetic potential, material throughput), industrialists' calling is to effect an orderly transfer of power.[1] If political leaders can, on occasion, do it (witness Nelson Mandela), certainly industrial leaders can. And it is not just a matter of will but of strategy. This chapter offers preliminary thoughts about such strategizing.[2]

## Imagining Company Exit amid Industry Decline

The external case for ending the fossil fuel era is straightforward: the seriousness of the lagged effects of current combustion make accelerated phase-out rational—rational, that is, from the perspective of global society, of present populations directly harmed by fossil fuels and of future generations. But what might the internal case be? Might there be a rational early exit for individual companies, or the premature decline of an entire industry sector such as exploration or refining or mountaintop removal? In this chapter, we explore the possibility of fossil fuel companies deliberately leaving their business. The emphasis, it should be noted, is on *possibility*; these are not predictions or prescriptions. This chapter aspires only to make conceivable (1) a planned premature decline of the industry itself, before it happens naturally, and (2) planned early exits of individual companies. Here, "premature" and "early" mean before oil producers, independent and national, are forced to exit for geologic, economic, political, or environmental reasons. Such deliberate measures would cut into the otherwise high-risk global bet now being effectively wagered by major players—producers, consumers, and governments—that a solution will be found, so business-as-usual continues. Put differently, how could a global mining operation be undermined before it's too late?

How might the fossil fuel industry itself unwind? Under what conditions would incumbent fossil fuel companies force the industry into an early decline, earlier than might otherwise happen were there no external consequences for the extraction and combustion of its offending substances? Could the industry decline prematurely while incumbent companies still meet their primary objective of maximizing shareholder value over a reasonable period of time, say, a few decades? Could a company lead the way by embracing a higher objective than maximizing shareholder value? What type of first-mover advantages might accrue to such a company? To entertain such questions is to begin to imagine and devise a politics for an early transformation of the industry, that is, one that may start one firm at a time but end the entire industry as we know it (box 12.1). Fossil fuel extraction will continue for a long time thereafter, we are assuming, just as extraction existed prior to the fossil fuel era.

But it will proceed on a drastically reduced scale, one that we hope fits within the regenerative capacities of relevant resources and waste sinks.

To imagine the fossil fuel industry's premature decline, it is first useful to consider reasons that we tend not to hear about firms' exit strategies and why they are difficult to plan. This opens the door for a special case, possibly an exceptional case: mining.

**Box 12.1**
Imagining the Exit

Imagining the end of the fossil fuel era requires, among other things, imagining the exit of fossil fuel companies, both private and state. This is hard to do—at least as hard as imagining the absence of a large percentage of current energy consumption (say, 80 to 90 percent).

So here is one trick: imagine a clothing company telling its customers to think twice before buying new clothes; or a health care company saying that healthy people require a healthy environment and the health care industry itself is harming that very environment. Then, putting aside any inclination, however warranted, to be cynical about advertising, consider these statements from Patagonia, a major clothing company, and Dignity Health, a large heath care company.

The culture of consumption ... puts the economy of natural systems that support all life firmly in the red. ...

[Patagonia wants] to do the opposite of every other business today. We ask you to buy less and to reflect before you spend a dime on this jacket or anything else.

Environmental bankruptcy, as with corporate bankruptcy, can happen very slowly, then all of a sudden. This is what we face unless we slow down. ...

Don't buy what you don't need. Think twice before you buy anything.[a]

Whether it's good soil, pure water, or clean air—our health is deeply connected to the health of our planet. Yet the very health care industry that's meant to heal us is a major contributor to environmental harm.

[Our environmentally friendly] actions may win us recognition, but they also create clout to take on bigger challenges [such as modernizing] the Toxic Substances Control Act in Congress.[b]

Now imagine the largest private oil company in the world for much of the nineteenth and twentieth centuries, Exxon Mobil, a company that has systematically denied climate change for years and funded deniers of many stripes. Imagine that some of the very scientists this company has funded say that climate change is real. Imagine further this company's CEO, one

**Box 12.1**
(continued)

of the most powerful people in industry worldwide, saying that climate change, an "important global issue," must be addressed, and proposing a specific measure to do just that:

There is another policy option that should be considered, and that is a carbon tax. As a businessman, it is hard to speak favorably about any new tax. But a carbon tax strikes me as a more direct, a more transparent, and a more effective approach.[c]

And, finally, consider that, according to a news report, Germany's giant electric power utility RWE recently lost 2.8 billion euros, its first loss in sixty years, admitting that it got its energy strategy wrong: it should have concentrated on renewables and distributed energy rather than fossil fuels. RWE's CEO, Peter Terium, admitted that the change in electricity markets, which has seen earnings from conventional generation gutted by the impact of solar and wind energy, was "unstoppable." "We were late entering into the renewables market—possibly too late." The trend of making less and less money on conventional coal- and gas-fueled power stations "will continue in the next few years and it is irreversible," said Terium. "Despite the difficulties that we face, our focus is on the future. We are, and will remain, the partner for the transformation of the energy system, and are orienting our operational business accordingly."[d]

Now for the really hard part in this envisioning exercise, one intended to elicit a sense of the possible. Imagine a fossil fuel company making a public statement like this:

Climate change is real. So are toxic buildup, soil loss, and freshwater depletion. Fossil fuel products and emissions are a major cause. The very environment the economy and this company depend on is being damaged—in fact, irreparably damaged. The culture of consumption puts the economy of natural systems at risk. It cannot continue.

And so we as a company cannot continue. The very industry that has powered the world is now a major contributor to rendering this world uninhabitable.

Our company as we know it will come to an end, as will the industry. And this end will be well before fossil fuels "run out." Capital availability, economic viability, and political opposition will be the proximate causes, geologic and biological forces the ultimate causes. Our scientists know this and our financial people know this.

We are now taking steps to protect our shareholders, our suppliers, and our customers as we all ease out of fossil fuel dependency. We want to do the opposite of every other business today.

We ask our customers to buy less oil and gas and to reflect on the uses they put them to.

**Box 12.1**
(continued)

We have achieved great efficiencies in our operations, but like the textile, health care, and electric power industries, we know it is not enough. We have the clout, built up over decades industrializing the world, to take on the much bigger challenge of "decarbonizing" the world, and we intend to use it.

Preposterous? Perhaps. But stranger things have happened, things that before they happened were universally seen as impossible and then, after they happened, as inevitable. It will probably take a first mover to head for the exit to get it all going, a company highly enlightened or farsighted (as most oil companies already are; twenty- or thirty-year time horizons are the norm, even fifty years) or one just willing to face the facts, facts that have already motivated financiers, insurers, water and land managers, farmers, forest firefighters, and, indeed, oil company engineers trying to navigate hostile environments, along with a host of others who are now dealing with the consequences of excess fossil fuel combustion. It may only take a company willing to do the calculations, with biophysical and social disruption factored in, to conclude, "Time's up."

Imagine all the exits. Imagine a post–fossil fuel world.

**Sources**

a. "Don't Buy This Jacket," advertisement, *New York Times*, November 25, 2011, A7. See also Common Threads Initiative, Patagonia, Patagonia.com.

b. "The Earth's Health Is Our Health," advertisement, *New York Times*, June 23, 2013, 7.

c. Rex Tillerson, CEO, Exxon Mobil, in a 2009 energy policy speech in Washington, D.C., quoted in Steven Coll, *Private Empire: ExxonMobil and American Power* (New York: Penguin Press, 2012), 535.

d. Rex Parkinson, "Germany: Decline of Fossil Fuel Generation Is Irreversible," *New Economy*, March 5, 2014, http://reneweconomy.com.au/2014/germany-decline-of-fossil-fuel-generation-is-irreversible-75224; Karel Beckman, "Exclusive: RWE sheds old business model, embraces transition," *EnergyPost*, October 21, 2013, http://www.energypost.eu/exclusive-rwe-sheds-old-business-model-embraces-energy-transition.

## Difficult to Exit

Firms exit an industry by closing, switching to another industry, or merging or being acquired.[3] But they do not generally declare their exit strategies publicly even when they build them into their overall strategies. One reason for the reticence is an aversion to sending a negative signal. In general, businesses must be positive: they must exude optimism, highlight expected gains, and downplay losses All this is needed to attract investors, talent, suppliers, and customers who want to be part of a successful operation. What's more, to even entertain the idea of going out of business publicly is to effectively plan for it; it becomes self-fulfilling as stakeholders react to the signal, and competitive and market forces take over.

Some of the factors that affect a company's decision to exit a declining or mature industry are the intensity of assets and technology, excess capacity, expectations of growth and profitability, and past performance.[4] When a firm does consider exit, there are considerable obstacles to planning and carrying it out. One is cost, to the firm and to broader society. Such planning would divert resources from the firm's core function of producing goods and services and rewarding shareholders. Planning an exit is to incur a cost, however intangible. Firms act to minimize costs, not increase them. Investors look for competitive returns—what is unlikely when plant and equipment are being considered for liquidation, especially when there is a small resale market, or none at all, due to excess capacity. If cost prevents decision makers from considering exit plans, so do expectations of future levels of profitability. If companies forecast a negative scenario for future earnings, they are more likely to consider exiting the industry. Successful companies that forecast a positive scenario for the foreseeable future, however, tend to dismiss exit in the short term. In addition, ongoing high levels of income are seductive, distract managers, and blind them to trends that may otherwise be obvious. These economic factors encourage a firm to stay in the industry even when its activity is no longer profitable,[5] even when anticipating the decline makes it easier to dispose of assets.[6]

Another difficulty in shutting down operations is the disruption it would cause in employees' lives and entire communities. This is an important concern of firms that create value not just through economic wealth but also through the wealth of relationships among a full range of stakeholders.

Another obstacle is temporal distance. An exit may be perfectly rational at the level of the firm but beyond the time frame of key decision makers, especially managers and board directors. Short-term thinking is by far the rule, not the exception, and economic wealth maximizing is the prevailing mandate of professional managers. A firm's decision makers may also reason that the firm's responsibility ends when operations end. Abandoned properties are what governments handle, along with picking up trash and clearing sewer lines. When an oil rig blows or a chemical truck overturns, government comes in. This is what governments do and what corporations pay taxes for.

Another reason exit is difficult is that some managers may believe that if an exit were necessary, the market or government would take care of it. When demand or capital dries up, or supplies of raw material or of labor tighten, the firm will end operations at some point. Alternatively, when the political and legal systems find the operation unsupportable or illegal, laws and regulations will hasten its demise. Either way, managers may decide, there is no reason to explicitly develop an exit strategy; the external environment will take care of the end game.

What these considerations suggest is that a firm's exit and an entire industry's demise may be perfectly rational from an investment perspective (scarce capital is being tied up in the declining industry, say) or a societal perspective (industry is exporting costs to the rest of society, for instance), and yet not on the table. Firms won't talk about it, commentators won't consider it, and business analysts won't study it. Yes, exits happen all the time. But that's just part of the business cycle. What's interesting (and profitable) is the fast-growing new ventures and the blue-chip cash cows (that enable the new ventures). Exit just happens. Here, in the twenty-first century, one exit in particular can hardly happen fast enough. The common-good value of ending the fossil fuel era requires accelerated exit. Among the conditions would be long-term thinking, a sense of high purpose, and actual experience with or understanding of others' exit.

Curiously, the one industry whose major firms often meet these three conditions is the mining industry, or at least one aspect of the industry. In mining, all concerned interests—explorers, investors, miners, managers—know the well or mine will play out, and when it does, the company will pack up and move on, to the next pool, to the next seam. It may take years or decades, but eventually the firm leaves. For centuries there has always been a next source. And at one level, there still is for fossil fuels (1 trillion barrels burned; another trillion of conventional to go, 4 or 5 trillion; some say 18 trillion of unconventional left).[7] But at another level, there is not a next source because waste sink filling is the issue, not resource depletion. What's more, there is no available next source when energy return approaches one (chapter 2), technological limits are reached (no device can resist melting tundra and shifting sea ice), capital is constrained (when capital shifts to even more pressing needs), and, quite possibly, public resistance becomes unbearable for leaders (part

2). Nevertheless, mining companies have plenty of exit experience at limited scales both spatial and temporal—leaving their wells and fields, departing hostile countries, shutting down unproductive investments, vertically deintegrating from the production chain. In principle, these practices could serve as precedents guiding the closure of the fossil fuel industry.

## Exits

### Wells and Fields

Every oil and gas company eventually abandons its wells and fields. It may take years or decades, but eventually the economic returns do not justify continued pumping. The costs and the sheer difficulty of continued pumping, not to mention the decline in petroleum quality, all point to an exit well before depletion is complete (if that were physically possible). Worldwide, recovery rates average between 30 and 50 percent, which is to say at least half of known reserves remain in the ground after extraction.[8] In some cases, drillers will return to a closed well when prices rise high enough. But it still takes a considerable investment to revive a well. Even then, again, some half stays in the ground.

In short, at the scale of a specific mine or field, an oil and gas company plans not just extraction but exit. And the exit calculation is fundamentally the same as the entry calculation: How does one maximize the net benefit stream? At what point do marginal costs begin to exceed marginal benefits, appropriately discounted?[9] If such a calculation can be made at the scale of wells and fields, then in principle, it can be made at larger scales too, up to and including the entire enterprise.

It is hard to imagine that no oil and gas company has thought about that final calculation, maybe even made that calculation. Certainly firms have made it with respect to mergers and acquisitions. Oil historian and consultant Daniel Yergin describes the "death of a major" in the 1980s, what had been inconceivable for nearly a century: "Gulf Oil, one of the Seven Sisters ... [was] built by the Mellon family on the basis of [the] discovery of Spindletop in 1901, and had grown into a major American institution and a global company." It was then taken over by T. Boone Pickens and Chevron. "The Gulf board had decided that the prudent course was to accept Chevron's all-cash offer," writes Yergin. "The shareholders would be better off. And that was the end of Gulf Oil." One Gulf executive recalled his shaken world: "I never had any intention but that Gulf would be here for ever and ever. It had been my whole life, my whole career. To think that it wasn't going to be there any more got to me."[10]

Now an entire industry must begin thinking that it isn't going to be there forever and ever. For those pushing an early exit, all it may take is a calculation, a communication to shareholders (a primary concern of responsive executives and boards), and an orderly exit plan. According to one former head of a major oil company, that plan could start with the simple step of ending exploration. Returns from existing wells and a phased-in liquidation would go to shareholder payouts, and in twenty to thirty years, the prevailing time horizon of the industry, the company is out. Shareholders are happy and operators have a clear goal: liquidate optimally. Many of the maddening uncertainties (especially political) are removed, and the public relations unit spins the company for being a leader in corporate social responsibility.[11]

### Country Departure

In the middle of the twentieth century, major oil companies left entire countries, coming back later, if at all, as "service contractors." Those exits were involuntary host government expropriations driven by nationalist sentiment, but they nevertheless involved give-and-take negotiations, often with active government involvement. In fact, many oil countries learned after the experiences of Mexico and Iran that outright nationalizations did not serve their national interest. Through a process called participation, exporting countries bargained for sovereign control while oil companies shifted to distribution and sales.[12] What's more, in at least one case—Shell Oil leaving apartheid South Africa—the exit was voluntary.

Although the context is far different now, such early exits can be seen as setting a precedent for voluntary, seemingly premature exit, Ecuador being a case in point (chapter 5).

### Unproductive Investments

Here oil companies and countries find that their initial expectations, often boosted by those seeking the next big find (chapter 3), aren't met. Exxon, for example, put some billion dollars into shale oil in the US Western Slope, only to abandon the entire operation due to insurmountable technical and economic problems. The boom from exploration and initial development went bust in a matter of days.[13] The fracking boom in the United States and elsewhere is showing similar signs despite, at the time of this writing, rampant boosterism (chapter 11).

## Vertical Deintegration

In the late twentieth century, major oil companies began selling off their gas stations in the United States. Now they are getting out of refining. If distribution is followed by extraction and finally exploration, a pattern emerges: the exit, intentionally planned or not, begins downstream (retail) and moves up (toward production). Vertical integration made sense in the oil business since the days of Rockefeller and Standard Oil, of Marcus Samuel and Royal Dutch Shell, characterized by endless new discoveries and steady, if not increasing, economic returns. But discoveries peaked in the 1960s. The industry is still immensely profitable (see below), but it may well be that people in the business, people professionally accustomed to looking down the road fifteen to twenty years, even forty to fifty years, see the writing on the wall: there will be an oil industry but nothing like that of the twentieth century; production will drop off drastically for a host of reasons in the twenty-first century. Time to get out.

## Investing Away

Yes, oil and gas companies bring in huge revenues and can be immensely profitable. In *Fortune*'s 2011 ranking of top global corporations, for example, eight of the top dozen were oil and gas companies (box 12.2). In 2008 Exxon was making profits at the rate of some $1,400 a second.[14] It's likely such returns will continue for some time, even if renewables kick in, capital becomes increasingly tight, and resistance to extraction increases (all of which this study presumes is likely). This prediction seems to be upheld by stock markets if we can presume investors' calculations include a time horizon of, say, a decade or two. Continued capital outlays, even with flat or declining production returns, suggest that the industry itself expects many years, even decades of substantial returns.

The strategic question, then, is, as is always the case for a profitable firm, where to put those earnings. From an exit perspective (as opposed to a maximize-shareholder-value perspective), it is to ask if investing elsewhere could make sense under some set of conditions, internal to the firm and external. The choice is roughly threefold: make payouts to shareholders, reinvest in current and new oil and gas operations, and invest elsewhere. The last one includes what we will call investing away, that is, away from fossil fuels and toward a new order. To be clear, investing away is not just investing outside the industry; it is deliberately investing in a way that presumes the demise of the industry and the rise of other energy systems. The choice to invest away is a

**Box 12.2**
The World's Largest Corporations

| Rank | Company | Revenues ($ millions) | Profits ($ millions) |
|---|---|---|---|
| 1 | Walmart Stores | 421,849 | 16,389 |
| 2 | Royal Dutch Shell | 378,152 | 20,127 |
| 3 | Exxon Mobil | 354,674 | 30,460 |
| 4 | BP | 308,928 | -3,719 |
| 5 | Sinopec Group | 273,422 | 7,629 |
| 6 | China National Petroleum | 240,192 | 14,367 |
| 7 | State Grid | 226,294 | 4,556 |
| 8 | Toyota Motor | 221,760 | 4,766 |
| 9 | Japan Post Holdings | 203,958 | 4,891 |
| 10 | Chevron | 196,337 | 19,024 |
| 11 | Total | 186,055 | 14,001 |
| 12 | ConcoPhillips | 184,966 | 11,358 |

*Note:* Notice that while the biggest company is retail, Walmart, the three most profitable here are independent oil companies, Shell, Exxon, and Chevron, with a national oil company close behind, China National Petroleum. *Source:* Fortune 500 Ranking, CNN Money, 2014. http://money.cnn.com/magazines/fortune/global500/2011/full_list.

function of future expectations and the company's operational definition of core mission. So, for instance, if the company expects that the fossil fuel era will continue for a long time, that climate disruption and other environmental problems will be solved by new technologies, and that new finds will get harder for geologic and political reasons, then it would pay out enough to keep shareholders from fleeing and otherwise reinvest heavily in production, even if production returns are declining. This arguably is the dominant industry position now and implies a core mission of "powering the world."

But if the company expects that fossil fuel availability and use will decline, possibly precipitously, and, at the same time, that other social forces will pull capital elsewhere—for example, into defensive measures against effects from past environmental abuses, such as extreme weather as a result of a century's loading of atmosphere and oceans with carbon dioxide, or maintenance of nuclear waste sites, a premise of this study—then the company will liquidate, make payouts to shareholders, or look for alternative investments. Among the company's strategic questions is where to invest outside the industry. If its core mission is defined in narrow terms—"we just produce oil and gas" or "we just make money for our shareholders"—then it simply invests for maximum financial return on investment. Grocery retail, fast food, land in Africa, high-technology gadgetry: it makes little difference. But if the company's core mission is (or was) of a higher order—to industrialize and modernize the world, say, or to defend free peoples from tyrants, to generate wealth and alleviate poverty and illness—then it will invest away.

Among the criteria for investing away would be economic activities that are congruent with (1) the broadest and highest expression of the company's core mission, especially perhaps as it emerged in the early days of the company's history or when the company's existence was threatened; and (2) plausible directions that relevant societies will take in the future, in a transition unprecedented in terms of human population size, consumption rates, fossil fuel dependency, the virtual absence of resource or waste sink frontiers, and bite-back from past environmental abuses. A company that foresees a straight swap out of fossil fuels and swap in of renewables would invest in solar, wind, and biofuels. A company that foresees a societal embrace of simple living would invest in do-it-yourself businesses. Wherever it invests, as long as the investments are congruent with the company's past and its vision of the future, the company will be, in effect, making its exit; it will be pursuing its core mission while no longer being a major oil and gas company. It may retain select fossil fuel production options, but it otherwise becomes a twenty-first century holding company, holding capital from the past—fossil fuel extraction—and allocating it to the future—self-renewing energy and material systems, or what we might call a sustainable economy.

The preceding rests on the key assumption that the company's strategic thinking is Janus-faced temporally—simultaneously looking into its past and asking what has been its higher mission and looking into the future and asking what kind of society it wants to be a part of (given the trends in resources, waste sinks, capital, and social organization). Arguably, few businesses are so inclined; they are busy pleasing consumers, regulators and investors, all of whom want product and revenue *now*. Long-term thinking is in short supply in the modern economy. But if any industry has a history of long-term thinking (longer than, say, the next quarterly earnings or longer than the election cycle), then it is the most capital intensive of industries, the one that often must wait many years, even a decade or more for its returns—oil and gas. The majors have long operated in terms of decades. National oil companies, for all the pressures of filling state coffers, would presumably have a similar long-term orientation or could if restive publics demand responsible corporate citizenship. What's more, oil and gas companies, for all the abuse leveled at them (recall that John Hofmeister, a former CEO of Shell Oil-US, titled his book *Why We Hate the Oil Companies*), have seen themselves as forces for good in a modern, progressive world, a world that at times has been regressive or totalitarian or terrorist.

Now, in the twenty-first century, "good" is a very different thing, intimately tied to the need to make the transition out of the very substances that allowed the fossil fuel industry to be a force for good. "Good" now is the *common good* when everyone, high consuming and low, (but especially high consuming), must figure out how to live in a world full of human activity and impact, a world without resource frontiers, a world with waste sinks filled. This is a world that did not exist in the first century of the fossil fuel era. It is a world in which, we contend, the fossil fuel industry has a role to play: pushing the end of the fossil fuel era.

If these speculations and propositions are at all on the mark, we might wonder why no one is talking about it. Why is there no public discussion of the fossil fuel exit? The simple answer is that few in the business—energy or manufacturing or financial—look that far down the road. Those who do tend to see the same road of continuous production. But the deeper answer is that it would be in no business's interest to lay out the exit strategy. In an economic system where key actors, namely, investors, are skittish about bond markets, sovereign funds, and a host of debt-related issues (where debts are in the trillions of dollars), such talk could be self-fulfilling, if not hastening. It could cause great harm with economic and political disruption. And yet, not to act is likely to cause even greater harm. So, as the coeditors have argued throughout this book, this is a historical moment with great potential and great risk. In this chapter we argue that the role of the fossil fuel industry is to accelerate the transition and do it with foresight and planning; it is *to seek a twenty-first century common good by effecting a gradual, minimally disruptive exit.*

## A General Strategy

The first step in strategizing an early exit is to entertain the possibility that the industry is already on the verge of its decline. However intentional or inadvertent company plans may be, what can be certain is that, like all else in the fossil fuel business since its earliest days, nearly everything firms do for external perception is strategic: predictions of the future are optimistic, projections of demand are increasing, and return on investment is endlessly positive. Given that, what might their final strategy be?

At one level, as noted, the game is the same whether for entry or exit: a firm maximizes returns for its owners. To do that, it must convince owners, customers, and their legal and military enablers—governments—that the end game is a long, long time away, if ever. Otherwise each firm plans its exit before the industry as a whole plays out its end game. And for this game, the cards are played very close to the chest. For a firm to reveal its hand would only hasten others' premature exit.[15] Again, every move, every statement, every study is strategic, aimed not at descriptive accuracy but at actors' behaviors, at the rules of the game, at public expectation.

At another level, the rational calculation would be that a company continues to produce as long as returns exceed its capital's next-best alternative employment. Production may decline, prices drop, revenues fall off, and net energy may even go negative (for instance, coal is used to make liquid fuels at a net loss in energy but a gain in usability), but if returns are positive, the company stays in the game. And the game continues as long as key players don't get spooked—as long as investors, politicians, regulators, and consumers are doing well (or at least see no alternative). For these decisions—to keep producing, investing, and buying the product—the time frame is short, say, a few years.

Attracting investors and customers when all signs point to continuous growth in supply (twentieth-century conditions) is, however, far different from keeping investors and customers in the game when all signs point to energy descent and industry decline (twenty-first-century conditions).[16] So at another level, the game now is fundamentally different, especially if the time frame extends into many years or even decades. How do fossil fuel companies—private and state—keep these actors in the game long enough to extract maximum net gains before exiting? The rational calculation shifts to strategic maneuver. How does that look now, as the game is currently played?

First, they deny the end. Firms dismiss, indeed ridicule, any suggestion that the game may be coming to an end. They highlight new finds and new technologies and claim there are vast untouched supplies. They downplay spills and blowouts, injuries and deaths, calling them "accidents," the cost of running an industrial empire, or the stuff that just happens. They ignore health effects and threats to ecosystems and the climate, calling them too "uncertain" or solvable by new technologies. They wave off repressive government practices, calling them just "politics." Above all, they emphasize the inevitability of fossil fuel use for decades, indeed, centuries to come. New finds are routinely trumpeted as "the next Saudi Arabia" (e.g., the North Caspian, Alberta's tar sands, North America's shale gas, Brazil's deep-sea Frade field) and new technologies "the answer" to environmental and social objections (hydrogen fuel cells, corn ethanol, hydraulic fracturing, carbon capture and sequestration, geoengineering). The boosterism (chapter 3) these days is unrelenting.

As the industry thus blows smoke in its opponent's eyes,[17] its firms must still plan their exits. Again, the rational calculation would be to strip off the least profitable operations. Vertical integration in mostly national or regional markets made sense in the heady days of going up the production curve, indeed since John D. Rockefeller first attempted to stabilize the industry by standardizing the product (hence, Standard Oil).[18] Going down the curve in a volatile, largely uncontrollable global market is a different story for the industry. The majors had a taste of that in the 1960s and early 1970s culminating in the Arab oil embargoes. But this is different. It is not a shift in corporate power, a turn in operations. It is getting out with the least disruption, the best returns, the highest sense of purpose, and the greatest legacy possible.

So now it makes sense to cast off unprofitable or marginally profitable (or expectedly unprofitable) operations. Likely candidates include downstream operations where competition is greatest and monopoly rents least, notably retail. In industry lingo, downstream operations are called "disposal," getting rid of that which is "produced." Disposal is what anyone can do—run a gas station, for instance. By contrast, what only oilmen can do is explore and produce. Indeed, as in the late 1900s, the majors sold off gas stations right and left. They also got out of shipping. Moving upstream, refining would be next, and indeed that appears to be happening. In fact, the last refinery built in the United States was in 1976. And one of Europe's biggest refineries recently announced it was selling (if it can find a buyer). Going yet further upstream, if a company's production returns on exploration and R&D investments decline, that company may well be planning an exit. After all, if it brings on no new wells, all else equal, it can only pump from existing wells. Then it is done. This would be rational from a shareholder standpoint as the firm's limited resources could be diverted from costly and doubtful exploration and R&D to dividends. And it would be strategic because the firm would be saying one thing—drill, pump, deliver—and doing another—withdrawing.

Finally, profitable withdrawal requires a liquidation strategy. That would rest on secrecy, what the industry is well accustomed to (chapter 3): early exiting firms will get the best prices selling off plant and equipment; to do that, the company once again must appear to be doing just the opposite: operating in the business for a long, long time.

Two additional considerations round out this general exit strategy. First, although we have constructed this proposition in terms of independent oil companies (IOCs; e.g., Exxon, Royal Dutch Shell, BP), the national oil companies (NOCs; e.g., Saudi Aramco, Gazprom, Statoil, Pemex), which hold some 90 percent of reserves and produce 65 percent of the world's oil and gas, may have their own reasons to plan an exit. First, their core mission is presumably to advance the nation's interest. At some point, perpetuating the fossil fuel era is in no one's interest. In international affairs, no country wants to be a pariah state. They too must respond to capital markets (or get allocations from their governments, the very entities that would otherwise be seeking revenues and that must themselves enter capital markets). Perhaps most significant, many of the NOCs rely on the IOCs for current technologies and technological innovation. The major exception is Statoil. It has considerable technological and financial capacity on its own accord. But this is precisely the NOC most subject to public pressure, pressure that we saw in chapter 10 seems to be increasing, and it is pressure from a public that prides itself on environmental values.[19]

Second, although a major hindrance to exit would appear to be capital markets and their imposition of short time frames, an independent, publicly traded company could escape such pressures by going private. Some of the largest companies in the world are privately held or nearly so (e.g., Cargill, Bechtel, Bosch), and there appears to be a trend to go private for companies that have the capital and whose mission requires insulation from investor pressures. Going private could be the necessary first step in an exit strategy.

To resist exit would presume a narrow financial motivation among key actors such as executives, board members, and investors—getting the highest, fastest financial return on their investments. Relaxing this assumption and expanding the motivational range to a broader set of stakeholders, and consequently allowing for longer-term returns makes the internal case (internal to the industry) even stronger for the rationality of early exit. For instance, a company's historical core mission could have been "powering industrial society" or "building a great nation" or "heating, feeding, and transporting millions." Such notions may have made sense in the nineteenth century and maybe well into the twentieth. But the transition of the twenty-first century is fundamentally different. It requires a company to update its mission to something like "living in a biophysically constrained world" or "securing necessities under conditions of recurrent weather extremes and financial crashes" or "transitioning out of fossil fuel dependency" or "creating a just and sustainable world" or "empowering communities to live in place." Such a revision of core mission may be unimaginable for hard-core return-maximizing business leaders and their compatriots in business schools and the business press. But if our investigations into the industry through interviews, histories, news accounts, and memoirs are indicative, motivations for creating shared value are prevalent, especially among senior members concerned about their personal and professional legacy and junior members entering the industry during its twilight years, expecting less than lifetime employment and hoping to make a difference in the industry quite unlike that of their swashbuckling "big find" predecessors. To work for an oil and gas company, even a coal company, whose explicit mission is to phase out its most famous products—gasoline, natural gas, petrochemicals—and phase in the components of a sustainable world, is something that even some financiers could get on board with. One means would be to reposition investment earnings, to "invest away" from fossil fuels and toward a just and sustainable world.

How would it end for a firm, for the industry? A reasonable guess is that it would end like other mining operations and financial bubbles end: for a while, maybe a long while, everything looks very good to investors, tax collectors, advertisers, and consumers. And then suddenly it stops. And everyone is surprised.

If these conjectures are close to the mark, if the dominant, century-long narrative of vast reserves and fossil fuel centrality is a strategic move, rhetorical devices to keep an inherently dangerous and finite game going, then it is likely the end game will come suddenly and perhaps much sooner than the general public, including experts and commentators in academe and the media, would believe. But it may still not be soon enough. Other actors may have to nudge the industry, goad it, or provide normative cover for doing what is still unthinkable in the modern, industrial, growth-obsessed, consumerist, fossil-fueled order: leave fossil fuels, the great bulk of them, in the ground.

## Easing the Exit

Assuming no oil company steps up to invest away, what can other actors do to prod an early exit? Put differently, how could a global mining operation be undermined before it's too late? Recall that "early" means before oil producers, independent and national, would be forced to exit for geologic, economic, political, or environmental reasons.

The obvious counterstrategy to business as usual is to expose the ruse of business as usual, which the climate change, environmental justice, and peak oil communities are in effect doing. A measure of their success would be the vehemence with which the industry and their enablers (politicians, media, think tanks) attack their efforts. On this score, the climate change community has mounted the most serious threat, even though it is indirect: the inadvertent waste products of the industry's activities—carbon dioxide, methane—could render the planet uninhabitable. The climate community's call for limits on emissions could have put a serious crimp in the fossil fuel game, endangering not just profits but the rhetorical ploy that this game requires—that it will continue for a long, long time, and that it must. But at least for the time being, the climate change community's efforts have been stymied, their politics rendered impotent by technological blinders (chapter 1). So what about the other two communities?

The environmental justice movement has never been able to mount a serious challenge. Its efforts have been largely out of sight and out of mind with respect to centralized decision making, both corporate and governmental. That may be changing with the subnational and transnational resistance occurring on a number of fronts. In fact, a major purpose of the cases of part 2 and chapter 11 is to demonstrate that resistance may well be mounting, gaining a foothold along with questions of justice and a decentralized, sometimes transnational politics.

While much empirical work needs to be done to establish such a fundamental political shift, it does appear that twentieth-century politics are giving way to a yet-to-be-defined twenty-first century politics (chapters 1 and 13). If the biophysical basis of twentieth-century politics—easy-to-get, easy-to-use, highly concentrated energy with significant costs easily displaced—engendered industrial dominance and a geopolitics of bipolar and unipolar hegemony, then the withdrawal of easy energy will result in something fundamentally different. What those politics will be, the authors of this book have only been able to hint at. One element on the material side, however, will be an increasing delegitimization of fossil fuel (and other mineral) extraction and the revalorization of fossil fuels as special (chapters 1, 3, and 4). Another, we have argued in this chapter, is the foresight of industry players themselves, using their power to effect a positive transformation of the industry and, consequently, a positive transition of society. Again, empirical work is crucial to test these propositions. But the argument is clear regarding the energetic sources of political power. The twentieth-entury source was concentrated energy—fossil fuels. As their availability declines and its dependent industries and companies with it, so too will the associated economic and political power. And because that power has been largely one of domination, "power over," it is reasonable to surmise that the balance will shift to "power with," the power of collaboration and living with natural systems. Twenty-first-century power will be that which is grounded in place, dependent on land and water and other regenerative systems. Those so dependent will increasingly build and defend such resources because their livelihoods and lives quite literally depend on it. Those who have been subject to external agents' extractive pressures will build a qualitatively different power base one of living in place. Their "power with" derives from the legitimacy of actually practicing sustainable resource use, legitimate because the 21$^{st}$-Century imperative now is living within our means and for the indefinite future. From practice to power, these are this book's last topics.

## Notes

1. For discussion of power, especially as it derives from concentrated energy—fossil fuels—and plays out as concentrated economic and political power, see chapters 1, 3, and 13.

2. A note on terminology may be useful for readers coming from different disciplinary and professional backgrounds. *Energy descent* is a term people in the energy policy, peak oil, and transition fields use to refer to the historical trend, only beginning to play out now, of less and less available—repeat, available—energy. (See chapter 2.). *Industrial decline* is one operational consequence of energy descent, one especially pertinent to the fossil fuel industry but also to fossil fuel–dependent industries such as automotive and petrochemical. Another operational consequence is *localization*, which we largely do not engage in this book; see, however, Raymond De Young and Thomas Princen, *The Localization Reader: Adaptations for the Coming Downshift* (Cambridge, MA: MIT Press, 2012). Although *industrial decline* is used in the business strategy literature to refer to the last stage of the life cycle of an industry (from embryonic to growth to maturity to decline), here we use it, in the first instance, as that which is driven by biophysical trends, a methodological premise of this book, and in the second, as that driven

by cultural norms and expectations. *Company exit*, then, is an operational consequence of industry decline. It can be inertial, denying, and reactive (as seems to be mostly the case in the corporate world to date, at least the public face of the corporate world) or anticipatory, high-minded, and proactive (what we argue for in this chapter). So while *industry decline* and *exit* are commonplace in the conventional business world, the descent, decline, and exit that inform this chapter—with their potential for exiting companies to push industry decline and, as a result, accelerate the end of the fossil fuel era—are unique.

3. D. Greenaway, J. Gullstrand, and R. Kneller, "Live or Let Die? Alternative Routes to Industry Exit." *Open Economic Review* 20 (2009): 317–337.

4. K. R. Harrigan, "Exit Decisions in Mature Industries," *Academy of Management Journal* 25 (1982): 707–732.

5. K. R. Harrigan, "Deterrents to Divestiture." *Academy of Management Journal* 24 (1981): 306–323.

6. Harrigan, "Exit Decisions in Mature Industries" and "Deterrents to Divestiture."

7. A. R. Brandt and A. E. Farrell, "Scraping the Bottom of the Barrel: Greenhouse Gas Emission Consequences of a Transition to Low-Quality and Synthetic Petroleum Resources," *Climate Change* 84 (2007): 241–263.

8. E. Tzimas, A. Georgakaki, C. Garcia Cortes, and S. D. Peteves, *Enhanced Oil Recovery Using Carbon Dioxide in the European Energy System* (Peten, Netherlands: Institute for Energy, December 2005), 1–117, Report EUR 21895 EN.

9. "Appropriately discounted" may be a leverage point for nonindustry interveners, private or governmental. If a firm can be shown that future benefits are too optimistic—that is, the firm's future must be discounted more heavily—then it may choose to hasten its exit and reap early mover rewards.

10. Daniel Yergin, *The Prize: The Epic Quest for Oil, Money and Power* (New York: Simon and Shuster, 1991), 739, 740.

11. Confidential interview, 2010.

12. Yergin, *The Prize*, 583.

13. Ibid., 716.

14. Steve Coll, *Private Empire: ExxonMobil and American Power* (New York: Penguin Press, 2012), 502–503.

15. We deliberately use the word *game* because, in part, that is how business leaders so often talk about their work: We're in it for the game, to compete, to win. We also use it to connote strategic relations as opposed to the prevalent discourse that is often technical (When will oil production peak? What is the price point for a shift to renewables?) or moralizing (All they care about is short-term profits!). The politics of keeping oil in the ground is, we contend in this chapter, ultimately that of strategy; it occurs in a decision realm where there is no determinative solution, where "the game" is to arrange the board such that others find it in their interests to do one's bidding. "One" here are those actors promoting early exit, whether in or out of the industry.

16. For evidence of industry decline (although not stated this way) from within the corporate world and from the leading global business consulting firm, McKinsey Company, see R. Dobbs, J. Oppenheim, F. Thompson, M. Brinkman, and M. Zornes, *Resource revolution: Meeting the world's energy, materials, food, and water needs* (McKinsey Global Institute Report, November 2011), http://www. mckinsey.com/insights/energy_resources_materials/resource_revolution. For discussions of the twentieth and twenty-first centuries from a transition perspective, see chapter 1 of this book. For definitions of *energy descent, industry decline,* and *company exit,* see note 2 above.

17. Many of these tactics come right out of the tobacco industry's playbook. See, for instance, Allan M. Brandt, The *Cigarette Century: The Rise, Fall, and Deadly Persistence of the Product that Defined America* (New York: Basic Books, 2007).

18. Yergin, *The Prize.*

19. Kari Marie Norgaard, *Living in Denial: Climate Change, Emotions, and Everyday Life* (Cambridge, MA: MIT Press, 2011).

# 13

## On the Way Down: Fossil Fuel Politics in the Twenty-First Century

Thomas Princen, Jack P. Manno, and Pamela L. Martin

When we began this project, a serious proposal to deliberately keep fossil fuels in the ground was a rare sighting in environmental and energy fields. Certainly the idea was implicit: if we tax the stuff, or offset it, or promote wind and solar (and nuclear), then we just won't need to use fossil fuels, so of course they'll stay in the ground. To presume that fossil fuels will stay in the ground as a by-product of rational environmental and energy policies is, to put it bluntly, politically deficient. It presumes that technorationalism—cost–benefit analysis, efficiency measures, new technologies, designed markets—will eventually overcome the capabilities of entrenched fossil fuel interests.[1] It presumes a politics without power. It neglects the tremendous societal inertia compelling continued extraction, inertia driven by some combination of these:

- Physical infrastructure valued in the billions (if it's there we must use it)

- A pervasive belief among policy and business elites in the solvent of growth (growth is needed to solve problems) and technological determinism (human ingenuity and markets overcome resource scarcity)[2]

- Economic rents (i.e., unearned revenues) that go to those who find themselves sitting on reserves or are able to acquire rights to such reserves

- Monetary returns on monetary investments that continue to out-perform nearly all alternative long-term investments, even in declining fields, even in high-risk environments, even with a "carbon bubble," even with the prospect of energy return on investment (EROI) approaching minimum levels (chapter 2)

- Political power expertly employed writing the rules of the game, especially the game of gaining access with legal, diplomatic and military means (chapter 3)

A politically sufficient approach, one attuned to the realities of the twenty-first century, begins, we've argued in various ways in this book, with explicit attention to multiple sources of power, some material (natural resources, weaponry, financial capital, for example), some ideational (growth is necessary and good), some overt (military and economic), some hidden (in writing the rules of the game and gaining access and dumping externalities on the marginalized). Above all, it must be grounded in the biophysical and social, including the likelihood of catastrophic outcomes if business as usual continues with current patterns of extraction and injustice.

We say this, however, with a certain humility, knowing no better than anyone else how such societal inertia can be overcome, especially at the requisite rates to avoid environmental and social catastrophe. For insights, we have turned to those working on the ground to keep fossil fuels in the ground. It is their view of the world, their vision, their initiative, indeed, their courage, that gives us hope that the inertia of the fossil fuel juggernaut can be tamed. While we hesitate to say a keep-it-in-the-ground (KIIG) movement exists, something like it, a growing transnational network, does appear to be emerging, given the failures of top-down efforts to arrest climate change, to end toxic substance production, and to stop net freshwater drawdown and soil loss (both powered by fossil fuels).[3] More and more now, we hear phrases like "keep coal in the ground" and "leave the oil in the soil." If there is any group of peoples for whom this makes sense, and has always made sense, it is Indigenous peoples. At risk of overgeneralizing, they see long, penetrating tendrils from oil, coal, and gas mines across the land and into the water, and into themselves and other life forms. In two of our cases, the Yasuní-ITT Initiative in Ecuador and, in the United States, New York State's moratorium on hydrofracking for natural gas, Indigenous peoples from very different environments have organized against extraction (chapters 5 and 11). The commonality of their actions, indeed their language and moral commitments, remind us of the long history of Indigenous and other local peoples bearing the brunt of the downside of "progress"; they remind us who has been dominated, if not who is dominating.[4] Even in the Global North, in places like eastern Germany and Appalachian America, the downside is longstanding, continuing to this day (chapters 9 and 6). Now, as it becomes clear at a planetary level that everyone, dominators and dominated, is potentially on the downside, it seems others are beginning to listen, including powerful elites.

In part, this book does what all social science does: identify a problem, collect evidence, conceptualize, generalize, suggest policy recommendations, and offer caveats along the way about limited sample size, generalizability, and predictability. But we have attempted to do more in this book. First, we have taken inspiration from those on the ground who are trying to keep dangerous substances in the ground. In one way or another, everyone in this project has a deep connection to the peoples and places that bear the brunt of fossil fuel and mineral extraction. Our feet have touched the places where oil, coal, natural gas, gold, and uranium are taken from the ground. Our hands have shaken the hands of activists who fight every day, often risking their lives, simply for the right to live well in a healthy environment. Some of us have been involved in the movements, standing toe-to-toe with activists, for example, as Laura Bozzi described in chapter 6. Robin Broad and John Cavanagh led a delegation to El Salvador. James Goodman and Stuart Rosewarne and their colleagues have been active in anti-uranium movements. For two decades, Tom Morton has covered the former East Germany as a journalist and documentary producer. Helge Ryggvik has testified abroad about Norwegian oil policies and protested at home. Berit Kristoffersen has worked with communities in the far north of Norway, exposing how the state and the oil industry worked together to access oil fields in the Norwegian Arctic. Pamela Martin has canoed the upper Amazon, speaking with Indigenous peoples and activists about oil concessions in their home, the rain forest. In one such place, not long after Martin's visit, Indigenous peoples living in voluntary isolation in Yasuní National Park allegedly killed two Huaorani leaders. In response, Indigenous, environmental, and community leaders from all over the world called for a moratorium on oil drilling in the Amazon. Jack Manno navigated the Hudson River in New York State to commemorate the Two Row Wampum Treaty of 1613 between the Dutch government and the five nations of Haudenosaunee (Iroquois) peoples. Tom Princen has shared meals with farmers, nuns, and oil executives all trying to envision a postgrowth, post–fossil fuel world.

Second, as scholars we have taken on a dual task, one of theorizing (see below) and the other of serving, serving those most burdened by extraction and serving all of us on this planet as we transition out of fossil fuels. We ask, for instance, what worldviews inform the dominant paradigm of extraction and expansion, externalization and empire. What are the dominant arguments in framing this paradigm, and who are its key actors? What are the relevant sources and expressions of power? What is the essence of the politics of those fighting extraction and the paradigm? What institutional structures are emerging in this movement? What would "success" mean in the cases we encountered in part 2? Above all, we ask if the time has come to challenge the very legitimacy of fossil fuels in the current materialist, consumerist, growth-obsessed, debt-driven culture to entertain the proposition that some stuff is better left in the ground.

### Not Yet a Theory

We do not presume to have a theory of fossil fuel cessation or of the structure of post–fossil fuel societies. Such theories should be inductively developed, which necessarily will have to await a richer and more defini-tive set of cases where the reasons for and meanings of success and failure are evident and key variables allow comparative purchase (introduction to part 2).[5] Such theories, like so many others in the social sciences, would be retrospective and descriptive, not prospective and normative. That is, they would say only what is and how it is, not what is likely to come nor what *should* come given a set of objectives.[6]

Our approach is precisely a forward-looking, anticipatory exercise that presumes an objective of peaceful, democratic, just, and biophysi-cally sustainable transition. At this historical juncture, it has no definitive case studies, only what we presume are transitional instances of early movers, social pioneers, moral entrepreneurs. Our chapter authors' con-clusions are necessarily tentative, their lessons provisional.

As for theory building, then, we have presented a hypothesis-generat-ing, not hypothesis-testing, study. Because few in the relevant disciplines have asked questions of transition, of exiting fossil fuels, of pursuing the good life without fossil fuels, there is no theory to test.[7] More important, however, this is a normative study. It dares ask what decision makers *should* do given a set of conditions. Among those conditions are the end of cheap energy, ever increasing concentrations of greenhouse gases in the atmosphere and oceans with associated environmental threats, and concomitant social changes, from the nature of the corporation to the structure of the economy, including the financial system, changes in the meaning of nation and community, of progress and the good life. The "should" in this normative exercise derives not in the first instance from the authors' personal ideologies but from a common conviction that these changes are profound and, on the whole, unprecedented. What is more, we authors presume we are not alone in wishing the transition to be peaceful, democratic, just, and biophysically sustainable. We put these conditions and these normative positions front and center.

## Future Scenarios

A future-oriented, theoretical work requires realistic grounding (here, actual cases) and plausible scenarios of that future. Two recent efforts to formalize alternative scenarios for the future are based on realistic options for dealing with the environmental and resource challenges facing the world: the Tellus Institute's Global Scenarios for the Century Ahead: Searching for Sustainability[8] and the scenarios developed for the Millennium Ecosystem Assessment (tables 13.1 and 13.2).[9] Each study refers to similar uncertainties: a wide range of possible climate change feedbacks and ecosystem responses, whether and how human values may change, the potential for global cooperation or conflict, and other potential feedback loops that could seriously alter societal change. As a result, each posits four scenarios it considers most plausible. Both seriously question the sustainability of the business-as-usual scenario and suggest social change will happen; the question is, "Toward what end?" The similarity between these scenarios and others suggests that while the future may be unknowable, given energy and environmental trends, plausible scenarios are not unlimited.

The plausible scenarios follow along the lines of business as usual, with utopian and dystopian futures. They posit that there will be, as there

### Table 13.1
Tellus Scenarios

---

**Conventional Worlds**

Market forces: Business as usual. Global incomes; GDP and population grow. Profound inequalities. Conflicts over scarce resources. Collapse.

Policy reform: Government-directed reforms toward sustainability objectives. Serious reduction in GHG greenhouse gas emissions. Internationally agreed poverty-reduction strategies.

**Alternative Visions**

Fortress world: Authoritarian order imposed. Elites retreat to protected enclaves. Environmental degradation exacerbated.

Great transition: Values shift to a just, sustainable world. Human solidarity and environmental stewardship. Reduction in consumption through frugal lifestyles. Voluntarily reduced population pressures.

---

**Table 13.2**
Millennium Ecosystem Assessment Scenarios

---

Global orchestration: Economic cooperation, global growth, trickle-down benefits for environment and other public goods.

Techno-garden: Ecological engineering and biotechnology follow adoption of reforms based on natural capitalism, profits from mimicking efficiencies of natural processes.

Adapting mosaic: Managing socioecological systems through adaptive management. Free flow of information. More restricted flow of trade goods and services. Great regional variation. Local and regional comanagement.

Order from strength: Breakdown of global cooperation, authoritarian responses to social and environmental crises

---

always is, a struggle for the future among competing perspectives on justice, fairness, righteousness, and faith, as well as competing political and economic interests. Energy analysis and social theories can inform what is possible but not necessarily what is likely. Dystopian and utopian tendencies will emerge together, and the outcome may be a mix of both for a long time to come. While the state of economic disparity, global climate, biodiversity, water, and other issues all trend toward the dystopian, there are signs of growing social movements that lead to what the Tellus scenarios call the Great Transition: the shift from the industrial-growth society to a life-sustaining civilization.[10] In that transition to a post–fossil fuel world, there will necessarily be lower levels of consumption, and maybe less economic security. In an energy-constrained world, fairness in energy access will become increasingly difficult to define and achieve. For many—for example, the roughly 1 billion people who have no access to electricity—the lack of refrigeration and health care can be a matter of life and death. Such energy poverty will require significant expansion of decentralized, renewable sources of electricity at the same time the world as a whole faces unavoidable energy constraint and reduced material consumption.[11]

What these scenario exercises share is a view that social crises can be times of danger and opportunity, challenge and creativity. A peaceful transition requires committed work from millions of people involved in social change activism. In our own work with students and community groups, we find a general acceptance of fundamental change, of a "new normal," and a yearning to make sense of that change and do something about it. Many young people are looking for direction in their education and in choosing work that facilitates a positive transition. As instructors and social analysts, we expect that with the end of cheap energy, negative scenarios are likely unless a strong movement for social change can explain broadly what is happening, why it is happening, and show how it can be positive. A rationale, based on declining EROI and the threshold relation between energy consumption and well-being, is one part of the explanation (chapter 2). Another part is an ethic of fossil fuel use that privileges ending fossil fuel use (chapter 4) and a corporate strategy that rationalizes exit from within the industry (chapter 12). And another is a deliberate politics of resistance to powerful actors and restoration of communities and ecosystems.

### Creating the Conditions

With these theoretical caveats and future scenarios in mind, in this concluding chapter, we offer five propositions regarding the prospects of ending the fossil fuel era. These are necessarily tentative and provisional. At a minimum, we hope they will prompt constructive criticism, empirical exploration, and more theorizing. More ambitiously (and perhaps presumptuously), we hope these propositions will help guide those on the ground doing the real work of keeping fossil fuels in the ground. Notice that four themes intersect the five propositions: biophysical imperative, place, good life, and power.

### Proposition 1: Normative Shift
*A dominant cultural norm of endlessly growing material wealth will shift to a norm of "the good life," or something like it, as*

- *Resource constraints tighten,*
- *Negative feedback—biophysical and social—intensifies, and*
- *Disparities in resource access widen.*

Keep it in the ground (KIIG) transcends mere resistance when actors (1) construct their own, particularist notion of the good life situated in place and (2) recapture its energy, including its flow of life-supporting energy.

Resource constraints and biophysical feedback are amply discussed in the environmental sciences. What is rarely discussed there, however, is the politics of this normative shift (aside from decrying the "lack of political will" and the partisan bickering). By contrast, this book, with a focus on fossil fuels, has been about the social feedback that has taken two forms. One is the positive feedback of the past two centuries in the form of wealth and power that until recently has overwhelmed any checks on such feedback (see also figure 3.1 in chapter 3):

*cheap energy* → *easy wealth* → *political power* → *resource access* → *cheap energy*

That feedback, along with an ethic of total extraction (chapter 4), has brought the world to an ecological precipice, prompting the second form of social feedback—resistance and restoration. Here the politics comes from (1) those whose place and livelihoods are violated, (2) those who see calamity in the scientifically established trends, and (3) those who allocate capital and must choose between life-supporting investments (e.g., for food and housing) and extractive industry-supporting investments (e.g., pipelines, deepwater rigs).

Regarding violated places, the common factor North and South is violence—violence to the land, to people, and to people's spirit. The case of El Salvador and tragic loss of life in this struggle highlight the urgency of a peaceful transition (chapter 7). But Norway too, like so many other producing countries, has experienced dramatic loss of life (chapter 10). Those who extract from the places of other peoples show little reverence for anything but return on investment, talk about jobs and development notwithstanding. As a result, people everywhere, whether attached to such places as their home or as sacred places far from home, whether dependent or, in the grander scheme of planetary connection, interdependent with those places, they are trying to live a good life and allow others to do the same. These dynamics were perhaps most evident in the coal cases of Appalachia and eastern Germany (chapters 6 and 9). So in this book, we adopt a term that we see rapidly spreading through the Global South but applicable to the Global North as well, especially in those places where extraction is synonymous with destruction. That term is *good life—buen vivir* in Spanish, *sumak kawsay* in Quechua, and *Ganigonhi:oh* in Onondaga. We intend this term, as we believe its users do, to delimit an approach to natural resources that is not primarily extractivist, leaves intact natural systems and in fact enhances their ecological integrity, and encourages human flourishing without endlessly increasing consumption and waste deposition. This is the politics of restoration.

So what are the underpinnings of the good life? First, they are not merely assumptions on which to base a new theory or a personal philosophy. Rather, they are ways of organizing oneself and others to live well by living well within biophysical limits. Part of that organizing is to resist the depredations of others, of those whose work has no place, of those who have so successfully organized the modern world to extract and dump and always need another place to do it some more. From Ecuador to El Salvador, Germany to Australia, the United States to Norway, the cases in this book have been primarily about that resistance.

Second, the underpinnings of the good life are ways of constructing the "lifeblood of living in place." *Lifeblood* in this context is a term that indicates both the vital necessity of sustaining life-support systems and the absolute urgency of transitioning out of fossil fuels and endless extraction and expansion. Throughout this book, we have endeavored to show how fossil fuels and minerals have threatened life. In the cases of Ecuador and El Salvador, indigenous peoples and activists have been jailed and murdered. Broad and Cavanagh (chapter 7) told the story of Marcelo Rivera who was murdered for his work opposing gold mining and supporting clean water. Princen (chapter 3) described the horrific work conditions in the 1800s of children with coal strapped to their backs. Laura Bozzi (chapter 6) brings the coal story to the present with tales of communities forced to live where mountaintops are blown apart, rivers and valleys filled, streams and water systems polluted, species lost forever, and homes and communities destroyed. As Bozzi noted, despite all the promises, none of this has truly increased employment or income in these communities over the long term. And according to 2011 net savings data from the World Bank, African countries would have increased their gross domestic products more by leaving fossil fuels in the ground than by extracting them.[12] In these places, with these practices, there is no good life.

So *lifeblood* here indeed connotes life, unlike its contemporary use where oil is the lifeblood of modern society. Industrial lifeblood derives from the long-ago deaths of microorganisms; the contemporary death of peoples, communities, and ecosystems worldwide; and the future deaths and lives wasted with climate disruption, toxic buildup, freshwater contamination, soil loss, and the like. If the industrial metaphor is at all apt, it is to suggest that modern life is running on external life support, that this grand experiment is a one-time blip, that the body politic is utterly dependent on cheap, high-energy transfusions, and when they end, so does modern society. So in this book we ask what good modern society is if the good life amounts to little more than a proliferation of consumer "goods." What good is modern life when a minority concentrate wealth and power and the rest struggle to sustain themselves? Thus, we deliberately take back a perfectly good metaphor, lifeblood, but locate it in that which sustains, rather than destroys, life over time.[13]

Third, because lifeblood as fossil fuel leads inexorably and tragically to societal collapse, societies can evolve into better living by deliberately getting off fossil fuels and pursuing a life-affirming good life. Significantly, people worldwide are not waiting; they are already doing it (see part 2 and boxes 5.1 and 11.1). The cases in this study, we must be clear, are not meant to provide definitive models from which to replicate or scale up the transition. Rather, they suggest some of the contours of a route to the good life, one that includes using energy according to measures of long-term welfare (Manno and Balogh's welfare returned on investment, chapter 2) and living productively (Princen's criteria of livelihood in transition, chapter 4). Such movement requires that the opaque power of destructive industries (as seen in chapter 3 and throughout the cases) be exposed, its contradictions highlighted, its prerogatives denied. Doing so creates twenty-first century realist notions of power, sovereignty, and, ultimately, politics, and at all levels from the personal to the global.[14]

One voice for the good life and against the current destructive order is that of Indigenous peoples. Groups from the Amazon to the Arctic are working to change the politics of the international system, a politics that takes as given that unending growth and consumption are desirable and right. In chapter 11, Manno and Martin demonstrated the deep meaning of the good life and good mind through Haudenosaunee and Andean indigenous concepts of living in harmony with nature. Such pathways of place-based development and productivity are based on a cosmological vision that does not view time in a linear fashion but rather views ancestors as living among us and includes future generations as part of the system. Therefore, say Indigenous peoples around the world, we have to walk with the past and think of the future all while living in the present. To do that, extracting fossil fuels from sacred grounds and contaminating soils and water systems is ethically and spiritually wrong. Indigenous Amazonian peoples see a form of energy in all that surrounds them, an energy that people from a wide range of spiritual traditions take as sacred. Fossil fuels cannot duplicate this energy. In fact, extractive practices destroy such energy. This is the energy of regeneration, the only true lifeblood of a sustainable and moral system.

Indigenous peoples in Ecuador, New York State, and elsewhere are communicating such values to the rest of the world. Theirs is an effort to blend traditional and natural knowledge with economic, political, and ethical theories and systems to form a new pathway to a truer develop-ment. The concept of the good life in the Andes is one that requires dia-logue and evolution over time. This means that it has global applicability yet requires local implementation. The implementation of the good life is one that is not based on individual acts and the ability to dominate nature and exploit natural resources. Rather, it is based on community interaction, developed in dialogue through participatory institutions. Illustrative of this interaction is the cooperation among labor unions and Indigenous peoples in Australia (chapter 8) to protect their lands from uranium extraction and uphold a larger ethic that opposes the destructive uses of nuclear weapons.

In Ecuador, the National Development Plan was formulated through citizen institutions in each province of the country, explicitly to develop a plan for the good life—incorporating biodiversity, pluriculturalism, and participatory democracy within nature's limits—according to their local cultures and natural environments. In this process, the participants reframed concepts of development and economy to reflect nature as the "source of life," not the source of unlimited resources for insatiable consumption.[15] Thus, they placed the economy within nature rather than outside it. They did so as a means of deepening democracy and respect-ing all life forms and cultural and ethnic groups. Those who wrote the 2008 Ecuadorian Constitution that enshrines the good life and rights of nature worked closely with Indigenous groups; they also drew on the ideas of ecological economists.[16] Basing an economy on vital ecosystems, ecosystems that are currently threatened by human activities, and a view of the planet as cooperative and interconnected aligns with principles of the good life and of ecological economics. It does not, however, align with a vision of the world that is individualistic, treats nature as raw material, and externalizes the costs of resource use. Nor does it assume that security is based on conflict rather than cooperation.

A premise of this book is that the old route of overconsumption powered by fossil fuels cannot and will not be sustained. The Interna-tional Energy Agency's *World Energy Outlook 2012* announced that two-thirds of current fossil fuel reserves need to stay in the ground if we are to limit a rise in temperature to 2 degrees Celsius, which it predicts will occur in 2017 at the current rate of carbon emissions. They also note that such a leave-it-in-the-ground policy would increase efficiency and lower energy bills.[17] The *Economist* called these "unburnable fuel[s]."[18] As Ryggvik and Kristoffersen showed, some Norwegians are asking themselves whether their oil is "unburnable" too if Norwegians are to live up to their ideals (chapter 10). Peoples around the globe need no further proof of this premise. They are moving toward new ways of creating livelihoods and building an imaginative politics that sustains the planet rather than destroys it.[19] Even in the United States, the largest consumer of fossil fuels, movements to divest university endowments from the fossil fuel industry are organizing and getting tangible results. This indeed may have been unimaginable, or even naive, only a few years ago, but demonstrates the steady evolution toward a politics of ending the fossil fuel era.[20]

Conjoined with the indigenous perspective is the ethical. In chapter 4, Princen made the case that while fossil fuels have brought great things, including great power, political, economic, and military, they destroy, and their destruction remains long after the stuff is burned. The biophysical imperative is to transition out of these substances. The ethical imperative is to delegitimize humans' deeply problematic, century-long relationship to fossil fuels. In this study, we have found that the power of fossil fuels is anathema to the good life and that a different energy, a different form of power, is needed (see below).

**Proposition 2: Strategic Alignment**
*KIIG efforts will have a higher likelihood of success to the extent they*

1. *Explicitly reject the language and behaviors of reductionism, commoditization, and placelessness, and*

2. *Seek community-affirming strategies that are simultaneously biophysical, cultural, ethical, and spiritual.*

A politics of resistance creates conditions that hasten and accentuate negative feedback as a check on a couple of centuries of positive feedback (easy energy leads to easy wealth, which is used to gain access to more easy energy; see figure 3.1). For that, language—ideas, concepts, metaphors, goals, stories—and reconfigured notions of the common good will be key.

This proposition starts with the fact that as we saw in chapter 2, a compelling case for KIIG can be made on strictly physical grounds: when it takes as much energy to get energy and to respond to its collateral damage, it is energetically rational to leave the resource in the ground. Decision making in a commercial, growth-dependent order, by contrast, tends to follow a different logic, where physical parameters are less important than price, including the price of energy and the price of capital (interest). To move away from such a simplistic reductionism,

Manno and Balogh proposed adding social welfare: when more energy use begins to reduce welfare, the resource should be left in the ground (chapter 2). This formulation retains the physical, what ultimately will happen over the long term anyway, while making explicit that the energy issue is not about energy per se but about human welfare, now and into the distant future.

In chapter 3, then, Princen argued that a fossil fuel culture has been deliberately constructed by boosterism and powerful, self-reinforcing feedback loops of physical energy and political power. One implication is that if a growth-dependent order can be constructed, it can also be deconstructed and an alternative constructed. There was nothing inevitable about the origin of the current order and there is nothing inevitable about its continuation. A post–fossil fuel order is possible; in fact, for biophysical reasons, it is inevitable. And because the cases of part 2 illustrate how acts of courage mixed with visions of an alternative create movement in that direction, an accelerated end to the fossil fuel order is entirely possible.

Notice that this hopeful note—unlike so much of the discussion that begins with carbon and greening and politics as usual—is empirically grounded, both physical and behavioral. A distinguishing feature of the KIIG perspective is our actor lens, one that doesn't privilege conventional actors (nations, intergovernmental bodies, transnational corporations) when it comes to effecting fundamental cultural change. Rather, our lens, which combines the biophysical, cultural, ethical, and spiritual, allows us to see a "people's politics," one that privileges those who are grounded in place and inclined to fight to sustain that place. The more we look, the more we see such people, and not just in all the usual places—peasant communities, Indigenous peoples, environmental organizations. Rather, we see them in countries rich and poor, North and South, on hardscrabble farms and in corporate boardrooms. We hope this book will spur documentation of more.

So what will the alternative construction require? At a minimum, a new language. That language will not be of extraction and empire, not of efficient production and satisfied consumers, not of globalized commodity flows and high-tech hucksterism. Rather, it will be a language of regeneration and partnership, of sufficiency and a productive citizenry, of living in place and living within our means. And it will be a language of resistance because the extractors will not go away; they will just get more clever in appropriating the wealth, natural and cultural, of others. All this, then, takes us to those higher-order concerns, what defenders of the status quo do not want to speak about: questions of purpose, of humans' place in the world (in their multiple worlds), to values and worldviews, to issues of living well by living well within our means, to questions of ethical precept and spiritual belonging.[21]

So in chapter 4, Princen called for a fossil fuel policy of "start stopping" governed by three ethical principles: lifesaving (seemingly self-evident yet routinely violated under the contemporary fossil fuel ethic), transition out of fossil fuels (the counterintuitive notion that fossil fuel use is justified if it explicitly ends fossil fuel use, a point that boosters of renewables don't seem to appreciate), and livelihood (everyone is justified in maintaining their ability to self-provision but not to luxuriate and externalize true costs). Then in chapter 11, Manno and Martin proposed a societal goal of *buen vivir* or a good life and a personal and cultural strategy known as the good mind (*Ganigonhi:oh*), an alignment of one's rational mind with one's dependence on the natural world, an understanding that nonhuman beings are relatives, not resources and that all these relatives have their duties given them by the Creator. Essential to this spiritual imperative of maintaining a good mind is gratitude toward the earth, water, plants, animals, and all other elements of creation that carry out their duties.

The authors of this book have found that it is the deliberate, integrated, strategic combination of these elements—physical, social, cultural, ethical, spiritual—that generates the possibility of a positive transition out of fossil fuels. This finding stands in stark contrast to the culture of fossil fuels, where new technologies and new energy sources are simply presumed, where strategy is all about access (to more energy sources) and prying open others' lands and violating others' livelihoods, where business as usual continues unabated. The dominant fossil fuel strategy depends on the ease with which coal, oil, and gas can be extracted and transmuted into a single thing, energy, measured by a single numeraire with all its ramifications of wealth and power—namely, money. As resource analysts, policymakers, and, especially, the industry like to say, it's all about price: if the price is high enough, we'll get the stuff out; there's plenty of it. With sufficient revenues, we can do it safely and cleanly. Such reasoning and its accompanying rhetoric, we assert in this book, have run their course. Energy and money can no longer justify endless extraction and expansion. Their time's up; game's over. This is the twenty-first century. Better to get ready, and sooner than later. Better not to coddle any more the defenders and holdouts of the nineteenth and twentieth centuries. A post–fossil fuel order is coming, like it or not, ready or not (box 13.1).

Getting ready, we argue in this book, starts with the unapologetic rejection of easy answers and proceeds forthwith to a confrontation—ideational, biophysical, normative, rhetorical—with the defenders, with the dying powers. Simultaneously, it creates an alternative: one farm, one store, one factory, one town at a time.

An integrated strategy, grounded from the biophysical to the spiritual, is the antithesis of energetic and monetary reductionism. It entails a wholly different language, one fossil fuel defenders cannot speak. If pressed to demonstrate an ethic beyond total use (chapter 4) or a societal goal beyond increased consumption or a connection beyond commodity, they have no answer, their public relations arms rendered discursively mute.

To paraphrase philosopher Richard Rorty, fundamental cultural change occurs not when people argue better but when they speak differently. We have argued a lot in this book, convincingly, we hope. But

**Box 13.1**
What's It All About?

What's it all about?
It's not about climate change;
    it's about carbon.
It's not about carbon;
    it's about fossil fuels.
It's not about fossil fuels;
    it's about mining.
It's not about mining;
    it's about endless frontiers.
It's not about endless frontiers;
    it's about excess.
It's not about excess;
    it's about doubt.
It's not about doubt;
    it's about disconnection.
What's it all about?
    Reconnection.
    Wonder.
    Sufficiency.
    Renewal.

should this book have impact, it will be because it has enabled different speaking. It will aid those who understand that fossil fuels are incompatible with the good life in making their case and doing so without adopting the language of extraction, expansion, consumerism, commoditization, and commercialism. It will give comfort as they muster the courage to resist the depredations and create the alternatives. It will help legitimize the language of embedded systems, of limited growth, of political pluralism, of sufficiency, of NIMBY (not in my backyard) and NOPE (not on Planet Earth), of placefulness, of, in short, the good life.[22]

## Proposition 3: Multiple Sovereignties
*An exclusive sovereignty—of the state—excludes from decision making those grounded in place. Multiple sovereignties allow for the privileging of those who sustain life-support systems.*

During one of the workshops to write this book, Laura Bozzi told a story of standing in front of a line of coal miners with blue overalls who were supporting future mining projects. As part of an activist group protesting coal mining, she had to look these men and women in the eye and explain why she, an outsider, would be at all interested in Appalachia, including the livelihoods of these people and their environment. Bozzi's experience illustrates one kind of sovereignty as well as what we coeditors call the insider-outsider dilemma: although we feel an attachment to such places as Appalachia and the Amazon, few of us reside there. We use products and drive cars that use the fossil fuels mined in such places, breathe in the particles they emit, sometimes drink the water they pollute. Our dilemma is that we are all complicit yet hardly all responsible. The perpetuators of a fossil fuel order like to argue that unless one lives and works in the affected regions, one is not part of the equation. This argument relies on a convenient notion of sovereignty, one the extractors and financiers of fossil fuels themselves barely respect. One thing we coeditors have learned in this project is that yes, we are all complicit (as consumers) and we are all victims (breathing the air, adapting to climate disruption), but "we" are not all *equally* responsible.

The insider-outsider dilemma is a dilemma only because policymakers and citizens alike accept the conventional notion of sovereignty, one that speaks a language of inclusiveness and democracy (we're all in this together; we all vote; and we all make consumer choices). In the end, however, it is a sovereignty that privileges the powerful few and reinforces the status quo. It is a sovereignty that locates primary decision making among those who can bid the highest and best write the rules, people comfortably ensconced in places like Washington, DC, London, Brussels, Tokyo, and Beijing. By contrast, a notion of multiple sovereignties allows primary decision making among those who have pride of place—those with primary responsibility for a piece of land, a forest, a fishery, a water supply. Here solutions to problems are based on local dialogue.

A second way that sovereignty enters debates about fossil fuels and their by-products is territorial and proprietary: who has soil rights, subsoil rights, water rights, and clean air rights. In El Salvador, where massive gold extraction is still possible, long-time residents have learned that their de facto water rights, a basic necessity of life, are subject to appropriation by outside extractors. The question Salvadorans and others around the world are facing is whether state and market actors should have such rights, effectively denying residents their right to pursue a livelihood, protect their health, even, in some cases, protect their lives. The answer, the authors in this book sense, is increasingly no. But actors on the ground keep bumping up against a state-based notion of sovereignty that confers formal authority (yet dubious moral authority) to those who wish to extract from citizens' land. In much of the lesser-developed world, subsoil rights are state rights, which give states sovereign control over minerals and fossil fuels.[23] As Martin elucidated in chapter 5 on the Yasuní-ITT Initiative in Ecuador, the United Nations Programme for Reducing Emissions from Deforestation and Forest Degradation projects do not address these subsoil sovereignty issues, thwarting the desires of resident peoples to determine their own future, including leaving fossil fuels and minerals, so valuable to outside extractors and their political enablers, in the ground.

A third way sovereignty enters fossil fuel debates is, as Robin Broad and John Cavanagh in chapter 7 and James Goodman and Stuart Rosewarne in chapter 8 illustrated, local-national-global complexity. In the case of gold in El Salvador, local peoples are limited by the International Centre for Settlement of Investment Disputes, housed at the World Bank in Washington, D.C., and part of the 2005 Central American Free Trade Agreement, also located in Washington, D.C., to which El Salvador is a member state. Pacific Rim mining company is suing the El Salvadoran government for apparent losses to their investment to the tune of $77 million in compensation. Here, a global institution and a multinational corporation, neither rooted in El Salvador, can challenge Salvadoran citizens' right to their land and resources, denying livelihood by denying both local and national sovereignty. Similarly, Goodman and Rosewarne told the story of the Australian Labour Party, whose platform since the late 1970s was "keep uranium in the ground," but has now changed so as to compete with other uranium exporters, like the United States, on the global market, all this despite cries from civil society about safety, pollution, and sovereignty on sacred Indigenous lands. El Salvador is near a ban on gold mining, and the Australian government is experiencing resistance in nearly every mining area of the country. Such resistance suggests that reclamation of local sovereignty is possible.

These challenges to and prospects for local community sovereignty lead to the final and perhaps most significant part of the sovereignty equation: displacement and localization. Indigenous peoples have suffered grave displacement, including loss of territory and entire tribes. In the case of oil extraction in Ecuador's northern Amazon, the Tagaeri and Taromenane tribes have been forced to move deeper and deeper into the forest, losing their hunting grounds and way of life in the Yasuní National Park. This and more has happened despite a legally mandated "intangible zone" that is supposed to protect them and their land. Pablo Fajardo, attorney for local peoples in the $19 billion lawsuit against Chevron Texaco, tells the tragic story of the murder of his brother and constant threats to his life and that of his family members.[24] The Taromenane and Tagaeri peoples and others of the Amazon and the Haudenosaunee peoples of New York State struggle to protect their lands from hydrofracking activities and live within their cultural and spiritual norms. Indigenous and community activists in El Salvador and Australia as well face living with polluted air and water and increased health risks. Their sovereignty is always subordinate to state and transnational sovereignty.

In sum, the insider-outsider dilemma, differentiated soil and subsoil rights, and questions of displacement raise crucial questions about competing sovereignties and people's rights to decide how to live on the land they depend on. The contemporary international system recognizes only state-based sovereignty, with preference given to the highest bidder, which is rarely local. Fossil fuel resistance and efforts to pursue livelihood reveal the need for an expanded notion of sovereignty, one that is multiple yet grounded in life-support systems, in the rights of communities to live in healthy environments with high-integrity ecosystems as the basis of their economies. Thus, confronting multiple sovereignties forces citizen and leaders alike to question the international system and its privileging of mining over sustaining, the distant over the residential, the commoditized over the subsistent, the placeless over the place based.

It forces a reexamination of people's place on the earth and the prerogatives of powerful actors. It forces everyone to ask whether the good life is possible in the current order, whether a pattern of extraction that is ultimately destructive can ever be legitimate.

## Proposition 4: Grounded Discourses

*In the competition among local, national, and international sovereignties in a resource-constrained world, sovereignty can no longer privilege state and capital prerogatives, including the prerogative to extract, externalize, and expand. Instead, in an ecologically full world, sovereignty must privilege those who are grounded in place and depend directly on life-sustaining resources.*

Some fifteen years ago, political scientist Karen Litfin wrote *Ozone Discourses*, a book that incisively engaged a topic that could hardly be more abstract, more globalist, more esoteric, more removed from everyday life. Stratospheric ozone, essential for ultraviolet protection, was being depleted by chlorofluorocarbons from refrigerants and aerosols; a global agreement was needed. Now Litfin is writing about ecovillages and building such a village herself, from the ground up.[25] Many of the authors in this book have made similar shifts, away from the abstract and idealized to the grounded and pragmatic, away from the globalist and placeless to the localist and placeful, away from the "realist" and materialist to the ethical and spiritual. We still write, conceptualize, and theorize, but we seek grounding when so much of the elite discourse is, indeed, abstract.

In part, the grounding we seek is literal—being attentive to what lies below our feet. In that, our journey through fossil fuel's cultural dominance, its physical pervasiveness, and its widespread destruction has taken us to countergrounding—grounding not in "energy" or commodity prices or rules and regulations. Rather, our journey has taken us, often as not, to water and soil, to, as Aldo Leopold put it so many decades ago, "the land." There we find that what is life-taking is the punctures and gashes and burns in the ground, wounds that bleed toxic fluids and belch disruptive gasses.

Our grounding is also in everyday life, in the workings of families and communities, in regions and nations, where people strive to live well by living well within their means. Here there is life giving. These are people who live in place, in *their* place, the place they depend on, the place they defend and keep faith in. When others take, it is these people's places they take from, exhibiting little reverence for such places. The takers do so in the name of growth or efficiency or modernization or development or, of all things, saving the environment. Their discourse includes place but it is in fact others' place and resources, others' jobs and consumption. Grounded discourse, by contrast, is about (1) multiple sovereignties where the place based is privileged; (2) power where "power with" is privileged—that is, the power derived from place is primary, from neighbors secondary, and from distant extractors tertiary, if at all; and (3) authority, real authority, moral authority, what can only originate in the place based (box 13.2).

## Proposition 5: Twenty-First-Century Realism

*Realism in the twenty-first century means that real security starts with biophysical and social limits. It then extends to the injustices of*

---

**Box 13.2**
Tactics for Dismissing "Keep It in the Ground"

Resources can be used slowly, conservatively, or not at all. But the dominant approach is total use (chapter 3). Anything contrary to using resources as fast and as thoroughly as possible, including keeping fossil fuels in the ground, is, in mainstream discourse, ignored at best, dismissed and ridiculed at worst. Among the most common tactics for dismissal and ridicule are the following. Notice how selective their users are in dealing with facts and how certain facts are ignored—for example, the impossibility of infinite material growth on a finite planet; the reality, in the fossil fuel era, of wealth for the few, environmental burden for the many.

### Narrow the Scope

Pipelines, labor leaders stated, "are a low carbon emissions method of transporting oil and gas"[a] compared with truck, rail, and tanker ship transport. Pipelines thus have fewer emissions and are good for the environment, as well as construction jobs.

*Explanation*: "The environment" in question here is the land immediately adjacent to the pipeline—that which can be polluted from leaks. "Low emissions" is an efficiency claim, one of the standard tactics for narrowing the scope. Here transport efficiency—that is, emissions per unit energy per mile—is compared to other modes of transport. The real issue is the assimilative capacity over time of waste sinks such as the atmosphere and oceans. What matters environmentally is total loading over time. If all else is equal—that is, if there is no reduction in fossil fuel use as a direct consequence of choosing a pipeline over a truck—that one more pipeline results in more emissions. That is, an additional pipeline makes for more total loading—not less.

**Box 13.2**
(continued)

## Claim Inevitability

"I believe the oil transferred from Canada is going to make it to some final destination no matter what we do in the United States."[b]—Cecil Roberts, president, US United Mine Workers, 2013

"Like it or not, fossil fuels are going to remain the world's dominant energy source for the foreseeable future."[c]—Joe Nocera, columnist, *New York Times*, 2013

"That oil [in Alberta's sands] will be sold, if not to you [Americans], then to someone else. That is not meant as a threat. It is just a fact."[d]—Joe Oliver, minister of natural resources, Canada, 2013

*Explanation*: A claim of inevitability is a prediction of a particular sort: it is an extrapolation from past practices suffused with wishful thinking. An extrapolation assumes that external conditions will remain the same and no new ones of significance will emerge. The trajectory of the past (here, oil always gets to its destination) will be maintained because the conditioning forces (e.g., demand, industrial necessity, strategic imperative) will be the same. The extrapolation ignores the fundamental uncertainty of the future, especially with irresolvable complexity, limited predictability, surprise, and rare events.

Part of the wishful thinking is, naturally, self-interested. Part is a lack of imagination, especially of what can happen and what should happen given certain biophysical and social realities (chapters 1 and 2).

## Claim Endless Abundance

Canada "has the resources to meet all of America's future needs for imported oil."[e]—Joe Oliver, minister of natural resources, Canada, 2013.

"These shale assets are forever," said Ralph Eads III, vice chairman of Jefferies & Company, a Houston investment bank. "They are going to produce for a hundred years."[f]

"... This transformation [in U.S. oil and gas production] could make the U.S. the world's top energy producer by 2020, raise more tax revenue, free us from worrying about the Middle East, and, if we're smart, build a bridge to a much cleaner energy future. All of this is good news, but it will come true at scale only if these oil and gas resources can be extracted in an environmentally sustainable manner. This can be done right, but we need a deal between environmentalists and the oil and gas industry to lock it in—now."[g]—Thomas Friedman, columnist, *New York Times*, 2012.

*Explanation*: Endless abundance is a call for investment, for limitless growth, for progress forever, for government to get out of the way (with

**Box 13.2**
(continued)

regulations), for government to get into the act (with subsidies and protections for investors).

## Use Doublespeak

Rather than "constrain" natural gas to domestic use, writes Michael A. Levi, a senior fellow for energy and the environment at the Council on Foreign Relations, exports should be promoted because "exports would likely reduce global greenhouse gas emissions." Exports provide incentives to extract and trade. "Allowing natural gas exports while protecting the environment and low-income consumers is the right way to go."[h]

*Explanation*: Trade is presumed good for everyone, so more trade must be better for everyone. This includes the poor because trade lowers prices. Consequently, this claim has it, the more "good" fossil fuels that are traded, the less fossil fuel emissions there will be, the better prices consumers will get, and the resulting surplus value can be used to "clean up the environment." This assumes a swap-out mechanism: the more natural gas, the less coal and oil. None is offered aside from price, which is presumed to take care of problems, even if market failures—monopoly power, incomplete information, or externalities, for example—are notoriously huge.

## Call Them Names

James E. Hansen, former director of NASA's Goddard Institute for Space Studies and perhaps the most famous of climate scientists, has taken an "utterly boneheaded" move by becoming a climate activist, according to Joe Nocera, a regular columnist in America's paper of record. What's more, Hansen's logic of social change is "backward."[i]

*Explanation*: If your view of the world conflicts with another's, rather than examining and being explicit about each other's underlying assumptions, call that person names, this tactic has it. Better yet, demand that the other person be logical, as if there was a single logic in the world. Here the critic's logic is strictly economic, which is to say individualized and devoid of any sense of power: "The emphasis should be on demand, not supply," writes Nocera. "If the U.S. stopped consuming so much of the world's oil, the economic need for the tar sands would evaporate." Yes, and if people just cooperated there would be no conflict in the world.

**Box 13.2**
(continued)

---

### Sources

a. Unnamed labor leaders quoted in Steven Greenhouse, "A.F.L.-C.I.O. Backs Keystone Oil Pipeline, If Indirectly," *New York Times*, February 28, 2013.

b. Cecil E. Roberts, president of the United Mine Workers in the United States, quoted in ibid.

c. Joe Nocera, "How Not to Fix Climate Change," *New York Times*, February 19, 2013.

d. Quoted in Joe Nocera, "Canada's Oil Minister, Unmuzzled" *New York Times*, Aril 24, 2013.

e. Quoted in Joe Nocera, "Canada's Oil Minister, Unmuzzled" *New York Times*, Aril 24, 2013.

f. Quoted in Clifford Kauss and Eric Lipton, "After the Boom in Natural Gas," *New York Times*, October 20, 2012.

g. Thomas Friedman, "A Good Question," *New York Times*, February 25, 2012.

h. Michael A. Levi, "The Case for Natural Gas Exports," *New York Times*, August 16, 2012, A19.

i. Joe Nocera, "How Not to Fix Climate Change," *New York Times*, February 19, 2013.

---

*concentrating benefits among the few and displacing costs to the many over long periods of time. Realist politics accepts the limits and reverses the injustices.*

Self-described "realists" in the study of international relations have prided themselves on facing the world as it really is, on not being idealistic. Placing security at the center of international life, they are tough on the bad guys, supportive of the wealth generators. If there has been a fossil fuel dimension to such "realism," it has been to analyze the role of oil in international politics and prescribe (without idealism) its free flow. Claiming to eschew utopian visions and reject the possibility of human perfection, even sustained cooperation, realists say they deal with the world as it is.[26] When, by contrast, the world is seen from the perspective of biophysical systems, at once resilient and fragile, let alone from the perspective of the rights and livelihoods of those peoples disrupted by extraction, realism takes on new meaning. No longer can realism privilege states as supreme agents, power as economic and military prowess, and security derived from fear of insecurity in a chaotic world.[27] Rather, from the biophysical and local perspective, where overshoot is entirely possible and security is a function of fertile soil and drinkable water, people living in place are the primary agents, that is, agents of sustainable living. The "realism" of the nineteenth and twentieth centuries, centered on the security of the state, gives way to the realism of the twenty-first century, centered on the security of peoples grounded in place. This is certainly not utopian.

Twenty-first century realism does have a security dimension. But what the chapters and cases of this book have uncovered is that not all securities are equal; the state tends to protect the economic security of select actors, those who extract "efficiently" and return revenues, who have the ear of legislators, who amass great wealth and have no qualms about turning that wealth into great power. In the process, farmers, shopkeepers, Indigenous peoples, and community activists face cyanide and arsenic poisoning, increased levels of cancer from petroleum extraction, while weather extremes around the world threaten everyone's livelihoods and lives. For example, chapter 8 examines the Australian Labor Party's move to extract uranium from Indigenous communities and other places and export it, which not only endangers local people's health, but threatens the security of all on the planet. Promises to trade only "good uranium" (used for energy, not weapons) like Norway's claim to "clean oil" (chapter 10) ring hollow. The realism here is in facing the reality that there are some substances humans cannot handle and rightfully should stay in the ground.

So twenty-first century security is, in the first instance, that of individuals and communities. The rights and well-being of interdependent people, especially those grounded in place, are what add up to vibrant, integrated communities and, eventually, nations. In *The Justice Cascade*, political scientist Kathryn Sikkink describes the devolution of international human rights laws to domestic courts and the increased vigilance of these courts toward improved standards of human rights all over the world.[28] Martin found that this is very much what is happening in Ecuador (chapter 5). Ryggvik and Kristoffersen suggest it may be happening among oil-affected fishing communities in Norway (chapter 10). Lines of sovereignty blur in these cases, or when mercury travels from Chinese coal plants to the US coastline and fossil fuel emissions everywhere lower the air quality for everyone as the planet warms. Similarly, Princen demonstrated in chapter 3 that the culture of fossil fuels pervades all societies and classes. Even the Appalachian activists adhere to the long cultural foundations of coal mining in their region (chapter 6). While these lines are invisible, their impacts are tangible, leading to the extinction of entire Indigenous tribes in the northern Ecuadorian Amazon and the loss of human biodiversity.[29]

When faced with the lived struggles of those who want to keep fossil fuels in the ground, a twenty-first century vision of the world is logical, indeed unavoidable. No longer can high-consumption societies (the Global North) afford to continue with politics as usual, a politics that places states and corporations at the center of power in an economic system based on limitless growth, and do so at the same time its biophysical and cultural foundations are crumbling. The old way of drawing the globe without regard for ecological systems, people, and species living within those systems cannot continue. In the twenty-first century, that way is anything but realist. The politics of ending the fossil fuel era unavoidably focuses attention on the emerging politics of resident peoples. Theirs is a sovereignty—and an economy—based on biophysical and social limits.[30] This is twenty-first century realism.

As argued in chapters 1 and 4, the ethical argument must be part of this new realism. While end-of-pipe emissions mitigation and adaption are convenient (chapter 1), justifying such marginal approaches becomes unfathomable in light of monumental biophysical trends. Projections of fossil fuel use into 2050, presumably with never-ending state subsidies, all the while climate scientists tell us that temperatures and increased emissions put the planet in imminent danger, cannot be realistically supported, only wished for. Moreover, ecologists tell us that biodiversity has declined precipitously to the point of irreversible harm, including acidification of oceans to the point of changing the chemistry needed for marine life.[31]

## Power With

All of this—questions of the good life, language, sovereignties, securities, and twenty-first-century realism—leads to a topic few energy analysts, technocrats, or even environmentalists wish to engage: power. Central to the current distribution of "goods" (delivered energy, consumables, services) and "bads" (degradation, displacement, injustice) is indeed power, what we put up front in the introduction to this book and now we end with.

What we have found in this study is that power looks different through the lens of fossil fuel extraction, manufacture, distribution, and marketing. It looks different when so much is concentrated—the energy sources themselves, the wealth, the economic and political power. It looks different when so much is invisible—the fuels themselves, the organizing, the politicking. It looks different when the lives of resident peoples trying to maintain their livelihoods and soldiers trying to keep the stuff flowing around the world count as much as the sovereignties of states and the rights of access of globalized mining companies, state and nonstate.

So what is that difference? One way to put it, following Hannah Arendt and others,[32] is that it is the difference between power over and power with. Power over in the international system depends on the threat of violence facilitated by advanced weaponry and cheap energy. Some individuals and communities necessarily lose as the state system is shaped according to state power. Power with is facilitated by widespread legitimacy, state and nonstate, including the public sense that state leaders and their corporate brethren are serving the public first, not themselves. Power with starts with the power of regenerative systems, biophysical and social, a power that is inherently respectful of limits, of individuals and of the importance of community, again both social and biophysical. As political theorist Les Thiel puts it, "Humans require community to become unique individuals. Shaping this community and ordering the interactions of the unique individuals composing it is the stuff of politics."[33] All politics, Thiel reminds us, has power at its core. And twentieth-century politics, due in part to cheap energy—cheap economically, energetically, and environmentally—had power over at their core. Twenty-first-century politics, due to the end of cheap energy and the beginnings of extreme conditions such as sea-level rise and mass migration, demands power with at its core.

Another way to understand the fossil fuels lens on power is the normative sway of easy energy and endless growth. In chapter 3, Princen showed how the boosterism of frontier settlement led to pervasive cultural norms such as total resource use, progress, consumer sovereignty, and, as a result, a mining economy that knows no bounds. The norm of cheap energy has become so embedded in American society, and now in others, that the entire economic system, indeed an entire way of life (chapter 11), is based on it. What's more, the invisible power of the fossil fuel industry results in the invisibility of impacts and even of the fossil fuel industry itself. Access goes to those who can most expeditiously get the stuff out and fill state coffers, all regardless of local resistance, health risks, or even deaths to residents. The case of El Salvador is particularly illustrative of this expression of power, one that privileges extractive industry over citizen livelihood (chapter 7).

A third way to see power through fossil fuels is through multiple sovereignties and multiple securities. Here, power over locates primarily in state authority and the privileging of the extractive and the expansionist, power with in the authority of those grounded in place, those who help themselves and help others live within their means. Market and military power may determine the structure of the state system, all in the name of security—protection of the state and its citizenry. Yet the cases in this book suggest another kind of security, one built on the protection of life systems, on living with nature, not taking from it. This is the security of livelihood in the context of limited resources and limited regenerative capacity of ecological systems. The Yasuní-ITT Initiative (chapters 5 and 11) with its global trust fund is one mechanism designed for such security, to protect one piece of forest habitat so as to move Ecuador a step toward its goal of a post–fossil fuel economy. Power derives from a combination of normative innovation and organizing from the local to the international. Security lies in that which protects cultural and biological diversity, economic prosperity in that which secures livelihood over the long term.

To entertain the idea of deliberately and planfully ending the fossil fuel era is to find power in the land and the people of a place. And it is to counter the power of endless extraction and deposition. The dramatic violence of state-to-state conflict, terrorism, and subnational conflicts may pervade the news, but "slow violence," that of ecological degradation, disease, and persistent contamination, is actually the norm for many societies, from the Amazon to Appalachia, from Central America to the Arctic, from Australia to western Europe.

This book, then, has sought to bring a particular kind of realism to bear on the politics of a world order dominated by cheap, concentrated energy, of resisting the depredations thereof, of imagining an industry exit, and of constructing alternative visions of the good life where living within our means, biophysical and social, is an ethical standard. By *realism*, we mean political analysis where power is central, but not the power typically found in contemporary study of international relations, that which derives ultimately from military and economic sources at the level of the state. For the purpose of understanding unsustainable trends and the phase-out of fossil fuels, we have focused on the power in the fossil fuel complex—independent and national oil companies and their enablers in finance and government—a power that for more than a century has overwhelmed localities, regions, nations, and the international system as a whole. The case studies in this book give a taste of that power, largely as seen through the eyes of those who are dumped on, whose land and livelihoods have become sacrifice zones for this last desperate phase of the fossil fuel era. Many more examples exist around the world. So it is through this resource-based analysis of power, at once ecological and ethical, that we have highlighted power as it originates in, interacts with, and reconstructs natural systems—including pools of petroleum and veins of coal, fertile soil, and recharging aquifers. We have argued that the fossil fuel basis of power has nearly spent itself, not geologically or economically, but ethically. It is no longer legitimate, no longer net beneficial, not in the long term. Much as feminist scholars uncover the hidden in the very real politics of gender, the personal as political, we have endeavored to uncover the hidden in the very real politics of concentrated energy and its relations with both the powerful and the marginalized, "the energetic as political."[34] This is the politics that undergirds both fossil fuel dominance and the struggles against that dominance. Some of these struggles are of those who depend directly on the ecosystems within which they live; others are from afar, who see the injustices both local and global. The normative aim of this project, then, has been to imagine a politics—and the power sources to go with it—that would accelerate the end of the fossil fuel era, closing out the nineteenth and twentieth centuries and, finally, bringing in the twenty-first century. Managing emissions won't do; the power of the fossil fuel complex is upstream where the rules of the game are written, capital is amassed, technological experiments are conducted, and wealth is accumulated. The end of the fossil fuel era starts with sources of power—energetic, economic, and institutional on the one hand, and place based, ecological, and spiritual on the other. For that, there is no better policy direction than to deliberately and gracefully leave fossil fuels in the ground.

## Notes

1. For prominent examples of such technorationalism, see, for economic arguments, Nicholas Stern, *The Economics of Climate Change* (Cambridge: Cambridge University Press, 2007); for efficiency solutions, Raimund Bleischwitz and Peter Hennicke, eds., *Eco-Efficiency, Regulation, and Sustainable Business: Toward a Governance Structure for Sustainable Development* (Cheltenham, UK: Edward Elgar, 2004), Livio D. DeSimone and Frank Popoff with the World Business Council for Sustainable Development, *Eco-Efficiency: The Business Link to Sustainable Development* (Cambridge, MA: MIT Press, 1997); Amory Lovins and Rocky Mountain Institute, *Reinventing Fire: Bold Business Solutions for the*

*New Energy Era* (White River Junction, VT: Chelsea Green, 2011); and for an alternative technology (fourth-generation nuclear power), Pushker A. Kharecha and James E. Hansen, "Prevented Mortality and Greenhouse Gas Emissions from Historical and Projected Nuclear Power," *Environmental Science and Technology* 47 (2013) 4889–4895, doi:10.1021/es3051197.

2. Sheila Jasanoff, "The Essential Parallel between Science and Democracy," *SEED*, February 17, 2009, 1–4.

3. Leah Temper, I. Yánez, K. Sharife, O. Godwin, and J. Martinez-Alier (coord.), "Towards a Post-Oil Civilization: Yasunization and Other Initiatives to Leave Fossil Fuels in the Soil," *EJOLT Report*, no. 6 (May 2013).

4. Christopher Lasch, *The True and Only Heaven: Progress and Its Critics* (New York: Norton, 1991); Michael Greer, "Progress vs. Apocalypse: The Stories We Tell Ourselves," in *The Energy Reader: Overdevelopment and the Delusion of Endless Growth*, ed. Tom Butler, Daniel Lerch, and George Wuerthner (Sausalito, CA: Foundation for Deep Ecology, 2009), 95–101; Donald N. Michael and Walter Truett Anderson, "Now That 'Progress' No Longer Unites Us," *New Options*, November 24, 1986, 1–2.

5. Paul F. Steinberg and Stacy D. VanDeveer, eds., *Comparative Environmental Politics: Theory, Practice, and Prospects* (Cambridge, MA: MIT Press, 2012).

6. On the failures of forecasting, especially in the energy arena, and the prospects of normative scenario building, see Vaclav Smil, *Energy at the Crossroads: Global Perspectives and Uncertainties* (Cambridge, MA: MIT Press, 2003) 121–80.

7. For exceptions to this general paucity of studies on societal transition (as opposed to energy or technological transition) see Butler, Lerch, and Wuerthner , *The Energy Reader*; Raymond De Young and Thomas Princen, *The Localization Reader: Adaptations for the Coming Downshift* (Cambridge, MA: MIT Press, 2012); Jörg Friedrichs, *The Future Is Not What It Used to Be: Climate Change and Energy Scarcity* (Cambridge, MA: MIT Press, 2013); J. Grin, J. Rotmans, J. Schot, F. Geels, and D. Loorbach, *Transitions to Sustainable Development: New Directions in the Study of Long-Term Transformative Change* (New York: Routledge, 2010); Mark Swilling and Eve Annecke, *Just Transitions: Explorations of Sustainability in an Unfair World* (Tokyo: United Nations University Press, 2012); Paul Raskin, Tariq Banuri, Gilberto Gallopin, Pablo Gutman, Al Hammond, Robert Kates, and Rob Swart, *Great Transitions: The Promise and Lure of the Times Ahead* (Boston: Stockholm Environment Institute—Boston and Tellus Institute, 2002); Temper et al., "Toward a Post-Oil Civilization."

8. R. A. Rosen, C. Electris, and P. D. Raskin, "Global Scenarios for the Century Ahead: Searching for Sustainability," Tellus Institute Report, http://www.mdpi.com/2071-1050/2/8/2626/pdf.

9. Millennium Ecosystem Assessment Scenarios Working Group, *Scenario Assessment, 2005*, http://www.maweb.org/en/Scenarios.aspx.

10. For such signs, see among others, Paul Hawken, *Blessed Unrest: How the Largest Movement in the World Came into Being and Why No One Saw It Coming* (New York: Viking, 2007).

11. See Practical Action, "Poor People's Energy Outlook 2013," accessed September 20, 2013, http://cdn1.practicalaction.org/docs/ppeo-2013-practical-action .pdf#page=11; David Korten, "Religion, Science, and Spirit: The Sacred Story of our Time," *YES Magazine,* January 17, 2013, http://www.yesmagazine.org/happiness/ religion-science-and-spirit-a-sacred-story-for-our-time.

12. As stated in Temper et al., "Toward a Post-Oil Civilization."

13. Matthew Huber, *Lifeblood: Oil, Freedom, and the Forces of Capital* (Minneapolis: University of Minnesota Press, 2013).

14. Cynthia Enloe makes this point regarding feminism as well, arguing that it "sees" multiple forms of power below the simple realism paradigm of international relations—for example, to the average worker, wife, and mother in the streets. See the video of Enloe at http://www.e-ir.info/2013/03/13/interview -cynthia-enloe/.

15. Republic of Ecuador, *National Plan for the Good Life 2009–2013,* 24, http:// www.politicaeconomica.gob.ec/wp-content/uploads/downloads/2012/09/Plan -nacional-del-buen-vivir-resumen.pdf.

16. Mario Melo, "How the Recognition of the Rights of Nature Became a Part of the Ecuadorian Constitution," in Global Exchange, the Council of Canadians, and Fundacion Pachamama, *The Rights of Nature: The Case for a Universal Declaration for the Rights of Mother Earth,* (San Francisco: Global Exchange and Council of Canadians, 2001), 83–84.

17. International Energy Agency, *World Energy Outlook 2012,* https://www.iea .org/newsroomandevents/pressreleases/2012/november/name,33015,en.html.

18. "Unburnable Fuel," *Economist,* May 4, 2013, http://www.economist.com/ news/business/21577097-either-governments-are-not-serious-about-climate -change-or-fossil-fuel-firms-are.

19. Thomas Princen, Jack P. Manno, and Pamela Martin, "Keep Them in the Ground: Ending the Fossil Fuel Era," in *State of the World 2013: Is Sustainability Still Possible?* ed. Worldwatch Institute (Washington, DC: Island Press, 2013), 161171.

20. Bill McKibben, *Oil and Honey: The Education of an Unlikely Activist* (New York: Macmillan, 2013).

21. For examples of such perspectives, see Peter G. Brown and Geoffrey Garver, *Right Relationship: Building a Whole Earth Economy* (San Francisco: Berrett-Koehler, 2009); Herman E. Daly and Kenneth N. Townsend, eds., *Valuing the Earth: Economics, Ecology, Ethics* (Cambridge, MA: MIT Press, 1993); Jerry Mander, *In the Absence of the Sacred: The Failure of Technology and the Survival of the Indian Nations* (San Francisco: Sierra Club Books, 1999); Thomas Princen, *Treading Softly: Paths to Ecological Order* (Cambridge, MA: MIT Press, 2010).

22. We thank Raymond De Young for introducing the term *placefulness,* implying something richer and deeper than the two-dimensional *place based.*

23. Kathleen Lawlor, Erika Weinthal, and Lydia Olander, "Institutions and Policies to Protect Rural Livelihoods in REDD+ Regimes," *Global Environmental Politic* 10, no. 4 (2010): 1–11.

24. Pablo Fajardo, interview with Pamela Martin, Quito, Ecuador, 2009.

25. Karen Litfin, *Ecovillages: Lessons for Sustainable Communities* (Cambridge, UK: Polity Press, 2013); Karen Litfin, *Ozone Discourses* (New York: Columbia University Press, 1994).

26. Reinhold Niebuhr, *The Irony of American History* (Chicago: University of Chicago Press, 2008).

27. John Mearsheimer, "Why We Will Soon Miss the Cold War," *Atlantic Monthly* 266, no. 2 (1990): 35–50; Kenneth N. Waltz, "Structural Realism after the Cold War," *International Security* 25, no. 1 (2000): 5–41.

28. Kathryn Sikkink, *The Justice Cascade: How Human Rights Prosecutions Are Changing World Politics* (New York: Norton, 2011).

29. Simon Romero and Clifford Krauss, "Ecuador Judge Orders Chevron to Pay $9 Billion," *New York Times*, February 14, 2011, http://www.nytimes.com/2011/02/15/world/americas/15ecuador.html?_r=0.

30. Raymond De Young and Thomas Princen, *The Localization Reader: Adaptations for the Coming Downshift* (Cambridge, MA: MIT Press, 2012).

31. Stephanie Mills, "Peak Nature?" in *The Post Carbon Reader: Managing the 21st Century*, ed. Richard Heinberg and Daniel Lerch (Healdsburg, CA: Watershed Publishers, 2012), 97–115.

32. Hannah Arendt, *The Human Condition* (Chicago: University of Chicago Press, 1959); Amartya Sen, "The Living Standard," in *Ethics of Consumption: The Good Life, Justice, and Global Stewardship*, ed. David A. Crocker and Toby Linden (Lanham, MD: Rowman & Littlefield, 1998), 287–311; Leslie Paul Thiele, *Indra's Net and the Midas Touch: Living Sustainably in a Connected World* (Cambridge, MA: MIT Press, 2011).

33. Thiele, *Indra's Net and the Midas Touch*, 172.

34. Cynthia Enloe, *Bananas, Beaches, and Bases: Making Feminist Sense of International Politics* (Berkeley: University of California Press, 2000), 195.

# Contributors

**Stephen B. Balogh** examines food and energy flows in urban ecosystems at SUNY College of Environmental Science and Forestry. He has published papers and book chapters on the importance of high-quality energy to societal development, the nexus of energy and the economy, the energetic efficiency of agricultural systems, and trade-offs in creating an urban green economy.

**Laura A. Bozzi** works for a West Virginia nonprofit organization on issues of environmental law and sustainable economic alternatives. She holds a PhD in forestry and environmental studies from Yale University.

**Robin Broad** is a professor of interdisciplinary international development at American University. She focuses on the local environmental, social, and economic impacts of globalization, as well as the social movements challenging it. Her publications include *Development Redefined: How the Market Met Its Match* (Paradigm) and *Global Backlash: Citizen Initiatives for a Just World Economy* (Rowman & Littlefield).

**John Cavanagh** directs the Institute for Policy Studies in Washington, D.C. His books on the global economy and global corporations include *Alternatives to Economic Globalization* (Berrett-Koehler) and *Global Dreams* (Simon and Schuster). He began working on these topics in the late 1970s as an international economist for the United Nations.

**James Goodman** researches social movements and globalization, focusing on global justice and climate change. He lectures at the University of Technology, Sydney, and is coauthor of *Justice Globalism: Ideology, Crises, Policy* (Sage) and *Climate Upsurge: The Ethnography of Climate Movement Politics* (Routledge).

**Berit Kristoffersen** explores issues of oil, fisheries, coastal tourism, and political autonomy at Universitetet i Tromsø, the Arctic University of Norway. She is a postdoctoral fellow at the Arctic Encounters Project there and coeditor of *Climate Change, Ethics and Human Security,* published by Cambridge University Press.

**Jack P. Manno** writes about sustainability, ecological economics, and Indigenous values at SUNY College of Environmental Science and Forestry. He is the author of *Privileged Goods: Commoditization and its Impacts on Environmental and Society* (Taylor and Francis).

**Pamela L. Martin** examines issues of sustainability, energy, and rights at Coastal Carolina University. She is the author of *Oil in the Soil: The Politics of Paying to Preserve the Amazon* and *Introduction to World Politics: Conflict and Consensus on a Small Planet* (both published by Rowman and Littlefield).

**Tom Morton** is a professor of journalism at the University of Technology, Sydney, and director of the Australian Centre for Independent Journalism. He is a radio documentary producer whose recent work includes *In King Coal's Kingdom* (ABC Radio National) and *A German Reunion/Deutschland eine Wiederbegegnung,* an international coproduction about the twentieth anniversary of German reunification.

**Thomas Princen** explores ecological and economic sustainability at the University of Michigan, focusing on social organizing principles (e.g., sufficiency), overconsumption, norms of resource use, and transition. He is the author of *The Logic of Sufficiency* and *Treading Softly: Paths to Ecological Order* (both published by MIT Press).

**Stuart Rosewarne** explores the ecologically contradictory tendencies of capital accumulation, developing a critical engagement with radical ecology, socialist ecological theory, and, more particularly, the political force of social movements. Based in the Department of Political Economy at the University of Sydney, he is coauthor of *Climate Action Upsurge: The Ethnography of Climate Movement Politics* (Routledge).

**Helge Ryggvik** is an economic historian working on energy-related issues at the Centre for Technology, Innovation and Culture at the University of Oslo. He is the author of *Til Siste Dråpe, Om Oljens Politiske Økonomi* (*To the Last Drop: About the Political Economy of Oil*) (Ascehoug) and *Olje og Klima* (*Norwegian Oil and the Climate*) (Gyldendal).

**Adele Santana** holds a PhD in business environment, ethics, strategy, and public policy from the University of Pittsburgh. She researches corporate social and economic sustainability at Sonoma State University. Her interests focus on corporate strategies for creation of shared value and on business design and leadership for a healthy world.

# Index